Industry and Tertiary Sectors towards Clean Energy Transition

Industry and Tertiary Sectors towards Clean Energy Transition

Editors

Claudia Toro
Chiara Martini

MDPI • Basel • Beijing • Wuhan • Barcelona • Belgrade • Manchester • Tokyo • Cluj • Tianjin

Editors

Claudia Toro
ENEA—Italian National Agency for New Technologies, Energy and Sustainable Economic Development, Energy Efficiency Department
Italy

Chiara Martini
ENEA—Italian National Agency for New Technologies, Energy and Sustainable Economic Development, Energy Efficiency Department
Italy

Editorial Office
MDPI
St. Alban-Anlage 66
4052 Basel, Switzerland

This is a reprint of articles from the Special Issue published online in the open access journal *Energies* (ISSN 1996-1073) (available at: https://www.mdpi.com/journal/energies/special_issues/industry_and_tertiary_sectors_towards_clean_energy_transition).

For citation purposes, cite each article independently as indicated on the article page online and as indicated below:

LastName, A.A.; LastName, B.B.; LastName, C.C. Article Title. *Journal Name* **Year**, *Volume Number*, Page Range.

ISBN 978-3-0365-4903-3 (Hbk)
ISBN 978-3-0365-4904-0 (PDF)

Contents

About the Editors

Claudia Toro

Claudia Toro (Ph.D. in Energy Engineering and MSc in Mechanical Engineering) is a researcher at the Energy Efficiency in Economic Sectors technical unit of the Italian National Energy Efficiency Agency—ENEA since 2018. At ENEA, she deals with energy audits, sectorial analyses, and the development and analysis of tools to promote energy efficiency at small and medium-sized enterprises. Previously, she was a researcher of the National Research Council and a post doc at the Department of Mechanical Engineering of the University of Rome 'Sapienza'.

Chiara Martini

Chiara Martini (Ph.D in Economics and Quantitative Methods and MSc in Environmental Economics) is a researcher at the Italian National Energy Efficiency Agency—ENEA since 2012, and currently works in the Energy Efficiency in Economic Sectors technical unit. At ENEA, she deals with energy audits, sectorial analyses of energy efficiency measures, and the development and analysis of tools to promote energy efficiency at small and medium-sized enterprises. Previously, she was a researcher at the Department of Economics of the University of Rome 'Roma Tre' and was seconded at the Organization of Economic Cooperation and Development, in the Centre for Tax Policy and Administration, Tax and Environment Unit.

energies

MDPI

Editorial

Special Issue "Industry and Tertiary Sectors towards Clean Energy Transition"

Chiara Martini and Claudia Toro *

Energy Efficiency in the Economic Sectors Laboratory (DUEE-SPS-ESE), Energy Efficiency Unit Department, Italian National Agency for New Technologies, Energy and Sustainable Economic Development (ENEA), 00123 Rome, Italy; chiara.martini@enea.it

* Correspondence: claudia.toro@enea.it

Abstract: The Special Issue "Industry and Tertiary Sectors towards Clean Energy Transition" is focused on technical, financial and policy-related aspects linked to the transition of industrial and services sectors towards energy saving and decarbonisation. These different aspects are interrelated, and as such, they have been analysed with an interdisciplinary approach combining economic and technical information. Collecting and analysing quantitative data would allow researchers to better understand the clean energy transition process, and how the international and national regulatory and policy framework are contributing to it. The papers within this Special Issue focus on energy efficiency and clean energy key technologies, renewable sources, energy management and monitoring systems, energy policies and regulations, and economic and financial aspects.

Keywords: energy efficiency in economic sectors; clean-energy technologies; energy policies and regulations; financial instruments; decarbonisation; renewable energy sources

Citation: Martini, C.; Toro, C. Special Issue "Industry and Tertiary Sectors towards Clean Energy Transition". *Energies* **2022**, *15*, 4166. https://doi.org/10.3390/en15114166

Received: 30 May 2022
Accepted: 1 June 2022
Published: 6 June 2022

1. Introduction

The global economy should undergo an epochal and radical change in the next few decades to combat climate change. Clean-energy transition, the shift from the use of non-renewable energy sources to renewable sources, is part of the wider transition to sustainable economies using renewable energy, the adoption of energy-saving measures and green technologies development. This is a long and complex process, but it will allow us to safeguard the health of the environment in the long-run. The European Union is among the leading major economies in this process. In December 2019, with the European Green Deal, the objective of achieving a climate-neutral EU by 2050 was endorsed, alongside the target of reducing net greenhouse gas emissions by at least 55% by 2030 compared to 1990 levels. Later on, Member States' presented Long-Term Strategies featuring national energy scenarios together with sectoral targets, which should be monitored to ensure that national contributions are consistent with the achievement of the European reduction path. The Commission presented its 'Fit for 55 package' in July 2021 to bring EU legislation in line with the 2030 goal. To reach these long-term targets, the contribution of everyone is required, from individual citizens to large multinationals, passing through SMEs. In this sense, national and international policies play a key role in paving the way for clean energy transition.

According to the well-known energy-efficiency gap, current energy-efficiency technologies may not be adopted due to different barriers; thus, several public policies exist to enhance and sustain their implementation. This Special Issue focuses on technical and policy-related aspects linked to the transition of industrial and services sectors towards energy saving and decarbonisation. These different aspects are interrelated; as such, they could be better analysed with an interdisciplinary approach, such as one combining economic and technical information.

Quantitative analysis of the main trends in the development and application of energy-efficiency technologies in different productive sectors can usefully complement policy monitoring and long-term policy planning. Collecting and analysing quantitative data would allow researchers to enhance their understanding of the clean-energy transition process, as well as how the regulatory and policy frameworks contribute and what improvements are required. The analysis focuses on energy efficiency and clean energy key technologies, renewable sources, energy management and monitoring systems, energy policies and regulations, and economic aspects.

2. A Short Review of the Contributions in This Issue

The articles included in this Special Issue address the topic of industry and tertiary sector energy transition from different perspectives. Some of these are purely related to technology development and analysis, while others focus on energy-efficiency policies and regulations. The role of different economic sectors in the clean-energy transition process is also analysed. What clearly emerges from the collected contributions is that different technologies and sectors could all play a significant role in the process, each with different impacts, and that energy-efficiency measures are critical for raising awareness and action, providing a useful information base to monitor policy outcomes and plan further actions.

The increased attention on energy efficiency, both at the national and international levels, has fostered the diffusion and development of specific energy consumption benchmarks for most relevant economic sectors. In this Special Issue, several articles examine energy consumption and the energy efficiency potential at sectoral level [1–4]. Energy audits (EAs) provide comprehensive information about the energy usage in a specific facility, identifying and quantifying cost-effective energy performance improvement actions (EPIAs). The crucial role of these tools in clean-energy transition is remarked by the European Energy Efficiency Directive (EED, Directive 27/2012), which introduces an obligation to implement EAs (art. 8). At member-state level, the database associated with mandatory energy audits could represent an important information basis to develop in-depth studies on energy consumption, energy performance indicators (EnPIs) and EPIAs. Basing on the database provided by mandatory EAs for large and energy-intensive enterprises in Italy (Legislative Decree 102/2014), Bruni et al. [4] developed a methodology to obtain energy consumption and energy-performance indicators, whereas Herce et al. [2] define a set of indicators to analyse EPIAs and the link between them and energy-consumption monitoring. The studies at sectoral level use the EAs information basis in different ways, and from different perspectives; moreover, two sectors are analysed using the methodology described in [4].

Two methodological studies were developed to fully exploit the database provided by the obligation to carry out an energy audit, enforced in Italy since 2014. Awareness of energy efficiency and sectoral benchmarking represents the first necessary step for companies to move towards energy transition. The novel methodology to assess energy performance indicators of productive and economic sectors presented in [4] could be potentially applied to all production sectors, providing key information needed to characterise various production processes from an energy perspective. Their paper provides details of the statistical method developed and a validation example on the NACE 23 division "Manufacturing of other non-metallic mineral products", with a focus on the cement industry.

Energy transition can only become a reality if everyone is involved: when energy efficiency is concerned, this implies that EPIAs are introduced in all sectors, reflecting the saving potential and specific conditions. The implementation of monitoring tools and energy-management systems (EnMSs) supports companies in their long-term energy-efficiency strategies and in the analysis of the effectiveness of EPIAs. Herce et al. [2] analyse the link between EnMSs (specifically ISO 50001) and EAs in the EED Article 8 implementation in two industrial and two tertiary sectors in Italy. Moreover, the impact of company size, energy-monitoring systems, and EnMSs on planned and/or implemented EPIAs is analysed. The findings show that, despite the complexity of the variables involved in the energy-efficiency gap, indicators such as "energy savings per company" and "EPIA per site" are higher in enterprises with an EnMS and monitoring system.

Both studies show how an obligation could become an opportunity at a twofold level: at company site level, EAs allow companies to better understand their energy-consumption structure and identify which EPIAs are most suited and where; meanwhile, at policy-making level, since the availability of reliable energy-consumption and saving information enables policy makers to better plan and monitor the strategies to reach long-term energy and environmental targets.

As far as studies at sectoral level are concerned, four single sectors are examined, one in the tertiary sector and three in the industrial sector [1,3–5]. In the second group, the refining sector is analysed; it is peculiar due to its key role in energy production [5].

Despite the high energy-consumption of hospitals and health structures, scientific literature lacks the presence of adequate energy-performance benchmarks, especially relative to the European context. Thus, Dadi et al. [1] aimed to define energy-benchmark indicators for the Italian private healthcare sector. EnPIs are calculated by considering the global energy-consumption of the different sites, based on the methodology developed in [4], and the sector's relevant variables are also employed. The results obtained are compared with those provided from the methodology adopted by the Environmental Protection Agency. In this way, the reliability of the proposed methodology could be validated, as well as the validity and future usability of the calculated indicators.

Looking at industries, the methodological contribution by [4] analyses cement as a case study, presenting results in terms of specific indicators based on an energy source. General results, methodological insights and validation of the proposed case study are discussed. The foundry industry is one of the most energy-intensive sectors; consequentially, many companies are trying to increase their energy efficiency. Choosing the most appropriate technological solution is a difficult task for several reasons, such as the high number of energy-saving technologies proposed by manufacturers and the literature, as well as rapid technological advances. Leoni et al. [3] investigated opportunities for reducing the energy consumption of Italian foundry companies and presents a list of available technological solutions validated by experts. Implemented and planned interventions were extracted from the EAs database, and the advantages of each technological solution were studied. It emerged that companies are strongly investing in increasing the efficiency of auxiliary systems such as compressors and motors. Petroleum refinement is very important in the European economy, and the continuous increase of energy efficiency is a key topic for this sector. Herce et al. [5] analyse ten Italian refineries based on mandatory EAs and public data, evaluating the primary, thermal and electrical specific energy-consumptions. Some insights into the impact of refined products mix and Nelson Complexity Index in energy consumption are also presented, together with an overview of EPIAs. This work presents a first step for the benchmark of Italian refineries.

In terms of specific technologies, those analysed by the contributions in this Special Issue mainly refer to energy use and electricity generation, namely waste-heat recovery [6,7], electricity storage and renewable electricity production [8,9] covering both energy efficiency and decarbonisation dimensions.

Waste-heat recovery is one of the most promising options for improving the efficiency and sustainability of industrial processes. Although it is abundantly available and technologies for its exploitation are consolidated, the implementation rate of waste-heat recovery interventions is still low. Besides technical, economic, financial and regulatory factors, the lack or incompleteness of information concerning the material and energy flows within the companies, the types and characteristics of waste-heat sources and possible sinks for their internal or external reuse is another barrier. Giordano and Benedetti [6] proposed a methodology to systematic identify and characterize low-temperature waste-heat sources and sinks in industrial processes, which was based on the data gathered from the analysis of EAs carried out by large and energy-intensive enterprises in Italy. In order to demonstrate its feasibility, the methodology was applied to the Italian dairy sector due to its large energy-consumption and enormous potential for utilisation of low-temperature waste-heat sources.

Waste-heat recovery also has great potential in a productive context completely different from the Italian one: the Algerian economy. Hydrocarbons represent more than 90% of exports and natural gas power plants produce approximately 90% of electricity. However, the ambitious governmental program launched to foster renewable energy and energy efficiency reflects the commitment to exploit the existing potential. In this context, reliable and time-efficient optimisation tools are needed, considering technical, economic, environmental and safety aspects. Redjeb et al. [7] built a mathematical tool capable of optimising both steam and organic Rankine units. The tool could perform single or multi-objective optimisations of the steam Rankine cycle layout and of a multiple set of organic Rankine cycle configurations. To show the tool's potentialities and improve awareness of waste-heat recovery in bio-gas plants, the authors selected an in-operation facility as test case.

Another very important aspect of the current energy scenario concerns the operation of electric power-systems. This is becoming increasingly difficult, as the peak load demand is growing continuously, and the daily and annual load factors are worsening. One countermeasure to overcome these problems is a study of the operation method of electric power systems, including novel energy-storage systems such as secondary batteries, superconducting magnets (SMES) and flywheels, which have demonstrated astonishing improvements lately. In general, the cost of power generation can be reduced if the energy-storage system is charged during the off-peak time interval and discharged during the peak time interval. To promote the commercialisation of electrical energy-storage systems, an assessment of their environmental issues is essential, particularly in terms of CO_2 emissions. Tae et al. [8] tackled an evaluation method for CO_2 emission based on an optimal algorithm to identify the best-mix solution of power sources to reduce potential adverse environmental impacts of electricity generation.

Arena et al. [9] focused on photovoltaic electricity production and predictive maintenance, which has received increasing attention and is considered fundamental in industrial applications. In fact, it contributes to guaranteeing healthy, safe and reliable systems and avoiding breakdowns that could potentially lead to a whole system shutdown. The paper focuses on a use case of robust anomaly detection applied to an Italian solar cell production plant in Catania. They considered a Monte-Carlo-based pre-processing technique as a valid alternative to other common methods due to several advantages, such as outlier replacement and the preservation of temporal locality with respect to the training dataset. After pre-processing, the authors trained an anomaly detection model based on principal component analysis and defined a suitable key performance indicator for each sensor in the production line based on the model errors. The algorithm allows anomalous conditions to be isolated by monitoring the above-mentioned indicators and virtually triggering an alarm when exceeding a reference threshold. Testing it on both standard operating conditions and an anomalous scenario was successful, anticipating a fault in the equipment and demonstrating robustness to false alarms.

After contributions on the state of the art of sectoral energy consumption, energy-efficient and low-carbon technologies and technological assessment solutions, a final contribution provided an overall picture synthesising the general trend of green-technology development in the European Union. Technology is one of the main drivers in the clean-energy transition.

The European Union has recently approved an ambitious unilateral mitigation strategy known as the European Green Deal, leading the way in the negotiation process under the Paris Agreement. Caravella et al. [10] presented a novel approach based on the analysis of patent data related to climate change and mitigation technologies. At the global level, the pace of generation of new green technologies as measured by patent data has slowed in recent years. Moreover, the current EU technological positioning with respect to green areas appears to be problematic in terms of technological sovereignty, with serious risks of potential technological dependencies from other countries. Given the ambitious environmental targets in the EU, and the radical technological shift required to achieve them, additional and directed investments should be enhanced further.

Author Contributions: C.M. and C.T. contributed equally to this work. All authors have read and agreed to the published version of the manuscript.

Funding: This research received no external funding.

Conflicts of Interest: The authors declare no conflict of interest.

References

1. Dadi, D.; Introna, V.; Santolamazza, A.; Salvio, M.; Martini, C.; Pastura, T.; Martini, F. Private Hospital Energy Performance Benchmarking Using Energy Audit Data: An Italian Case Study. *Energies* **2022**, *15*, 806. [CrossRef]
2. Herce, C.; Martini, C.; Salvio, M.; Toro, C. Energy Performance of Italian Oil Refineries Based on Mandatory Energy Audits. *Energies* **2022**, *15*, 532. [CrossRef]
3. Leoni, L.; Cantini, A.; De Carlo, F.; Salvio, M.; Martini, C.; Toro, C.; Martini, F. Energy-Saving Technology Opportunities and Investments of the Italian Foundry Industry. *Energies* **2021**, *14*, 8470. [CrossRef]
4. Bruni, G.; De Santis, A.; Herce, C.; Leto, L.; Martini, C.; Martini, F.; Salvio, M.; Tocchetti, F.A.; Toro, C. From Energy Audit to Energy Performance Indicators (EnPI): A Methodology to Characterize Productive Sectors. The Italian Cement Industry Case Study. *Energies* **2021**, *14*, 8436. [CrossRef]
5. Herce, C.; Biele, E.; Martini, C.; Salvio, M.; Toro, C. Impact of Energy Monitoring and Management Systems on the Implementation and Planning of Energy Performance Improved Actions: An Empirical Analysis Based on Energy Audits in Italy. *Energies* **2021**, *14*, 4723. [CrossRef]
6. Giordano, L.; Benedetti, M. A Methodology for the Identification and Characterization of Low-Temperature Waste Heat Sources and Sinks in Industrial Processes: Application in the Italian Dairy Sector. *Energies* **2022**, *15*, 155. [CrossRef]
7. Redjeb, Y.; Kaabeche-Djerafi, K.; Stoppato, A.; Benato, A. The IRC-PD Tool: A Code to Design Steam and Organic Waste Heat Recovery Units. *Energies* **2021**, *14*, 5611. [CrossRef]
8. Tae, D.-H.; Lee, H.-D.; Shen, J.; Han, B.-G.; Rho, D.-S. Estimation Method of Greenhouse Gas Reduction for Electrical Energy Storage Based on Load-Leveling Application. *Energies* **2021**, *14*, 5492. [CrossRef]
9. Arena, E.; Corsini, A.; Ferulano, R.; Iuvara, D.A.; Miele, E.S.; Ricciardi Celsi, L.; Sulieman, N.A.; Villari, M. Anomaly Detection in Photovoltaic Production Factories via Monte Carlo Pre-Processed Principal Component Analysis. *Energies* **2021**, *14*, 3951. [CrossRef]
10. Caravella, S.; Costantini, V.; Crespi, F. Mission-Oriented Policies and Technological Sovereignty: The Case of Climate Mitigation Technologies. *Energies* **2021**, *14*, 6854. [CrossRef]

energies

Article

Anomaly Detection in Photovoltaic Production Factories via Monte Carlo Pre-Processed Principal Component Analysis

Eleonora Arena [1], Alessandro Corsini [2], Roberto Ferulano [3], Dario Alfio Iuvara [1], Eric Stefan Miele [2], Lorenzo Ricciardi Celsi [3,*], Nour Alhuda Sulieman [4] and Massimo Villari [4]

1 Enel Green Power S.p.A., Contrada Blocco Torrazze sn, Zona Industriale, 95121 Catania, Italy; eleonora.arena@enel.com (E.A.); dario.iuvara@enel.com (D.A.I.)
2 Dipartimento di Ingegneria Astronautica, Elettrica ed Energetica, Sapienza Università di Roma via Eudossiana 18, 00184 Roma, Italy; alessandro.corsini@uniroma1.it (A.C.); ericstefan.miele@uniroma1.it (E.S.M.)
3 ELIS Innovation Hub, via Sandro Sandri 81, 00159 Roma, Italy; r.ferulano@elis.org
4 Dipartimento di Scienze Matematiche e Informatiche, Scienze Fisiche e Scienze Della Terra, Università di Messina, Piazza Pugliatti 1, 98122 Messina, Italy; nosulieman@unime.it (N.A.S.); mvillari@unime.it (M.V.)
* Correspondence: l.ricciardicelsi@elis.org

Abstract: This paper investigates a use case of robust anomaly detection applied to the scenario of a photovoltaic production factory—namely, Enel Green Power's 3SUN solar cell production plant in Catania, Italy—by considering a Monte Carlo based pre-processing technique as a valid alternative to other typically used methods. In particular, the proposed method exhibits the following advantages: (i) Outlier replacement, by contrast with traditional methods which are limited to outlier detection only, and (ii) the preservation of temporal locality with respect to the training dataset. After pre-processing, the authors trained an anomaly detection model based on principal component analysis and defined a suitable key performance indicator for each sensor in the production line based on the model errors. In this way, by running the algorithm on unseen data streams, it is possible to isolate anomalous conditions by monitoring the above-mentioned indicators and virtually trigger an alarm when exceeding a reference threshold. The proposed approach was tested on both standard operating conditions and an anomalous scenario. With respect to the considered use case, it successfully anticipated a fault in the equipment with an advance of almost two weeks, but also demonstrated its robustness to false alarms during normal conditions.

Keywords: anomaly detection; principal component analysis; Monte Carlo simulation; PV cell production line; predictive maintenance

Citation: Arena, E.; Corsini, A.; Ferulano, R.; Iuvara, D.A; Miele, E.S; Ricciardi Celsi, L.; Sulieman, N.A; Villari, M. Anomaly Detection in Photovoltaic Production Factories via Monte Carlo Pre-Processed Principal Component Analysis. *Energies* **2021**, *14*, 3951. https://doi.org/10.3390/en14133951

Academic Editor: Lyes Bennamoun

Received: 14 May 2021
Accepted: 25 June 2021
Published: 1 July 2021

1. Introduction

In recent years, predictive maintenance has been receiving an ever increasing attention and has been considered fundamental in industrial applications. In fact, it contributes to guaranteeing healthy, safe and reliable systems, as well as to avoiding breakdowns that could potentially lead to a whole system shutdown.

As known, the main benefit of Principal Component Analysis (PCA) lies in its capability to reduce the dimensionality of data by selecting the most important features that are responsible for the highest variability in the input dataset. Namely, PCA allows to concentrate the analysis on a compressed version of the original dataset without compromising the reliability and the robustness of a predictive model. Among other factors, a key quality in PCA is the inherent capability of processing large multivariate datasets as customary in industrial equipment sensor networks. As a result, PCA formed a field of choice in predictive analytics in several use cases, e.g., maritime and transport applications, as well as decision support systems in healthcare [1,2].

On the other hand, the well known disadvantage of PCA stems from the sensitivity to outliers in the data. In this respect, in the literature four known algorithms have been very recently devised in order to sort outliers' observations out, namely the spherical principal component based algorithm, PCA based on robust covariance matrix estimation, robust PCA (ROBPCA) and the PCA projection pursuit algorithm [3].

To this end, based on measurements collected by the sensor network of a photovoltaic production plant, the paper proposes Monte Carlo (MC) simulation as the pre-processing stage to deal with outliers before applying PCA [4,5]. In this respect, the proposed approach is shown to be a valid alternative to relying on the classical Interquartile Range (IQR) method in order to omit outliers when applying PCA for anomaly detection purposes.

1.1. Related Works

Recently, the scientific community has devoted much attention to the use of data analytics and machine learning models in the operation domains, e.g., manufacturing and energy management. In particular, many applications have focused on predictive maintenance and anomaly detection [6–8].

In this context, industrial systems have adopted PCA for detecting anomalous scenarios in their operational processes. In particular, key performance indicators (KPIs) are usually defined starting from the PCA model in order to trigger alarms and prevent failures [9].

Many works focus on fault isolation techniques which are employed to classify different occurring errors and to isolate the system variables mostly affected by them [10]. Specifically, they often propose statistical methods for fault detection, like Hotelling T^2 or squared prediction errors Q [11,12].

Even though plenty of these works deal with error classification and isolation in the context of anomaly detection and predictive maintenance, other papers and practical experiments shed light on innovative strategies to pre-process the input data that will feed the predictive model. To this end, MC simulation has been largely applied for data pre-processing in order to define more robust models. For example, in [13] the authors process geodetic data by applying MC simulation to perform uncertainty modelling [14].

However, choosing the statistical method for MC simulation becomes difficult when the involved dataset is highly affected by the presence of outliers. In this respect, a robust estimation procedure has been investigated in [15]: The authors exploit the median since it provides an estimator with the highest breakdown point and it always guarantees a feasible solution for the considered optimization problem.

In general, MC simulation is used as a valid pre-processing strategy in order to successfully manage uncertainty with respect to experimental use cases in manufacturing and energy management, namely for predictive maintenance [16–19] or predictive analytics purposes [20].

Moreover, the number of data points sampled by MC simulation is another crucial parameter, since it could lead to inaccurate outputs [21]. This parameter is particularly challenging to optimize since it strongly depends on the use case and the quality of data. In [22] the authors test different MC simulations to determine the relationship between the sample size and the accuracy of the sample mean and variance.

Despite larger samples could provide for a better estimation of the input distributions, in [23] results demonstrated the need to restrict the number of MC runs to a number not greater than the sample sizes used for the input parameters, since a large number could be unnecessary or even harmful.

Despite the clear advantage of such approaches, they often still need to be validated in practice. So, to the best of the authors' knowledge, this paper proposes the application of MC simulation to a real photovoltaic production scenario, as an effective way to pre-process the data stream coming from the sensors deployed throughout the production site.

The related literature also reports pre-processing techniques for similar anomaly detection scenarios based on the IQR method (e.g., [24]), which, however, offers only the property of outlier removal and not the additional benefit of outlier replacement that is consequential to applying MC simulation, as further discussed in Section 3.

1.2. Paper Structure

The paper is structured as follows. Section 2 provides the use case description and problem setting. In Section 3 we explain our contribution in terms of exploiting MC simulation as an innovative approach to data pre-processing with respect to the considered anomaly detection and predictive maintenance application. Later on, in Section 4 we discuss PCA for anomaly detection. Section 5 presents the experimental setup and numerical results. Finally, Section 6 concludes the paper.

2. Problem Setting

Enel Green Power needs to implement, in the production line of sun cells in the 3SUN Factory, an artificial intelligence application capable of predicting faults relative to a piece of process equipment, the so-called Automatic Wet Bench (AWB) machine, for predicting any malfunctioning of the fans that ventilate the different stations within such machine. The data collected on the Manufacturing Execution System (MES) are fed as input to the predictive analytics engine in order to predict faults.

2.1. Use Case

In Figure 1 we show the process steps involved in the cell production. Each process equipment has a specific purpose: Raw wafers enter the first machine in the line, the so-called Wafer Inspection System (WIS), to check the quality of the input wafers; then, they are subject to texturization and cleaning through the AWB equipment; next, the Plasma Enhanced Chemical Vapor Deposition (PeCVD) equipment is used for the deposition of doped and un-doped layer of amorphous Silicon (aSi) on both side of the wafers. Then, the Physical Vapour Deposition equipment (PVD) is used for the sputtering process. Finally, the block formed by the Screen Printer, Tester and Sorter equipment are responsible, respectively, of collecting the electric charge of the cell (fingers) and to let the flow between one cell and the other (Bus Bar) in the assembled modules, testing the electrical I-V measurements of the cells and classifying them depending on their perfomance.

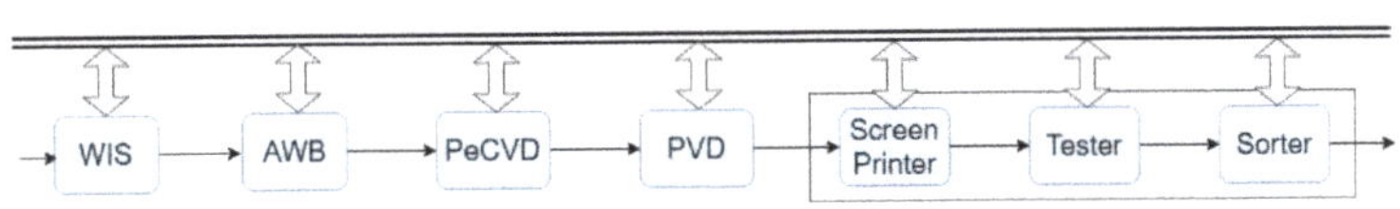

Figure 1. Photovoltaic cell production line in the 3SUN Factory.

The process equipment we refer to in this paper in order to predict the occurrence of faults is the AWB, where the wafers are chemically etched to roughen the surface to maximize the quantity of absorbed light and therefore the cell efficiency.

Along the production line, two parallel AWB machines are installed, each consisting of a loading station (the first one) and an unloading station (the last one) and, midway between the two, several stations where the chemical processes are performed. Within the AWB stage, the wafers are loaded onto specific containers called carriers, which move from one station to another until the process ends; the carriers do not enter in all stations but only some of them, as the same task can be carried out indifferently by one station or another, so that the carrier is moved by the automation system to the first available station that can carry out the required task.

More specifically, the stations composing the production line serve three main purposes: Pre-conditioning, texturing and cleaning. Each station is equipped with a sensor that records measurements when carriers enter and exit the station.

We now provide a brief description of the most frequently occurring fault inside the AWB and for which we design a suitable predictive analytics strategy. Such a fault is generally due to the malfunctioning of the fans that ventilate the different stations within each AWB stage.

For each AWB stage, there exist two drying tanks which must work properly in parallel and can never break down (not even alternatively), otherwise the AWB throughput would be halved, thus compromising the whole production line. Since the fault episode is generally preceded by the occurrence of anomalous vibrations, there is room for a suitable predictive analytics strategy aimed at anticipating the occurrence of the fault through the detection of such vibrations.

At a specific slot of time, an unexpected error may happen in one of its machines and block the production completely for several few days.

2.2. Sensor Measurements

The sensors mounted onto the production line stations measure several relevant parameters characterizing each station, such as station temperature, pump speed, flow speed, and ozone concentration level.

The measurements recorded by the sensors were collected only during the enter, exit and dosing phases of each carrier, thus leading to a non-constant sampling frequency. This produced many discontinuities of variable length in the sensor data streams, making standard time series analysis impossible. For this reason, the collected measurements were treated as an ordered set of samples rather than time series. In order to capture the time evolution of carriers going through a line, each sample is composed by the measurements coming from all the stations, collected during the enter, exit and dosing phases of a carrier.

Let k stations out of the total number N account for the main path drawn by a carrier entering the AWB stage to undergo pre-conditioning, texturing and cleaning. The remaining $(N-k)$ stations are parallel to the k principal ones and ensure the robustness of the whole AWB stage in the following way: If one of the k stations fails, there is at least a redundant station among the available $(N-k)$ that is properly working and can thus be entered by the carrier to undergo the whole production process.

For the sake of simplicity and without loss of generality, we assume to have k stations only, and we neglect the remaining ones. Each station contains m sensors. Each sensor measures the carrier up to t times.

The considered dataset collects the t measurements carried out by the m sensors in the k stations over n batches or carriers, assuming a batch to account for a couple of wafers flowing through the whole production line.

So we wrap all the available data into a structured dataset represented by a matrix X with n rows and $y := k \times m \times t$ columns.

As our approach is totally data-driven, without losing generality and for the scope of the model, hereinafter we assume $k = 7$ and $m = 6$. Moreover, we assume $t = 3$, because each sensor measures the carrier three times while it is inside the considered station.

3. Monte Carlo Based Pre-Preprocessing

In this section we illustrate a novel pre-processing approach based on Monte Carlo (MC) simulation and compare it with a commonly used method based on the Interquartile Range (IQR). This last is considered as a reference and the goal is to prove that our approach is a valid alternative to the IQR method. Since both these methods concern only the outlier removal phase, we also briefly describe the preliminary pre-processing steps required to standardize the data and handle missing values or flat signals.

3.1. Preliminary Data Cleaning

Independently on the method, a preliminary data cleaning and preparation stage is required before removing outliers. The following steps are applied:

- signal filtering when the missing values are above 5% of the total number of measurements. Above this threshold, data interpolation can lead to distortions so we preferred to discard the involved signals.
- linear interpolation of signals when the missing values are less that 5% of the total number of measurements.

- flat signals removal when the derivative is zero for at least 50% of the signal length since constant measurements do not provide any meaningful information.
- signal standardization in order to make the scales of the different signals comparable. This operation was achieved by subtracting the mean value and dividing by the standard deviation.

In the next sections we describe the reference IQR method, followed by the discussion of the proposed approach based on MC simulation.

3.2. *IQR Method*

The Interquartile Range (IQR) method is a simple but effective method used to identify outliers by isolating samples below the 25th percentile or above the 75th percentile [25].

3.3. *Monte Carlo Method*

In this paper we propose an innovative method for removing outliers based on MC simulation, which has been largely applied in other scenarios like estimation of sum, linear solvers, image recovery, matrix multiplication, low-rank approximation, etc. [26]. In our case, the idea is to generate new data points providing a more robust dataset by applying an estimator to random samples extracted from the original dataset.

By using the median estimator, there is no need to remove outliers from the raw data since this estimator is proved not to be affected by outliers [27].

Moreover, the size of the estimator dataset can be chosen arbitrarily, and can even be greater than that of the original one.

In the next sections we discuss the choice of the proper estimator, the number of samples used for MC simulation and the sliding window approach adopted to preserve the temporal locality of the sensor signals. Finally, we present the pseudocode illustrating the general pre-processing approach used to generate the new estimator dataset as input to the PCA model.

3.3.1. Mean Versus Median

The mean and the median are considered to be the most reliable estimators of the central tendency of a frequency distribution. Choosing the appropriate estimator is a challenging issue when using MC simulation since different results can lead to different correlations between signals, and thus different principal components when applying PCA. Let

$$x_i = (x_{p,z,w})_{\substack{p\,=\,1,\dots,k\\ z\,=\,1,\dots,m\\ w\,=\,1,\dots,t}} \tag{1}$$

denote the i-th row of the $n \times y$ data matrix X accounting for the measurement of sensor z during phase w in station p relative to batch i. In this way, each column f_j ($j = 1, \dots, k \times m \times t$) of X describes the temporal evolution of the measurements recorded by a specific sensor in a station during the processing of the batches.

Let $R^{IQR} = [r_{ij}^{IQR}]$ with $i, j \in \{1, \dots, n\}$, $i \neq j$, $r_{ij}^{IQR} = \frac{\sigma_{f_i f_j}}{\sigma_{f_i}\sigma_{f_j}}$ and $-1 \leq r_{ij}^{IQR} \leq 1$ denote the correlation matrix computed between the columns of the dataset resulting from the IQR pre-processing. Recall that $\sigma_{f_i f_j}$ denotes the covariance between the columns f_i and f_j, whereas σ_{f_i} denotes the variance of the i-th column.

Let $R^{MC,median} = [r_{ij}^{MC,median}]$ and $R^{MC,mean} = [r_{ij}^{MC,mean}]$ ($i, j \in \{1, \dots, n\}$, $i \neq j$), formulated as above, denote the correlation matrix computed between the columns of the dataset resulting from the median-based and the mean-based MC simulation pre-processing methods, respectively.

Let $\Delta := [\delta_{ij}] = R^{IQR} - R^{MC}$ account for the deviation between the two matrices, letting R^{MC} denote alternatively the correlation matrix relative to the median-based or the mean-based MC pre-processing method.

In order to evaluate which estimator suits our purpose best, we run the following statistical hypothesis test:

$$\begin{cases} H_0 : \delta_{ij} < \alpha \quad \forall i,j \\ H_1 : \delta_{ij} \geq \alpha \quad \forall i,j, \end{cases} \tag{2}$$

considering the difference between the correlation matrix computed after the application of the IQR method and the correlation matrix of the new dataset resulting from the previous section (that is, the MC dataset).

We can state that there exists a significance level α such that $\delta_{i,j}^{MC,median} < \alpha, \forall i,j$, and $\exists (i,j) : \delta_{i,j}^{MC,mean} \geq \alpha$, allowing us to choose H_0 only under the median-based MC method.

In particular, in the considered use case, the difference in the correlation matrices considering the median-based MC method is less than $\alpha = 6 \times 10^{-2}$ in absolute value and this proves to be a consequence of the median insensitivity to outlier observations.

3.3.2. Choosing the Size of the Monte Carlo Sample

Choosing the proper number of samples has a significant effect on MC simulation since it considerably improves estimation reliability. We recall that samples are chosen out of the data matrix X, where x_i, as defined in (1), represents a generic row of X accounting for the measurement of sensor z during phase w in station p relative to batch i.

Up to the authors' knowledge, the literature claims that increasing the sample size reduces the variance and decreases the noise of the simulation results method [28]. Calibrating the sample size depends on many factors such as dataset size, the pursued objective and the complexity of the phenomenon the designer is modeling [29]. Therefore, we have tested different sample sizes before defining a methodology aimed at finding a suitable number of samples for each round in MC simulation.

By comparison with the highly dispersed original dataset, by increasing the number of samples we obtain a proportional decrease in variance. The desired sample size will allow to remove only the outliers and at the same time preserve the rest of the information contained in the original dataset.

By excessively increasing the number of samples, the risk is that a significant part of the information is lost, thus affecting the accuracy of the PCA model.

In order to select the proper sample size for MC-based outlier removal, we evaluate the impact this parameter has on the PCA model.

To demonstrate that MC pre-processing is a valid alternative to the IQR-based pre-processing method, we compared the PCA models resulting from both approaches for different sample sizes, ranging from 1 to 100. In particular, we measured the proportion of the variance of the MC-PCA components that is explained by the IQR-PCA components in terms of R^2. In this way, high values of R^2 correspond to similar PCA models, thus confirming the equivalent performance of the two pre-processing methods.

From Figure 2, it is evident that by considering three samples we obtain the highest value of R^2 (around 97.5%), thus demonstrating that, by choosing the proper sample size, the MC pre-processing method achieves very similar results to those obtained by the IQR-based pre-processing method.

Figure 2 presents the results of the previous steps where it is experimentally proven that PCA with three-sample size has the best results.

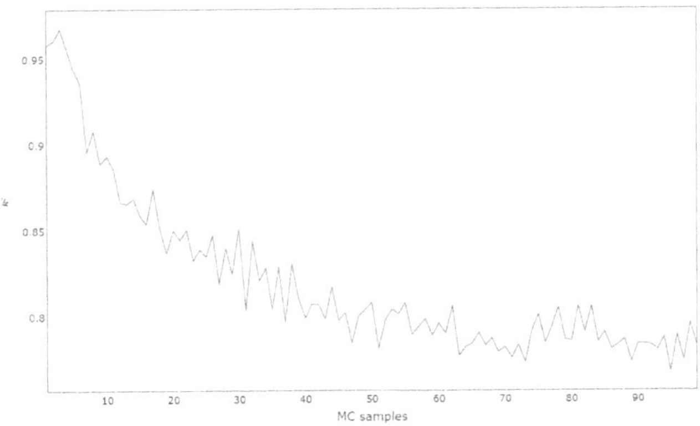

Figure 2. Testing R-squared for different sample sizes.

3.3.3. Preserving Trend Properties through a Suitable Choice of the Monte Carlo Sample

Since PCA is based on the linear correlation among variables, any trends intrinsic to the signals themselves will not be considered. For this reason, a random sampling among all the batches for the purpose of median computation may result in the loss of the temporal dependencies characterizing signals.

Therefore, we refined the procedure for the MC sample selection accordingly. In particular, for each batch in the original dataset we considered a time window centered around the batch itself. Samples considered for the median computation were therefore extracted inside such window, thus preserving the temporal locality among subsequent batches.

3.3.4. Pseudocode for the Pre-Processing Method Based on MC Simulation

The pseudocode reported in Algorithm 1 illustrates the steps required to generate a new estimator dataset by using a pre-processing procedure based on MC simulation, as proposed in Sections 3.3.1–3.3.3.

Algorithm 1 Pre-processing algorithm based on MC simulation

Input X: The original $n \times y$ data matrix
Output $\hat{X}$: The new estimator $\hat{n} \times y$ data matrix
Parameter $\hat{n}$: The size of the new estimator dataset
Parameter b: The number of samples considered for MC simulation
$i \leftarrow 0$
while $i < \hat{n}$ **do**
$idx \leftarrow$ generateRandomInteger$[b, n - b - 1]$
for j in range$[0, y - 1]$ **do**
$window \leftarrow X[idx - b : idx + b, j]$
$\hat{X}[i, j] = \leftarrow$ median($window$)
end
$i \leftarrow i + 1$
end

4. Principal Component Analysis for Anomaly Detection

Principal Component Analysis (PCA) is a well known method commonly used to reduce the dimensionality of a dataset, by transforming the original set of variables into a smaller one that still contains most of the information in terms of variance. In particular, it is a linear dimensionality reduction method based on Singular Value Decomposition (SVD) that projects the data on a lower dimensional space.

Being $\hat{n}$ the number of samples and let y the number of variables, the $\hat{n} \times y$ data matrix $\hat{X}$ is centered (by removing the mean of every feature) and SVD is applied on its covariance matrix, thus leading to a subset of orthonormal dimensions, namely the Principal Components (PCs) [30]. Since SVD computes PCs incrementally, their number depends on the pre-defined stopping criterion in searching for the next PC. A common strategy is to define the number of PCs as a function of the minimum variance information to be preserved with respect to the original dataset in order to compress the data sufficiently without loosing too much information.

In this paper we use PCA to perform anomaly detection. For this purpose, it is necessary to isolate a subset of data points associated with a normal behavior of the equipment. This subset is used as input to the PCA algorithm to compute a set of PCs considering as stopping criterion a high variance preservation (at least 90%). Having defined the $y \times z$ projection matrix Π composed by the z PCs, it is now possible to project each data point $\hat{x}_i$ on a lower dimensional space as:

$$c_i = \hat{x}_i \Pi \tag{3}$$

where c_i is the z-dimensional compressed version of $\hat{x}_i$. Then, we transform c_i back to its original space by multiplying it by the the inverse of the matrix Π (being Π orthonormal, the inverse coincides with its transpose), thus obtaining the reconstructed version of the input data:

$$\hat{x}_i' = c_i \Pi^T \tag{4}$$

Finally, we compute the reconstruction error of the sample $\hat{x}_i$ as:

$$e_i = |\hat{x}_i' - \hat{x}_i| \tag{5}$$

where the vector e_i contains the residual of every input feature. Since the model is trained on normal behavior data, the reconstruction error should be low for samples belonging to the same distribution. However, during an anomalous scenario, the error is expected to be high since the associated samples will deviate from such distribution. By considering these vectors as KPIs for the stations in the production lines, it is not only possible to detect anomalies when high errors occur, but also go back to the sensors mostly involved by inspecting the residuals of each single input feature.

Remark 1. *Thanks to the property of outlier replacement, to the median-based approach as introduced in Section 3.3.1, to the optimal choice of the sample size as described in Section 3.3.2 and to the preservation of any temporal dependencies characterizing the input signals as stated in Section 3.3.3, the proposed MC-based pre-processing approach turns out to be a robust alternative to IQR pre-processing. In fact, as it can be seen from the experimental results reported in Section 5, using median-based MC simulation in place of the IQR method for the pre-processing stage yields very similar results, although the number of PCs obtained when applying PCA after MC simulation is slightly higher than the number of PCs obtained when applying PCA after the IQR method.*

Remark 2. *The proposed pre-processing approach based on MC simulation is more adapt to the scenario of energy plants whose data require extensive cleaning. In this respect, if the input data are not cleaned enough, the IQR method, by isolating samples below the 25th percentile or above the 75th percentile, may end up removing a significant part of the original dataset, thus potentially compromising the quality of the subsequent data analytics task. Instead, MC simulation overcomes this obstacle by enabling the data scientist to tune the dimension of the dataset resulting from pre-processing according to the technical specifications of the considered task.*

5. Experimental Results of Anomaly Detection

In the experimental phase, we compared the results of the proposed anomaly detection approach considering both the IQR and MC pre-processing methods. In both scenarios, the relevant data were collected from the MES of the 3SUN Factory and a set of normal behaviour samples was defined for training the PCA model.

5.1. Training and Test Sets

According to the data format of matrix X specified in (1), we isolated a week of normal condition samples as training set, going from 8 July 2020 to 15 July 2020. This period was labelled as a period of standard operation by the operators working in the plant, together with other periods going from 1 November 2020 to 14 November 2020 and from 1 May 2020 to 8 May 2020, respectively, which we considered as test sets. The operators reported a fault in the plant on 4 July 2020, so we isolated 24 days of data before the fault as a further test set to see if the proposed model actually detects the anomaly, possibly in advance.

5.2. Pre-Processing Phase

Before the application of the anomaly detection approach based on PCA, we pre-processed the dataset as described in Section 3. In particular, 10 signals were filtered since they were completely flat, 12 signals were discarded since they presented an excessive rate of missing values, and eight signals were linearly interpolated. After this phase, the dataset counted 36 variables on which the two outlier removal methods were applied.

5.2.1. Outlier Removal Results

From the results it is evident that both the IQR and MC methods were able to filter outliers successfully. In Figure 3a, the original sensor signals are plotted in order to highlight the presence of outliers, while in Figure 3b,c, respectively, the pre-processed signals after the IQR and MC outlier removal methods are presented. It is important to notice that the IQR method does not handle the substitution of outliers (e.g., by interpolation) and it is limited to their identification and filtering. The MC method, instead, handles the presence of outliers by replacing all data points with the median over a sliding window, without requiring any additional substitution phase for the filtered values.

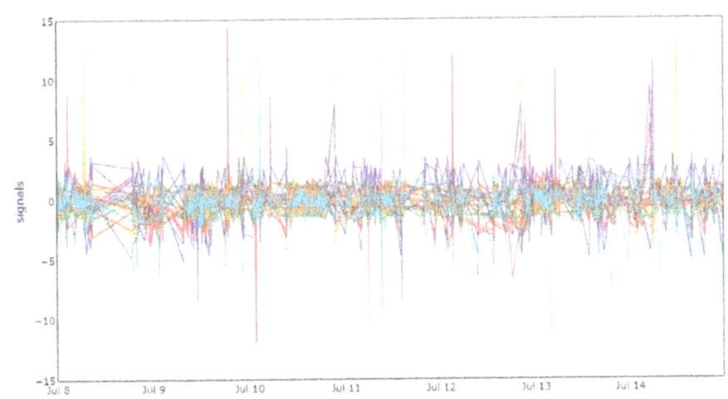

(a)

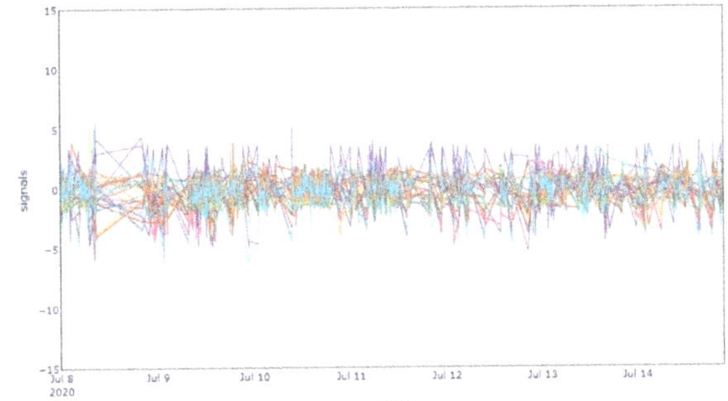

(b)

Figure 3. *Cont.*

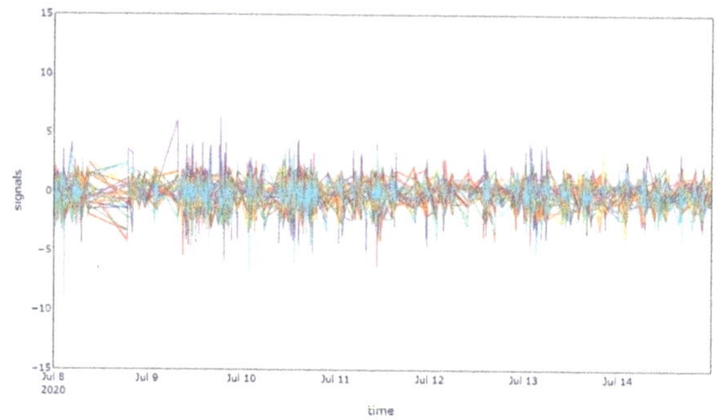

(c)

Figure 3. (**a**) shows the sensor signals without the removal of outliers, while (**b**,**c**) represent the signals over time after the IQR and MC methods were applied respectively for the outlier removal phase.

5.3. *Anomaly Detection Results*

The PCA algorithm was run onto the two scenarios, namely considering an IQR and MC pre-processing phase, by setting as stopping criterion a minimum of 90% of explained variance. In the case of IQR, the PCs computed by the PCA algorithm were 16, while using the MC method led to 19 new dimensions.

5.3.1. Testing in Normal Operating Conditions

The robustness of the anomaly detection model has been tested on normal behaviour conditions (Figure 4) in a period going from 1 November 2020 to 14 November 2020, namely on the data collected during the week following the training period. Figure 4a plots the reconstruction errors of the model without pre-processing, while Figure 4b,c display, respectively, the residuals considering IQR and MC for pre-processing. In all scenarios the reconstruction errors are never persistently exceeding a threshold of 20 units, which was taken as a reference considering the errors computed on the training data. In fact, the operating conditions are very similar to the normal behaviour period on which the model was trained and demonstrate that there are no substantial differences between the two pre-processing methods.

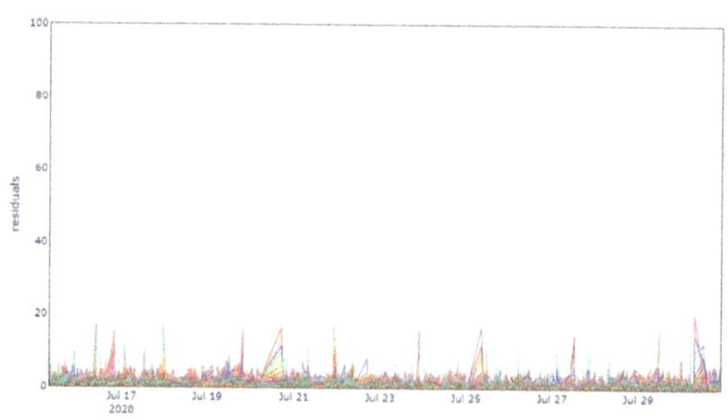

(a)

Figure 4. *Cont.*

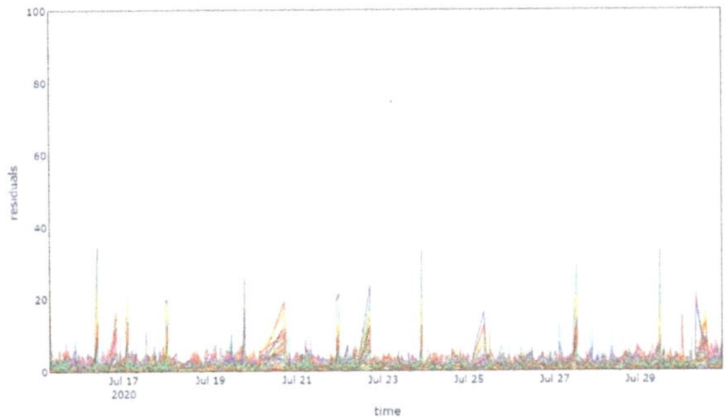

(**b**)

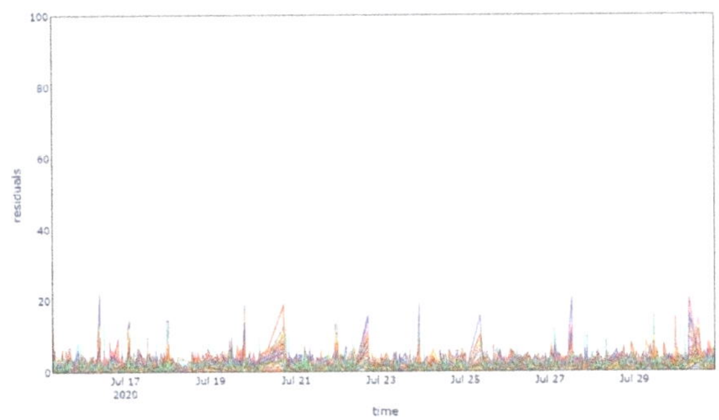

(**c**)

Figure 4. (**a**) shows the KPIs associated to all sensors without the removal of outliers in a normal operating condition period, while (**b**,**c**) represent the KPIs (5) over time after the IQR and MC methods were applied respectively for the outlier removal phase.

5.3.2. Testing in Anomalous Conditions

As a final step, we evaluated the model in a critical period going from 20 June 2020 to 8 July 2020, during which a technical problem led to equipment failure, as reported by the operators. Figure 5 shows the residuals of the model considering no outlier removal phase (Figure 5a), the IQR (Figure 5b) and the MC (Figure 5c) pre-processing methods. In proximity of the failure event (on 4 July 2020), the anomaly is detected by the residuals drastically exceeding the training reference threshold of 20 units, anticipated by another reconstruction error spike on 3 July 2020. Without outlier removal the residuals never persistently exceed the threshold in the period preceding the fault. When considering the IQR and MC methods, instead, residuals above 20 units are already frequent starting from 20 June 2020, anticipating the fault by more or less two weeks. As for the normal behaviour scenario, also in an anomalous period the two pre-processing methods demonstrated their similarity by achieving comparable results.

It is important to notice that it is possible to isolate the sensors of the stations that are mostly related to the anomalous conditions by inspecting the residual of each input feature of the model. In this anomalous period, stations 12 and 13 were isolated by looking at the large residuals two weeks before the fault. During the fault itself, instead, stations 19 and 20 were involved according to the model reconstruction errors.

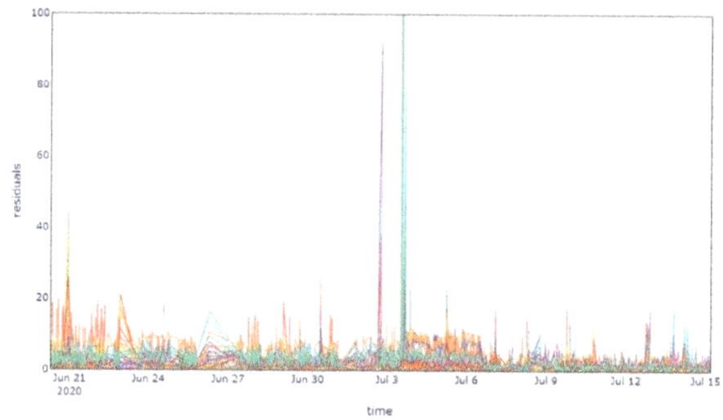

(**a**)

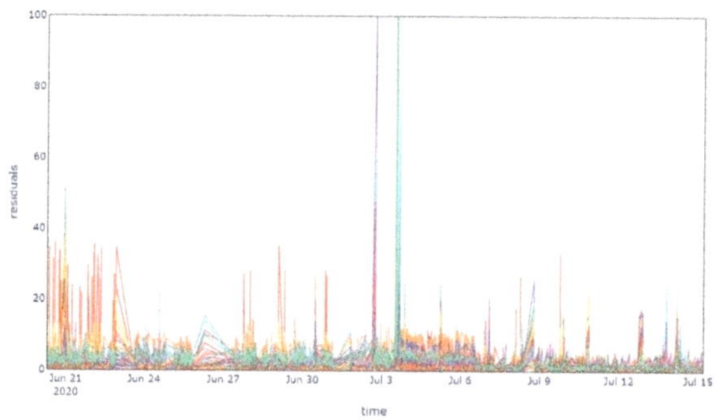

(**b**)

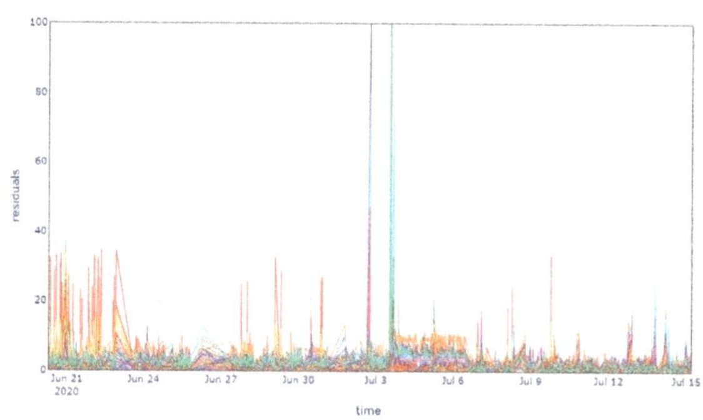

(**c**)

Figure 5. (**a**) shows the KPIs (5) associated to all sensors without the removal of outliers before and after the break, while (**b**,**c**) represent the KPIs (5) over time after the IQR and MC methods were applied respectively for the outlier removal phase.

6. Discussion

The proposed method for data pre-processing based on MC simulation exhibits the following features:

- preserving temporal locality with respect to the training dataset;
- outlier removal;

- outlier replacement, by contrast with traditional methods which are limited to outlier detection only (for example methods based on z-scores [31] or IQR techniques [32]).

As discussed in Section 3.3 and confirmed in [27], the median was chosen as the most accurate estimator in order to obtain a suitable dataset using Monte Carlo simulation to be provided as input to the PCA-based model. In particular, the median-based MC method proved to be more effective against outlier observations with respect to the mean estimator.

Moreover, we selected the optimal sample size for MC simulation by measuring the percentage of variance of the PCA components trained on the MC pre-processed dataset explained by the PCA components trained on the IQR pre-processed dataset in terms of R^2 due to many considerations in the literature which report pre-processing techniques for similar anomaly detection scenarios based on the IQR method [24]. This analysis led to an optimal value of three samples to be considered for the median computation. In particular, we adopted a sliding window sampling approach in order to preserve the temporal locality of subsequent batches.

From the results is Section 5.3 it is evident that the IQR and MC-based pre-processing methods produce similar results, demonstrating their capability to successfully deal with outliers. Nevertheless, they present substantial differences. In fact, a standard method like IQR is limited to isolating outliers and possibly remove them from the dataset. This is a limitation because filtered observations generate missing values which require a substitution algorithm (e.g., mean imputation [33], KNN [34], linear interpolation [35]). The MC method, instead, intrinsically deals with outlier substitution by computing the median of randomly selected points, thus generating a new estimator dataset with an arbitrary number of samples.

The PCA models for anomaly detection demonstrated their capability to successfully anticipate a fault in the equipment as shown in several other works and practical experiments [6–8]. In particular, two PCA models were trained, respectively, on the IQR and MC pre-processed datasets. Both models highlighted an anomalous condition almost two weeks before the equipment failure by producing KPIs (residuals) above a reference threshold which was used to discriminate between healthy and anomalous states of the equipment as done in [36].

Moreover, it is important to notice that, without any pre-processing, the algorithm is unable to detect the anomalies with such an advance and is limited to spotting only the occurrence of the actual fault, which is also detected by the IQR and MC approaches.

Both models were also tested in standard operating conditions in order to prove their robustness to false alarms. In fact, in normal conditions, the residuals of the models never exceed the reference threshold persistently.

Finally, by inspecting the residual of each input feature of the model, the proposed approach allows to isolate the sensors of the stations that are being subject to anomalous conditions.

The authors have selected a reference period in order to calculate the average downtime for the AWB stage of the production line shown in Figure 1, and then to compute an estimate of the AWB downtime reduction resulting from the adoption of our predictive model.

Considering that only 50% of the predicted machine-down events can be totally avoided—in fact, only in some cases it is possible to take advantage of scheduled preventive maintenances to repair the equipment in advance, the authors measured a reduction in AWB downtime by 0.55%. Assuming to extend the implementation of the predictive model to the entire equipment of the 3SUN production line (as shown in Figure 1), the authors expect an overall downtime reduction between 1% and 2%, which corresponds to an increase in the annual photovoltaic panels production in the order of approximately 1–2 megawatts.

7. Conclusions

In this paper, the authors have presented a use case of robust anomaly detection applied to the scenario of a photovoltaic production factory—namely, Enel Green Power's 3SUN solar cell production plant in Catania, Italy—by considering a Monte Carlo based pre-processing technique.

The proposed pre-processing algorithm demonstrated its ability to handle outliers like other standard methods, with the additional advantage of intrinsically dealing with outlier substitution and taking into account the temporal locality of subsequent samples.

After pre-processing, the authors trained an anomaly detection model based on Principal Component Analysis and defined a key performance indicator for each sensor in the production line based on the model errors. In this way, by running the algorithm on unseen data streams, it was possible to isolate anomalous conditions by monitoring the key performance indicators and virtually trigger an alarm when exceeding a reference threshold.

The proposed approach was tested on both standard operating conditions and an anomalous scenario. In particular, it successfully anticipated a fault in the equipment with an advance of almost two weeks, but also demonstrated its robustness to false alarms during normal conditions.

Finally, given the data-driven nature of the approach and its robustness to outliers and irregular sampling frequencies, this approach could be applied to multiple lines in the production plant. In fact, as future work, the authors look forward to testing the proposed method on multiple pieces of equipment in order to further validate its scalability.

Author Contributions: Conceptualization, E.A. and D.A.I.; Methodology, A.C., R.F., E.S.M., L.R.C. and M.V.; Writing—original draft, E.S.M., L.R.C. and N.A.S.; Writing—review & editing, E.S.M. and L.R.C. All authors have read and agreed to the published version of the manuscript.

Funding: This work was supported by ELIS Innovation Hub within a Joint Research Project with Enel Green Power S.p.A.

Conflicts of Interest: The authors declare no conflict of interest.

Abbreviations

The following abbreviations are used in this manuscript:

AWB	Automatic Wet Bench
CVD	Chemical Vapor Deposition
IQR	Interquartile Range
KPI	key performance indicator
MC	Monte Carlo
MEC	Manufacturing Execution System
PC	Principle Components
PCA	Principle Component Analysis
PeCVD	Plasma Enhanced Chemical Vapor Deposition
PVD	Physical Vapour Deposition
ROBPCA	Robust PCA
SVD	Singular Value Decomposition
WIS	Singular Value Decomposition

References

1. Wang, H.; Ni, G.; Chen, J.; Qu, J. Research on rolling bearing state health monitoring and life prediction based on PCA and Internet of things with multi-sensor. *Measurement* **2020**, *157*, 107657 [CrossRef]
2. Kimera, D.; Nangolo, F.N. Improving ship yard ballast pumps' operations: A PCA approach to predictive maintenance. *Marit. Transp. Res.* **2020**, *1*, 100003. [CrossRef]
3. Chen, X.; Zhang, B.; Wang, T.; Bonni, A.; Zhao, G. Robust principal component analysis for accurate outlier sample detection in RNA-Seq data. *BMC Bioinform.* **2020**, *21*, 1–20. [CrossRef]

4. Kim, S.; Hur, J. A Probabilistic Modeling Based on Monte Carlo Simulation of Wind Powered EV Charging Stations for Steady-States Security Analysis. *Energies* **2020**, *13*, 5260. [CrossRef]
5. Yoo, J.E.; Rho, M. Large-Scale Survey Data Analysis with Penalized Regression: A Monte Carlo Simulation on Missing Categorical Predictors. *Multivar. Behav. Res.* **2021**, 1–29. [CrossRef]
6. De Benedetti, M.; Leonardi, F.; Messina, F.; Santoro, C.; Vasilakos, A. Anomaly detection and predictive maintenance for photovoltaic systems. *Neurocomputing* **2018**, *310*, 59–68. [CrossRef]
7. Bashir, N.; Chen, D.; Irwin, D.; Shenoy, P. Solar-TK: A Data-driven Toolkit for Solar PV Performance Modeling and Forecasting. In Proceedings of the 2019 IEEE 16th International Conference on Mobile Ad Hoc and Sensor Systems (MASS), Monterey, CA, USA, 4–7 November 2019; pp. 456–466.
8. Bonacina, F.; Corsini, A.; Cardillo, L.; Lucchetta, F. Complex Network Analysis of Photovoltaic Plant Operations and Failure Modes. *Energies* **2019**, *12*, 1995. [CrossRef]
9. Zhou, F.; Park, J.H.; Liu, Y. Differential feature based hierarchical PCA fault detection method for dynamic fault. *Neurocomputing* **2016**, *202*, 27–35. [CrossRef]
10. Chen, Z.; Li, X.; Yang, C.; Peng, T.; Yang, C.; Karimi, H.R.; Gui, W. A data-driven ground fault detection and isolation method for main circuit in railway electrical traction system. *ISA Trans.* **2019**, *87*, 264–271. [CrossRef]
11. Harkat, M.F.; Kouadri, A.; Fezai, R.; Mansouri, M.; Nounou, H.; Nounou, M. Machine learning-based reduced kernel PCA model for nonlinear chemical process monitoring. *J. Control. Autom. Electr. Syst.* **2020**, *31*, 1196–1209. [CrossRef]
12. Bencheikh, F.; Harkat, M.F.; Kouadri, A.; Bensmail, A. New reduced kernel PCA for fault detection and diagnosis in cement rotary kiln. *Chemom. Intell. Lab. Syst.* **2020**, *204*, 104091. [CrossRef]
13. Niemeier, W.; Tengen, D. Stochastic Properties of Confidence Ellipsoids after Least Squares Adjustment, Derived from GUM Analysis and Monte Carlo Simulations. *Mathematics* **2020**, *8*, 1318. [CrossRef]
14. Zelditch, M.L.; Swiderski, D.L.; Sheets, H.D. *Geometric Morphometrics for Biologists*, 2nd ed.; Academic Press: San Diego, CA, USA, 2012; pp. 189–224.
15. Fang, X.; Zeng, W.; Zhou, Y.; Wang, B. On the total least median of squares adjustment for the pattern recognition in point clouds. *Measurement* **2020**, *160*, 107794. [CrossRef]
16. Parashar, S.; Swarnkar, A.; Niazi, K.R.; Gupta, N. Optimal integration of electric vehicles and energy management of grid connected microgrid. In Proceedings of the 2017 IEEE Transportation Electrification Conference (ITEC-India), Pune, India, 13–15 December 2017; pp. 1–5.
17. Fernando, T.M.L.; Marcelo, L.G.E.; David, V.M.H. Substation Distribution Reliability Assessment using Network Reduction and Montecarlo Method, a comparison. In Proceedings of the 2019 FISE-IEEE/CIGRE Conference—Living the Energy Transition (FISE/CIGRE), Medellin, Colombia, 4–6 December 2019; pp. 1–7.
18. Leger, G. Combining adaptive alternate test and multi-site. In Proceedings of the 2015 Design, Automation & Test in Europe Conference & Exhibition (DATE), Grenoble, France, 9–13 March 2015; pp. 1389–1394.
19. Kong, X.; Tong, X. Monte-Carlo Tree Search for Graph Coalition Structure Generation. In Proceedings of the 2020 13th International Congress on Image and Signal Processing, BioMedical Engineering and Informatics (CISP-BMEI), Chengdu, China, 17–19 October 2020; pp. 1058–1063.
20. Saracco, P.; Batic, M.; Hoff, G.; Pia, M.G. Uncertainty Quantification (UQ) in generic MonteCarlo simulations. In Proceedings of the 2012 IEEE Nuclear Science Symposium and Medical Imaging Conference Record (NSS/MIC), Anaheim, CA, USA, 27 October–3 November 2012; pp. 651–656.
21. Garcia-Alfonso, H.; Cordova-Esparza, D.M. Comparison of uncertainty analysis of the Montecarlo and Latin Hypercube algorithms in a camera calibration model. In Proceedings of the 2018 IEEE 2nd Colombian Conference on Robotics and Automation (CCRA), Barranquilla, Colombia, 1–3 November 2018; pp. 1–5.
22. Chen, Y.; Sun, R.; Borken-Kleefeld, J. On-Road NOx and Smoke Emissions of Diesel Light Commercial Vehicles–Combining Remote Sensing Measurements from across Europe. *Environ. Sci. Technol.* **2020**, *54*, 11744–11752. [CrossRef]
23. Heijungs, R. On the number of Monte Carlo runs in comparative probabilistic LCA. *Int. J. Life Cycle Assess.* **2020**, *25*, 394–402. [CrossRef]
24. Nair, P.; Kashyap, I. Hybrid Pre-processing Technique for Handling Imbalanced Data and Detecting Outliers for KNN Classifier. In Proceedings of the 2019 International Conference on Machine Learning, Big Data, Cloud and Parallel Computing (COMITCon), Faridabad, India, 14–16 February 2019.
25. Zwillinger, D.; Kokoska, S. *CRC Standard Probability and Statistics Tables and Formulae*; CRC Press: Boca Raton, FL, USA, 2000; p. 18, ISBN 1-58488-059-7.
26. Ji, H.; Li, Y. Monte Carlo methods and their applications in Big Data analysis. In *Mathematical Problems in Data Science*; Springer: Cham, Switzerland, 2015; pp. 125–139.
27. von Brömssen, C.; Röös, E. Why statistical testing and confidence intervals should not be used in comparative life cycle assessments based on Monte Carlo simulations. *Int. J. Life Cycle Assess.* **2020**, *25*, 2101–2105. [CrossRef]
28. Dongxiao, F.; Chuan, P.; Guoxing, Z.; Rui, Z.; Fang, L.; Zhenhua, D.; Hongliang, M. Research on Simulation Method for Reliability Prediction of Pyrotechnical System. In Proceedings of the 2020 11th International Conference on Prognostics and System Health Management (PHM-2020 Jinan), Jinan, China, 23–25 October 2020; pp. 556–559.

29. Meuleman, B.; Billiet, J. A Monte Carlo sample size study: How many countries are needed for accurate multilevel SEM? *Surv. Res. Methods* **2009**, *3*, 45–58.
30. Fadhel, S.; Delpha, C.; Diallo, D.; Bahri, I.; Migan, A.; Trabelsi, M.; Mimouni, M.F. PV shading fault detection and classification based on IV curve using principal component analysis: Application to isolated PV system. *Sol. Energy* **2019**, *179*, 1–10. [CrossRef]
31. Aggarwal, V.; Gupta, V.; Singh, P.; Sharma, K.; Sharma, N. Detection of spatial outlier by using improved Z-score test. In Proceedings of the 2019 3rd International Conference on Trends in Electronics and Informatics (ICOEI), Tirunelveli, India, 23–25 April 2019; pp. 788–790.
32. Vinutha, H.P.; Poornima, B.; Sagar, B.M. Detection of outliers using interquartile range technique from intrusion dataset. In *Information and Decision Sciences*; Springer: Singapore, 2018; pp. 511–518.
33. Donders, A.R.; Van Der Heijden, G.J.; Stijnen, T.; Moons, K.G. A gentle introduction to imputation of missing values. *J. Clin. Epidemiol.* **2006**, *59*, 1087–1091. [CrossRef]
34. Malarvizhi, M.R.; Thanamani, A.S. K-nearest neighbor in missing data imputation. *Int. J. Eng. Res. Dev.* **2012**, *5*, 5–7.
35. Noor, N.M.; Al Bakri, Abdullah, M.M.; Yahaya, A.S.; Ramli, N.A. Comparison of linear interpolation method and mean method to replace the missing values in environmental data set. In *Materials Science Forum*; Trans Tech Publications Ltd.: Zurich, Switzerland, 2015; Volume 803, pp. 278–281.
36. Parzinger, M.; Hanfstaengl, L.; Sigg, F.; Spindler, U.; Wellisch, U.; Wirnsberger, M. Residual Analysis of Predictive Modelling Data for Automated Fault Detection in Building's Heating, Ventilation and Air Conditioning Systems. *Sustainability* **2020**, *12*, 6758. [CrossRef]

energies

MDPI

Article

Impact of Energy Monitoring and Management Systems on the Implementation and Planning of Energy Performance Improved Actions: An Empirical Analysis Based on Energy Audits in Italy

Carlos Herce, Enrico Biele, Chiara Martini, Marcello Salvio and Claudia Toro *

Energy Efficiency in the Economic Sectors Laboratory (DUEE-SPS-ESE), Energy Efficiency Unit Department, Italian National Agency for New Technologies, Energy and Sustainable Economic Development (ENEA), C.R. Casaccia, Via Anguillarese 301, 00123 Rome, Italy; carlos.herce@enea.it (C.H.); enrico.biele@enea.it (E.B.); chiara.martini@enea.it (C.M.); marcello.salvio@enea.it (M.S.)
* Correspondence: claudia.toro@enea.it

Abstract: The implementation of monitoring tools and energy management systems (EnMSs) supports companies in their long-term energy efficiency strategies, and they are essential to analyse the effectiveness of energy performance improvement actions (EPIAs). The first fundamental step towards increasing energy efficiency is the development of energy audits (EAs). EAs provide comprehensive information about the energy usage in a specific facility, identifying and quantifying cost-effective EPIAs. The crucial role of these tools in clean energy transition is remarked by the European Energy Efficiency Directive (EED), which promotes the implementation of EAs and EnMS programmes. The purpose of this work is to better understand the link between EnMSs (specifically ISO 50001) and EAs in the EED Article 8 implementation in two industrial and two tertiary sectors in Italy. Moreover, the impact of company size, energy monitoring systems, and EnMSs on planned and/or implemented EPIAs is analysed. Our findings show that, albeit the complexity of the variables involved in energy efficiency gap, the "energy savings/company" and "EPIA/site" ratios are higher in enterprises with an EnMS and monitoring system. Thus, a correct energy audit must always be accompanied by a specific monitoring plan if it is to be effective and useful to the company decision maker.

Keywords: energy audits (EAs); energy management systems; energy performance improved actions (EPIAs); energy efficiency; manufacturing industry; tertiary sector

Citation: Herce, C.; Biele, E.; Martini, C.; Salvio, M.; Toro, C. Impact of Energy Monitoring and Management Systems on the Implementation and Planning of Energy Performance Improved Actions: An Empirical Analysis Based on Energy Audits in Italy. *Energies* **2021**, *14*, 4723. https://doi.org/10.3390/en14164723

Academic Editor: Mohamed Benbouzid

Received: 24 June 2021
Accepted: 29 July 2021
Published: 4 August 2021

1. Introduction

The Energy Efficiency Directive 2012/27/EU (EED) [1] (and the 2018/2002 directive amendment [2]) is one of the pillars of European legislation on energy. It is the regulatory framework to help the EU reach its energy efficiency targets (an increase of 20% by 2020 and ≥32.5% by 2030, relative to 1990 levels), and it is composed of a balanced collection of binding measures and recommendations. EED Article 8 is fully devoted to the promotion of cost-effective high-quality energy audits and the implementation of energy management systems. These are two crucial tools to evaluate the existing energy consumption, to identify all the opportunities to save energy, and to implement a continuous improvement on energy efficiency in the industry and in enterprises. The development of energy audits is the first step towards overcoming the main barriers to implementing energy efficiency actions [3].

The Italian government transposed the EED in 2014 and 2020 (by enacting Legislative Decrees 102/2014 and 73/2020, respectively), extending the obligation (from 5 December 2015) of carrying out mandatory energy audits at least every 4 years not only in large companies but also in a specific group of energy-intensive enterprises (mostly SMEs).

The Italian definition of large enterprise is a business organization that has more than 250 employees and has either an annual turnover exceeding EUR 50 million and/or an annual balance sheet total exceeding EUR 43 million. The size of the company is calculated, taking into consideration the activities of all the sites of the core company and partner/linked enterprises within the Italian territory. Other companies obliged to carry out energy audits are the energy-intensive enterprises (in Italian, "*Energivori*") subjected to tax relief in part of the purchased electricity and registered in the list of the Environmental Energy Services Fund (CSEA, a government agency on electricity). These companies present large energy consumptions (in absolute terms and relative to their internal costs), and they must be part of some specific industrial sectors (mainly Annexes 3 and 5 of EU Guidelines 2014/C 200/01 [4]). Enterprises that do not comply with the mandatory energy audits are subject to administrative and monetary penalties.

According to Article 8 of Italian Legislative Decree 102/2014, ENEA manages the Italian energy audit programme, including data gathering and subsequent sectorial analysis [5]. From the beginning of the programme (2015), ENEA has managed more than 25,000 EAs. The present work is focused on data gathered in relation to the first year of the second compliance cycle (2018). On 31 December 2019, 6434 enterprises were submitted to 11,172 energy audits of their production sites. Most of the EAs were related to the manufacturing sector (53%) with particular importance to the plastic (8%), iron and steel (9%), food (6%), textile (3%), and paper (2%) industries. More than 14% of the EAs were from the trade sector. In the second cycle, compliance cycle was observed in that more than 70% of the audits collected by ENEA presented data of energy consumption from specific monitoring systems.

The purpose of this research analysis is to evaluate the impact of energy monitoring systems and energy management systems on a company's propensity to plan and/or implement energy efficiency measures. In order to achieve this objective, energy audits in four different sectors in Italy were analysed to better understand the possible existing link between energy management and monitoring systems and mandatory energy audits in the EED Article 8 implementation. Moreover, it is important to note that Italian legislation includes the development of energy monitoring systems or plans and the implementation of energy performance improvement actions (EPIAs) according to the energy audits submitted to the national database. The identified sectors for analysis are two manufacturing industries and two branches of the tertiary sector, in order to provide us with insights from two different perspectives.

Previous related research focused on the problem of potential savings due to the implementation of EnMSs, but they were not linked to EAs. Commonly, the data used in research are based on voluntary surveys. Hence, the main novelty of our work is the high quality and amount of data analysed: more than 1600 EAs from more than 700 companies, including more than 1000 implemented and 4000 planned EPIAs. Moreover, specific data and analysis of small and medium enterprises are scarce. Finally, to the best of our knowledge, there is no empirical evidence of the impact of monitoring systems on the effective implementation of EPIAs. Hence, this work is an empirical demonstration of the impact of the promotion of EAs and EnMSs as a crucial part of energy efficiency policies.

2. Context

Energy audits (in Article 2 of EED, energy audit is defined as "*a systematic procedure with the purpose of obtaining adequate knowledge of the existing energy consumption profile of a building or group of buildings, an industrial or commercial operation or installation or a private or public service, identifying and quantifying cost-effective energy savings opportunities, and reporting the findings*" [1]) are the first step towards increasing energy efficiency within a firm and implementing an EnMS, such as ISO 50001. Energy-saving strategies cannot be implemented without having detailed and regular energy consumption data of a facility. Starting from the energy audit programmes, many studies, as analysed by Schleich et al. [6], refer to the residential sector, and only a few refer to enterprises. A recent study carried out by the EIB remarks that, for SMEs, the probability of investing in energy efficiency actions is 1.5 times greater for enterprises with an energy audit compared with those without one [7].

An energy management system (in Article 2 of EED, an energy management system is defined as *"a set of interrelated or interacting elements of a plan which sets an energy efficiency objective and a strategy to achieve that objective"* [1]) helps an enterprise build a structured process for monitoring its energy consumption and improve its internal efficiency through EPIAs. The adoption of an energy management system can lead to a reduction in energy consumption [8], gains in industrial productivity, and improvements in global enterprise performance, in addition to several other cobenefits positively affecting the overall company competitiveness [9,10]. Energy management is intrinsically connected to economic and environmental issues, but it could also lay the foundations of a comprehensive management system, which includes not only energy efficiency but also quality and environmental management, occupational safety and health, and other risk components [11,12]. However, instead of the multiple benefits of the adoption of energy efficiency strategies, there are multiple barriers involved in the energy efficiency gap that limit the implementation of EnMSs or EPIAs [13–15], or the adoption of the EnMSs in companies with implemented environmental management systems (EMSs) [16].

Regarding ISO 50001, Fiedler and Mircea, in their analysis [17], mentioned that cost saving is probably the key driver for most organizations adopting EnMSs and that certification may be useful for a company strategy and image. Fuchs et al. [18] conducted an analysis of the identification of drivers, benefits, and challenges of ISO 50001 through case study contents. The result was that the biggest motivations for ISO 50001 certification are: existing values and goals, cost savings, environmental sustainability concerns, government incentives or regulations, and gaining competitive advantage via visibility. These results are aligned with those of other works [19] and the 2015 AFNOR European survey "International survey energy management practices in ISO 50001-certified organizations". Another interesting analysis of the effectiveness of the ISO 50001 implementation shows a detailed framework analysis of gaps and potential improvements in order to boost the deployment of EnMSs [20].

McKane et al. [21], through the ISO 50001 Impacts Methodology, speculate both energy and nonenergy benefits. According to their analysis, considering a scenario by 2030 with 50% of the global enterprises under ISO 50001 management, the cumulative savings could reach nearly USD 700 billion, 105 EJ of primary energy, and 6.500 million tons of avoided CO_2 equivalent emissions.

An analysis based on a German energy audit national database [22] indicates that energy-intensive enterprises tend to prioritize energy efficiency projects compared with less energy-intensive ones. In terms of company size, larger companies are inclined to implement more energy efficiency measures than smaller ones. Similar empirical results were observed in Sweden [23] and Latvia [24]. Fleiter et al. [25] conclude that their result identifies high initial investment costs as the main barrier to the adoption of energy efficiency measures. Therefore, to accelerate the adoption of those measures, energy audit programmes should be supported by financing schemes. Moreover, they found evidence that higher satisfaction through energy audits increases the predisposition to implement suggested energy efficiency measures.

Italy is the third country in the world with the highest number of certifications in 2016 [26]. The main motivations for companies to implement an EnMS are, first, to increase competitiveness and, second, to reduce energy and costs [27]. Based on ENEA's analysis of the first obligation period data (started in December 2015) in the plastic sector, a relevant share of proposed interventions referred to ISO 50001 and monitoring systems (15% of 1051). A possible explanation for this relevant share is that the claimed payback time is lower than 2 years. This interesting payback period is confirmed in the energy audits presented for the ceramic sector, where on the same energy audit campaigns show an average payback lower than 1.5 years. A further confirmation of low payback periods for ISO 50001 is found in the FIRE-CEI-CTI survey carried out in 2016, where 70% of the participants declared a payback time lower than 3 years for ISO 50001 EnMSs and a return of investments in line with their expectations in 85% of the cases [28]. A report carried out by Accredia showed that the reason for certification is business strategy for 74% of the interviews, while only 10% is mainly for cost reduction [29].

3. Materials and Methods

From the preliminary analysis of the EED Article 8 implementation in the second obligation period (started in December 2019), the overall percentage of ISO 50001 sites amounted to 9% (about 1050 sites) of the total number of sites accomplishing their Article 8 obligation, while the overall percentage of sites with an installed energy monitoring system amounted to 70%. The number of certified ISO 50001 companies that presented EAs was 358, with 27% of them being SMEs [30].

The ISO 50001 EnMS standard includes the implementation of a monitoring system. However, it is important to note that the number of monitored sites is sensibly higher than the number of sites with certified EnMSs. Hence, the impact of both variables was analysed separately: the installation of an energy monitoring system only and the implementation of an EnMS (in particular, ISO 50001).

Implemented and identified EPIAs were analysed under companies that were ISO 50001 certified and had a monitoring system and were SMEs. It is important to note that the Italian manufacturing sector is dominated by SMEs [31]; therefore, class size was included in the analysis.

Additionally, a focus on *general EPIAs* was carried out. *General EPIAs* include capacitation of energy management, implementation of energy management systems, monitoring of energy consumption, extension and improvement of current management and/or monitoring systems, and other actions not strictly related to the production process or technical EE measures. The impact of the presence of an energy monitoring system on planned and/or implemented energy efficiency measures and on the corresponding savings was analysed.

A descriptive statistical analysis was developed based on both qualitative (number and type of EPIAs) and quantitative (energy impact of EPIAs) information. The database informing such analysis consists of all the implemented and identified EPIAs reported in the EAs uploaded until December 2019 on the website managed by ENEA (https://audit102.enea.it/) (reference database update 17 May 2020). It is worth specifying that each EA should include information on implemented and identified EPIAs, but this is not always the case. Moreover, information characterizing EPIAs could be incomplete, for example, regarding investment costs and achieved or expected energy savings.

Seven 4-digit NACEs were examined, covering 4 different sectors:

- Banks: K64.19—other monetary intermediation;
- Retail: G47.11—retail sale in nonspecialised stores (hyper- and supermarkets);
- Ceramics: C23.31—manufacture of ceramic tiles and flags and C23.32—manufacture of bricks, tiles, and construction products in baked clay;
- Plastics: C22.22—manufacture of plastic packing goods and C22.29—manufacture of other plastic products.

The analysed sectors were chosen based on their relevance in terms of both energy consumption and ISO 50001-certified companies. The energy audits in the sample reflected the number of obliged parties according to Article 8 of EED, which clearly differs by NACE sector. The two manufacturing sectors were dominated by SMEs, whereas the tertiary sectors were dominated by large companies. Consistently, in the tertiary sector a higher number of sites belonging to the same company were observed than in the industrial sector. An overview of the companies and energy audits analysed is presented in Figure 1.

Different NACE sectors have different patterns when looking at the share of total final energy consumption of companies that have an ISO 50001 certification and a monitoring system and that are defined as SMEs (Figure 2, which shows the share of SMEs, companies with a monitoring system and ISO 50001 certification in the final energy consumption of audited companies in 2019). In absolute value, the final energy consumption was relatively lower in the tertiary sector than in the manufacturing sector. In the tertiary sector, retail had a higher final energy consumption than banks, consuming 171 and 57 ktoe, respectively. In manufacturing, the total final energy consumption of the two NACE codes examined in the ceramic sector was double the consumption of the two NACE codes in the plastic sector (1100 vs. 577 ktoe).

The analysis of both implemented EPIAs (EPIAs, starting from here) and planned EPIAs covers, in addition to general EPIAs, also measures in technical intervention categories, such as pressure systems, heat recovery systems and thermal plants, inverters and other electrical machines and installations, transport, heating and cooling, and building envelope [32]. Measures in the categories of cogeneration and trigeneration and production from renewable sources were excluded from the analysis since they are associated with savings of primary energy [33].

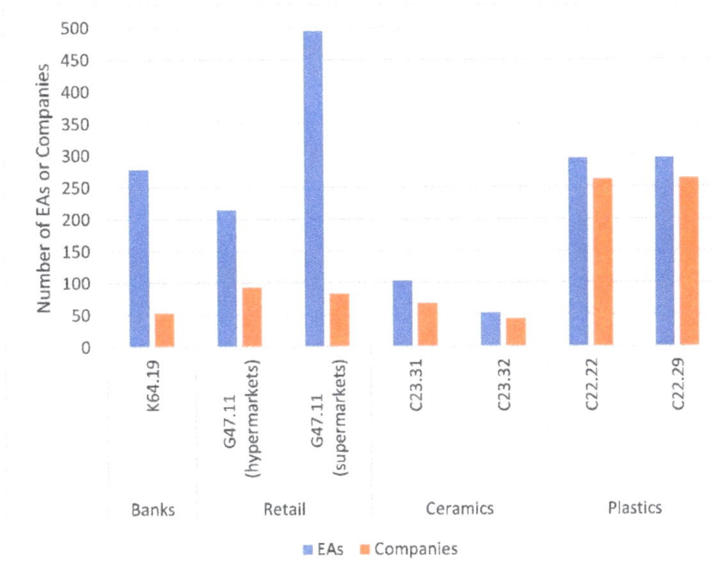

Figure 1. Number of EAs and companies by NACE code.

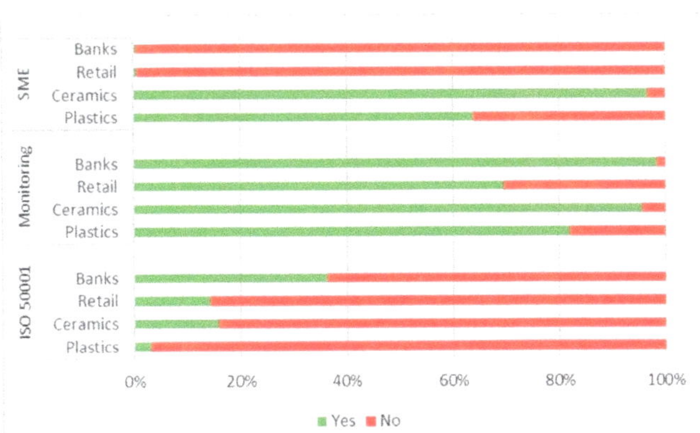

Figure 2. Share of the total final energy consumption by category.

In different NACE sectors, the number of EPIAs in enterprises that have an ISO 50001 certification and a monitoring system and are defined as SMEs is shown in Figure 3. The highest number of EPIAs was observed in plastics, with 558 implemented energy efficiency measures, followed by ceramics (218) and retail, with slightly lower numbers of measures (193). Banks had the lowest number of EPIAs (83). Clearly, this pattern is influenced by the number of EAs by sector; nevertheless, the number of EPIAs per site or per company could show different patterns by sector, as will be further investigated based on the indicators presented in next section. Regarding the total number of EPIAs, the share of measures reporting information on achieved energy savings was 53%, and this share varied by NACE sector, with retail having the highest share (85%). Figure 3 also shows the number of sites and companies that have an ISO 50001 certification and a monitoring system and are defined as SMEs: as anticipated, SMEs were absent in retail and very few in banks, so they were excluded from the analysis.

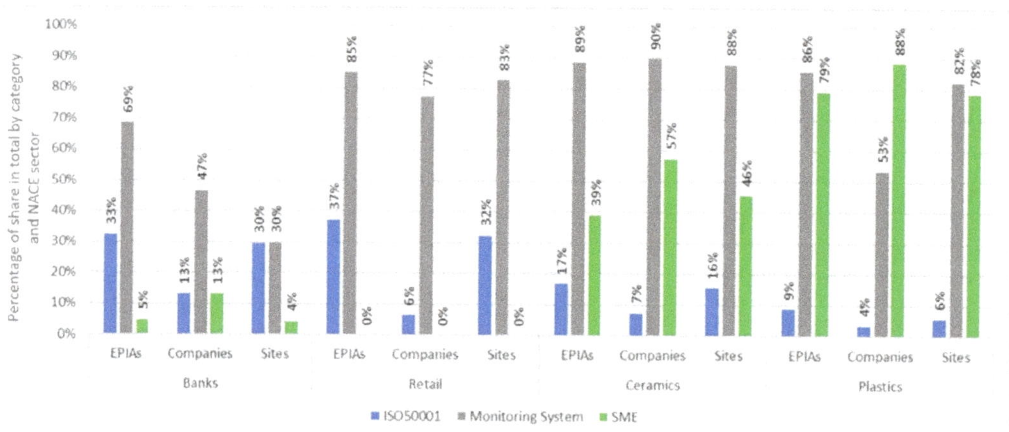

Figure 3. Share in total EPIAs by category.

In the following section, several indicators will be proposed, computing them also for general EPIAs (when available information allows):

- Number of EPIAs per site: it refers to all interventions, as well as those with no saving or investment information available.
- Energy saving per site or per company: it refers to final energy saving, and it is computed excluding sites without saving information.
- Saving: it is computed as the share of saving in total energy consumption of the relevant NACE code. Since the indicator includes only the available information on EPIA reporting savings, it represents a lower threshold for both achieved savings (EPIAs) and potential savings (identified EPIAs). In the second case, the potential nature of savings should be highlighted; namely, they are not likely to be achieved in full since companies would implement only part of the identified EPIAs and in different periods. These potential savings are not presented in this work, but they are employed in the calculation of the average cost effectiveness of the identified EPIAs.
- Investment per site: it is computed by excluding sites without investment information.
- Average cost effectiveness: it is computed as the average of the ratio between investment and saving calculated for each EPIA and identified EPIA, and it refers only to EPIAs including both figures. Such indicator is aimed at representing the cost of saving a toe of final energy and then the effectiveness of different NACE sectors in investing in energy efficiency.

- PBT: it represents simple payback time computed as the ratio between investment cost and energy saving expressed in economic terms. Such information is available only for identified EPIAs.

Payback time and cost effectiveness information does not include information on the effect of Italian incentive schemes on energy efficiency, such as tax deduction scheme for energy renovation, white certificates or regional funds, and tax relief for energy-intensive enterprises. Such incentive mechanisms are likely to have an impact on investment costs, each one in a different way, and then on both examined indicators. Access to each incentive scheme is likely to differ greatly by NACE sector due to different factors represented, for example, by the profile of energy consumption and the company dimension. Banks represent the NACE sector where heating and cooling and building envelope are the prevailing areas of intervention, and therefore, access to the tax deduction scheme is likely to be most relevant. This would pave the way to several insights in terms of investing behaviour and access to existing incentive mechanisms, but these are outside the scope of the present work.

The energy consumption and savings, the quality of data extracted from the energy audits, and the main economic indicators from implemented and planned EPIAs and *general* EPIAs are statistically analysed in Appendix A. Due to the variability of the terms of technology and the size of the EPIAs, the mean values of economic indicators are presented, but they are analysed qualitatively.

4. Results and Discussion

4.1. Ceramics and Plastics

The two manufacturing sectors evaluated in this study (plastic and ceramic) present some important insights in terms of EPIA distribution among the different categories analysed (ISO 50001-certified sites, sites with energy monitoring systems, and size class). As shown in Figures 1 and 2, the EA sample analysed from the plastic sector is dominated by small- and medium-sized enterprises in terms of both share of total final energy consumption and share of total EPIAs. In the ceramic sector, on the other hand, similar numbers of large and small enterprises operate, but the energy consumption share of large companies for the presented EAs is about 80%.

Around 40% of plastic manufacturing sites reported the implementation of any kind of EPIAs in the last 4 years, while this percentage reached 57% for the ceramic manufacturing sites. Thus, the implementation potential of EPIAs was still high in both sectors. The average number of EAs for plastic companies was 1.1 EAs, while it was 1.4 for ceramic companies. However, it is important to note that this number increased to 1.3 for plastic companies and 2.8 for ceramic companies if ISO 50001 certified.

Table 1 presents the impact of *general EPIAs* and the investment in plastic and ceramic manufacturing sites. Plastic and ceramic showed a similar distribution of EPIAs per site (2.35 and 2.42) and a ratio for "*general/total*" EPIAs (15% and 13%). In both cases, the ISO 50001-certified and monitored sites presented a higher degree of implementation of EPIAs per site compared with the sites without EnMSs or monitoring systems.

The number of implemented general EPIAs was very low for both sectors, and for ceramics, it was not possible to evaluate the related cost effectiveness for lack of information.

In Figure 4, energy savings per site, EPIAs, and companies in the plastic sector are presented. It is clear that the global energy savings (% compared with the total sector consumption in EAs) were higher in companies with ISO 50001 certification and monitoring systems compared with companies without these systems. Therefore, the use of EnMSs at the corporate level seemed to effectively increase energy savings. This effect was not observed if savings were evaluated at the site or EPIA level for ISO 50001-certified companies.

Table 1. Plastic and ceramic sector implemented EPIAs.

	IMPLEMENTED EPIAs	General EPIAs (%)	EPIAs per Site (#)	General EPIAs per Site (#)	General EPIA Savings (toe/site)	General EPIA Cost Effectiveness (EUR/toe)	Investment per Site (EUR)
22—Plastics	ISO 50001	6%	3.50	0.17	n.a	n.a.	675,910
	Not ISO 50001	16%	2.28	0.30	n.a.	n.a.	355,375
	Monitoring	16%	2.45	0.30	1.50	9956	456,916
	Not Monitoring	8%	1.90	0.18	0.94	n.a.	253,641
	Large Enterprise	12%	2.29	0.17	0.46	n.a.	497,732
	SME	15%	2.37	0.32	2.15	7387	376,688
	Total	**15%**	**2.35**	**0.28**	**1.41**	**7847**	**369,088**
23—Ceramics	ISO 50001	14%	2.64	0.36	n.a.	n.a.	733,731
	Not ISO 50001	13%	2.38	0.32	n.a.	n.a.	399,433
	Monitoring	11%	2.44	0.27	n.a.	n.a.	513,983
	Not Monitoring	32%	2.27	0.73	n.a.	n.a.	126,500
	Large Enterprise	14%	2.71	0.39	n.a.	n.a.	640,374
	SME	12%	2.07	0.24	n.a.	n.a.	221,288
	Total	**13%**	**2.42**	**0.32**	**n.a.**	**n.a.**	**466,292**

Similar trends for ISO 50001 companies were observed in ceramics, as shown in Figure 5. The number of sites without a monitoring system and including savings data was very low, and for this reason, it was not possible to evaluate properly the effect of the monitoring system on savings.

A comparison of cost effectiveness for the different categories analysed is reported in Figure 6. The average cost effectiveness of the implemented EPIAs in the analysed ISO 50001-certified plastic manufacturing site was higher than that of the noncertified sites, implying a worst performance in the former. This was mainly due to the fact that most of the interventions carried out in certified sites related to the replacement of process machinery (press, compressors, etc.) for which the main benefit lies in improving process productivity rather than energy efficiency. On the contrary, the average cost effectiveness for ISO 50001-certified ceramic manufacturing sites was lower than that for noncertified sites, showing a better performance in the former. In these sites, the most common interventions were related to the substitution or revamping of process machineries, installation of more efficient pumps and compressors, reduction of leaks, and energy consumption in intake ducts.

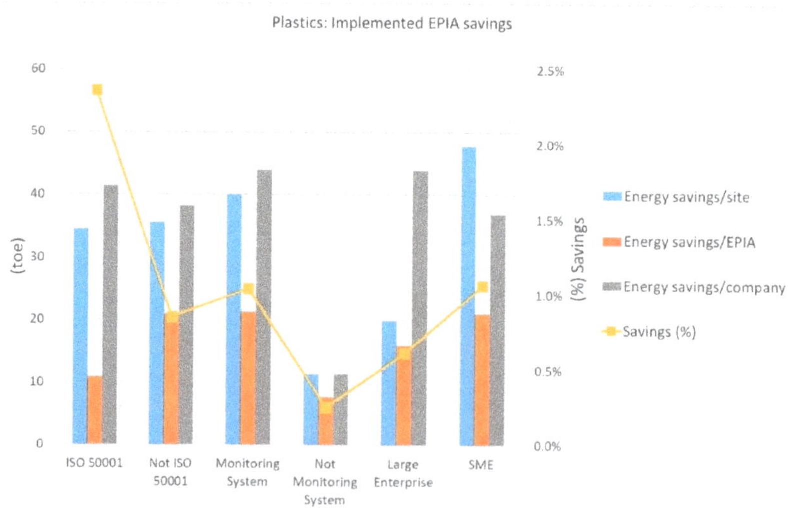

Figure 4. Plastics: implemented EPIA savings.

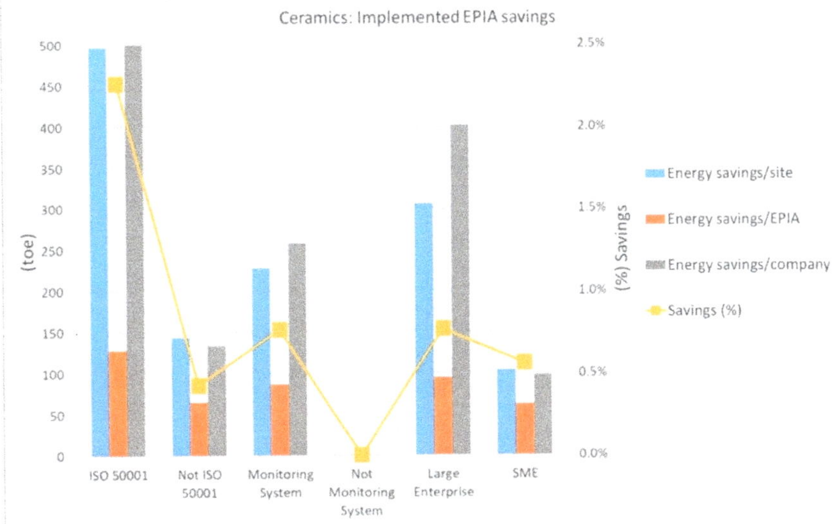

Figure 5. Ceramics: implemented EPIA savings.

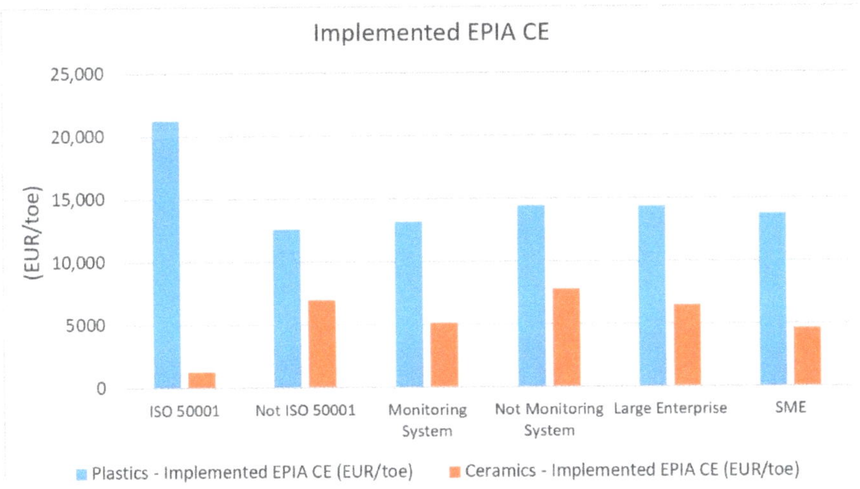

Figure 6. Implemented EPIA cost effectiveness (plastics and ceramics).

The main results of the analysis of the planned EPIAs for the plastic and ceramic sectors are shown in Table 2. A total of 2145 EPIAs were identified (excluding the integration of RES, 283 EPIAs, and CHP, 121 EPIAs), of which 17.7% were *general EPIAs* (mainly implementation of monitoring systems, EnMSs, and capacity training).

In the plastic sector, it seemed that in *general EPIAs* planned under ISO 50001, monitored and large enterprise sites presented lower CE, probably due to better understanding of energy savings and EE investments. On the contrary, CE for global EPIAs was higher for ISO 50001 and monitored sites due to the major share of process-related interventions (substitution of process machineries) planned in these sites. In ceramic sites with global EPIAs planned under ISO 50001, monitored and large enterprise sites presented lower CEs.

About 40% of the interventions in ISO 50001 sites were related to lighting, while general interventions were not considered. In ceramic production sites not subjected to monitoring, interventions on the lighting system prevailed (about 28%), while in the monitored sites, there was a prevalence of interventions concerning lighting (about 21%) and also compressed air (20%) and electric motors (18%).

Table 2. Plastic and ceramic sector planned EPIAs.

	PLANNED EPIAs	Companies (#)	Sites (#)	EPIAs (#)	General EPIAs (#)	EPIA Cost Effectiveness (EUR/toe)	General EPIA Cost Effectiveness (EUR/toe)	EPIA PBT (y)	General EPIA PBT (y)
22—PLASTICS	ISO 50001	17	22	57	7	8294	2804	4.0	1.8
	Not ISO 50001	470	513	1594	301	5929	3476	4.1	3.4
	Monitoring	329	371	1147	198	6438	3146	3.8	3.2
	Not Monitoring	158	164	504	110	5679	3739	4.4	3.7
	Large Enterprise	73	94	252	31	5417	1839	3.8	1.9
	SME	414	441	1399	277	6116	3657	4.1	3.5
	Total	**487**	**535**	**1651**	**308**	**6011**	**3303**	**4.1**	**3.3**
23—CERAMICS	ISO 50001	3	11	14	0	4699	n.a.	9.0	n.a.
	Not ISO 50001	101	131	480	72	5399	3691	3.9	2.2
	Monitoring	84	119	414	62	5245	3963	4.0	2.2
	Not Monitoring	20	23	80	10	6307	1859	4.4	2.0
	Large Enterprise	30	65	247	42	5153	5700	3.8	2.4
	SME	74	77	247	30	5640	2242	4.3	2.0
	Total	**104**	**142**	**494**	**72**	**5374**	**3691**	**4.1**	**2.2**

4.2. Banks and Retail

The two tertiary sectors evaluated (retail and banks) presented some important differences compared with the manufacturing ones. First, these sectors are dominated by large enterprises. The number of SMEs that presented EAs was very low (<5%), and the number of sites with implemented or planned EPIAs was lower than 2%. Hence, the analysis of class size in the tertiary sector was considered negligible. Second, these sectors are characterized by the clustering of multiple sites (supermarkets/hypermarkets and bank offices) with relatively low consumptions (240 and 200 toe/site for retail and banks, respectively). Therefore, the relative weight of *general EPIAs* induced a great impact in the different sites. Third, only a partial analysis of the results could be performed in these sectors due to missing information (specifically the savings of EPIAs in ISO 50001 banks). The impact of missing information on clusters of big companies was difficult to comprehensively analyse.

Only 18% of the sites reported the implementation of any kind of EPIAs in the last 4 years. Thus, the implementation potential of EPIAs was enormous in both sectors. Each retail company presented 4 EAs; meanwhile, each banking company had 5.2 EAs. However, it is important to note that this number increased to 11.6 and 20 EAs/company if there was ISO 50001 certification. Table 3 presents the impact of *general EPIAs* and the investment in tertiary sectors. Retail and banks showed a similar distribution of EPIAs per site (1.5 and 1.8) and a ratio for "*general/total*" EPIAs (37% and 31%). In both cases, the certified and monitored sites presented a higher degree of implementation of EPIAs per site compared with the sites without EnMSs or monitoring systems. However, the detailed distribution by EnMS and monitoring was very different. On the one hand, in the retail sector, the number of EPIAs per site was stable (between 1.3 and 1.7), and the *general EPIAs* were concentrated in the ISO and monitored sites. On the other hand, in banks there was a high variability in the number of EPIAs per site (from 1 to 4.1), and it was not possible to identify specific trends due to general EPIAs.

Table 3. Retail and bank sector implemented EPIAs.

	IMPLEMENTED EPIAs	General EPIAs (%)	EPIAs per Site (#)	General EPIAs per Site (#)	General EPIA Savings (toe/site)	General EPIA Cost Effectiveness (EUR/toe)	Investment per Site (EUR)
47—RETAIL	ISO 50001	81%	1.7	1.3	3.8	5791	19,653
	Not ISO 50001	12%	1.3	0.2	0.6	5926	142,402
	Monitoring	44%	1.5	0.7	2.1	5804	80,533
	Not Monitoring	0%	1.3	0	n.a.	n.a.	83,501
	Large Enterprise	37%	1.5	0.5	1.8	5804	81,819
	SME	n.a.	n.a.	n.a.	n.a.	n.a.	n.a.
	Total	**37%**	**1.5**	**0.5**	**1.8**	**5804**	**81,819**
64—BANKS	ISO 50001	19%	1.9	0.4	n.a	n.a	5016
	Not ISO 50001	38%	1.7	0.6	7.0	4640	34,270
	Monitoring	32%	4.1	1.3	10.0	5225	31,119
	Not Monitoring	31%	1	0.2	1.5	3982	34,537
	Large Enterprise	32%	1.8	0.6	7.5	4292	35,966
	SME	n.a.	n.a.	n.a.	n.a.	n.a.	n.a.
	Total	**31%**	**1.8**	**0.6**	**7.0**	**4640**	**32,690**

The lower cost effectiveness seemed to indicate that the EPIAs were implemented more efficiently in sites with energy management systems (in the retail sector). Moreover, the general EPIAs presented higher savings per site under ISO 50001 and monitoring systems. However, due to lack of information, these trends must be subsequently studied in other tertiary sectors.

It is worth noting that investments were strongly different between retail (81 k€/site) and banks (33 k€/site). Practically half of energy consumption in supermarkets was due to refrigeration [34]. Hence, a high number of technical EPIAs were related to the increase in efficiency of these systems and presented a relatively high cost compared with other technical EPIAs [35]. In banks, EPIAs were mainly related to non-residential uses of buildings (lighting, HVAC, and electric and electronic systems) in common with the retail sector [36,37]. The lower investment in ISO 50001 sites compared with noncertified sites could be explained by the clustering of the sites. Four certified companies reported 32% of sites with implemented EPIAs; hence, the relatively low investment by site was compensated by a high investment policy of ISO 50001 enterprises.

In Figure 7 are presented the energy savings per site, EPIAs, and companies in the retail sector. It is clear that the energy savings were higher in companies with ISO 50001 certification (110 toe/Co.) and with monitoring systems (97 toe/Co.) compared with companies without these systems (64 and 31 toe/Co., respectively). Therefore, the use of EnMSs at the corporate level seemed to effectively increase energy savings. This effect was not observed when savings were evaluated at the site or EPIA level. The global savings (compared with the total sector consumption) due to ISO 50001 or not due to ISO sites were very similar (0.9% and 1%). However, the impact on the use of a monitoring system significantly affected global saving, being that the sites monitored were responsible for at least more than 1.1% savings on global consumption, meanwhile nonmonitored systems had close to 0.6%.

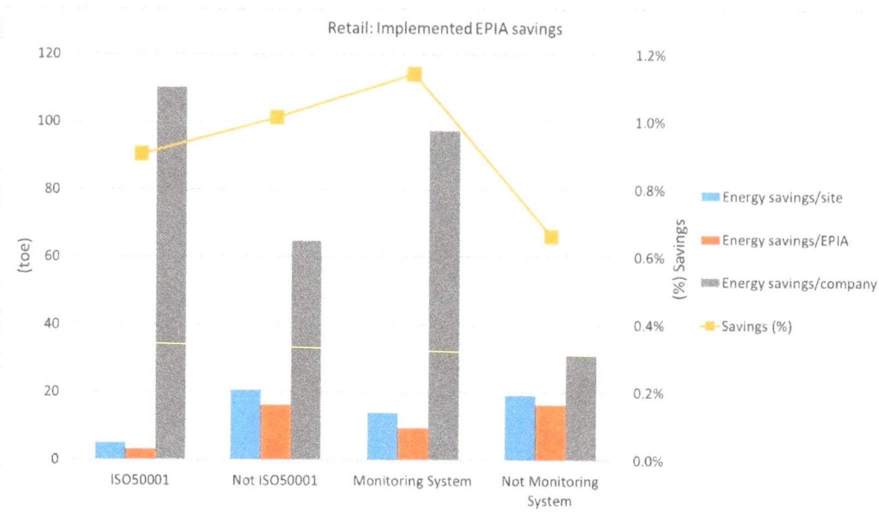

Figure 7. Retail: implemented EPIA savings.

The crucial impact of monitoring systems on energy savings was increased in the bank sector. The savings per site, EPIA, company, and globally were at least sensibly higher in monitored banks (21.7 toe, 13.4 toe, 86.9 toe, and 0.67%, respectively) compared with the nonmonitored ones (1.8 toe, 1.5 toe, 8.1 toe, and 0.33%) (see Figure 8). Unfortunately, the missing information on savings did not allow us to extend this study to ISO 50001 companies in the banking sector.

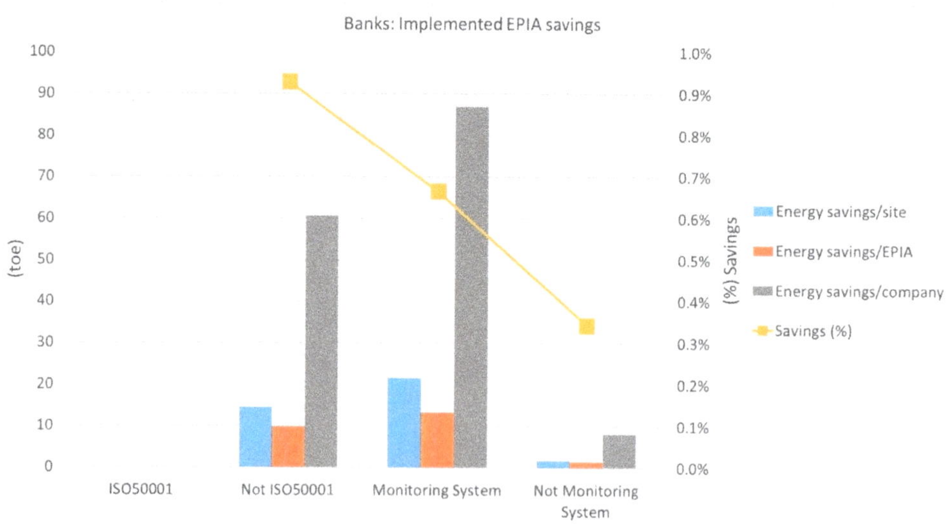

Figure 8. Banks: implemented EPIA savings.

In the tertiary sector, the EPIA cost effectiveness (EUR /toe) was aligned with the values observed in manufacturing (Figure 9). On the one hand, *general EPIAs* presented a lower CE than overall EPIAs. This means that the efficiency of the investment in *general EPIAs* was higher than in other measures. Hence, the promotion of these general practices (also promoted by the use of EnMSs) seemed to be convenient despite its limited impact (2.7 toe/site). On the other hand, CE spanned from 4000 to 10,000 EUR/toe as a function of the kind of EPIAs. From a general point of view, the CE of the refrigeration measures were higher for HVAC (medium CE) or lighting (low CE mainly promoted by the implementation of LEDs).

An analysis of planned EPIAs was carried out (see Table 4). A total of 1854 EPIAs were identified (excluding the integration of RES, 220 EPIAs, and CHP, 15 EPIAs), of which 17.4% were *general EPIAs* (mainly implementation of monitoring systems, EnMSs, and capacity training).

The CE of the identified *general EPIAs* was lower than that of the global EPIAs. This trend was similar to the values observed in implemented EPIAs. From a general point of view, it seemed that the global and general EPIAs with an ISO 50001 certification or a monitoring system presented lower CE, probably due to a better understanding of energy savings and EE investments. However, the specific CE by sector should be analysed with caution because it diverged from implemented to planned EPIAs, while in implemented EPIAs, CE was aligned between the two sectors, in the case of planned EPIAs, bank CE doubled retail CE. In any case, this trend was coherent with the lower PBT observed in the retail sector due to the intervention in refrigeration processes.

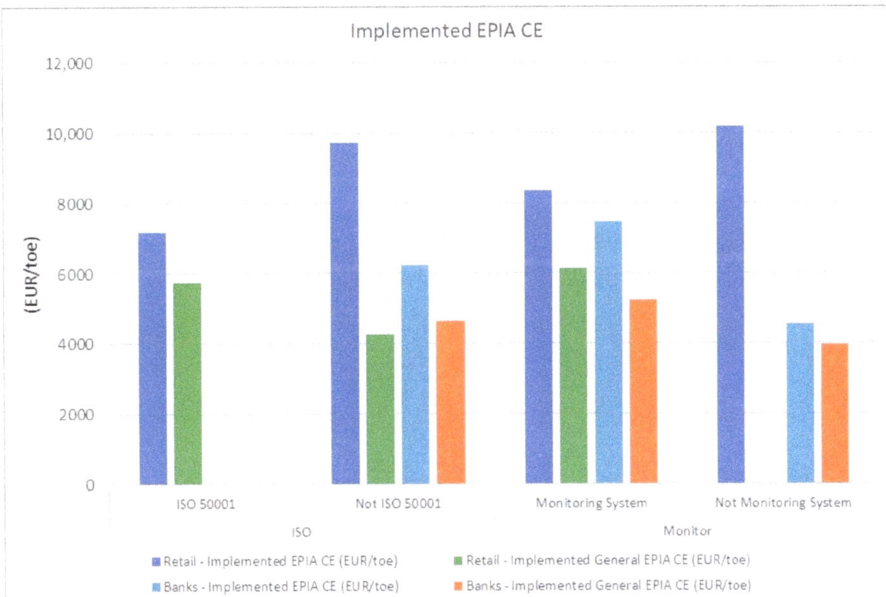

Figure 9. Implemented EPIA cost effectiveness (retail and banks).

Another interesting aspect was related to simple payback time (PBT). PBT was lower in *general EPIAs* than in overall EPIAs. This aspect was mainly due to the relatively low-risk investment associated with the general EE measurement [38]. Another important aspect was related to the lower PBT in the retail than in the banking sector. This fact can be due to several reasons. First, the technical refrigeration EPIAs (only in the retail sector) had a high impact on general site consumption, reducing the PBT. Second, the integration of energy-efficient technologies in supermarkets was usually incentivized by government legislation [39]. Third, banks' energy efficiency investments were supported by incentives related to non-residential buildings. These incentives were not considered in the EAs; therefore, PBT became longer [40].

However, the proposed EPIAs were not binding, and an analysis of the evolution of their execution should be carried out in order to increase the accuracy of this analysis. In any case, all the EAs were carried out by certified energy auditors and ESCOs; hence, all the information related to the proposed EPIAs was reasonable.

Table 4. Retail and bank sector planned EPIAs.

	PLANNED EPIAs	Companies (#)	Sites (#)	EPIAs (#)	General EPIAs (#)	EPIA Cost Effectiveness (EUR/toe)	General EPIA Cost Effectiveness (EUR/toe)	EPIA PBT (y)	General EPIA PBT (y)
47—RETAIL	ISO 50001	4	97	340	106	5474	3368	3.3	2.2
	Not ISO 50001	75	365	870	88	7782	5292	4.3	2.7
	Monitoring	42	334	882	169	7050	3968	3.8	2.4
	Not Monitoring	37	128	328	25	7280	5464	4.4	2.4
	Large Enterprise	75	457	1193	189	7072	3970	4.0	2.4
	SME	4	5	17	5	10,903	13,805	4.6	4.5
	Total	**79**	**462**	**1210**	**194**	**7111**	**4133**	**4.0**	**2.4**
64—BANKS	ISO 50001	2	40	123	36	18,478	6279	4.2	1.6
	Not ISO 50001	39	170	521	93	14,775	7318	8.8	4.9
	Monitoring	13	116	371	65	13,733	5875	7.2	3.8
	Not Monitoring	28	94	273	64	16,766	8023	9.1	5.0
	Large Enterprise	40	207	640	129	15,307	7256	8.1	4.5
	SME	1	3	4	0	1938	n.a.	1.9	n.a.
	Total	**41**	**210**	**644**	**129**	**15,201**	**7256**	**8.0**	**4.5**

4.3. Synthesis

The information presented can be summarized in a qualitative way in the following table, which includes information on both implemented and planned EPIAs: in Figure 10, green cells indicate that companies that are ISO 50001 certified and have a monitoring system or are defined as SMEs have better performance for each of the examined indicators; red cells, opposite results; and orange cells, mixed results.

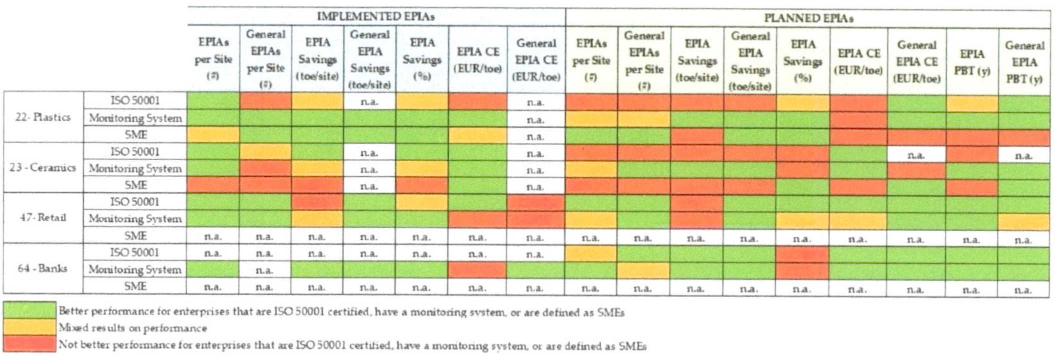

Figure 10. Impact of EnMS, monitoring, and SME class in implemented and planned EPIAs.

The results should be analysed while keeping in mind the sector-specific characteristics highlighted in previous sections, such as higher share of SMEs in the plastic sector, in terms of both total energy consumption and total EPIAs, or high concentration of multi-site companies in the retail and bank sectors. The results were also affected by the distribution of implemented and planned EPIAs among different technology and intervention domains.

Looking at the implemented EPIAs, having a monitoring system and being ISO 50001 certified had a positive impact on the global number of EPIAs in all the examined sectors (except for banks, where there was no information available on ISO 50001-certified sites). In all the sectors with available information (banks, retail, and plastics), having a monitoring system positively affected savings on total energy consumption and average savings from general EPIAs per site. In the two manufacturing sectors, monitoring systems also implied better cost effectiveness results.

Planned EPIAs showed mixed results when analysed in different sectors and by distinguishing by ISO 50001 certification, monitoring system, and class size. It should be considered that planned EPIAs were not binding and would deserve further analysis over time, in particular, relative to their implementation. The number of both global and general EPIAs had a slight tendency to be positively affected by having a monitoring system, which would require further investigation. The results seemed to be influenced by the specific intervention mix at the sectoral level, as described in previous sections. In general, monitoring systems seemed to have a positive impact on average savings when only general EPIAs were examined. To confirm this, the CE of general EPIAs was better in three out of the four sectors examined, and so was the average PBT of investments in general EPIAs. Finally, it is interesting to note that the average PBT was lower in all the analysed sectors for the monitoring system category.

5. Conclusions

In this work, the possible existing link between energy management and monitoring systems and energy audits in the EED Article 8 implementation in four different sectors in Italy was analysed. Additionally, an investigation on the impact of energy monitoring systems and an energy management system on planned and implemented energy performance improvement actions was developed.

The analysis showed that the manufacturing subsectors, plastics and ceramics, had a similar distribution of EPIAs per site (2.35 and 2.42) and a ratio for "*general/total*" EPIAs (15% and 13%). In both cases, the ISO 50001-certified and monitored sites presented a higher degree of implementation of EPIAs per site compared with the sites without EnMSs or monitoring systems. In the plastic sector, it was clear that the global energy savings (% compared with the total sector consumption in EAs) were higher in the companies with ISO 50001 certification and with monitoring systems compared with the companies without these systems. Therefore, the use of EnMSs at the corporate level seemed to effectively increase energy savings. This effect was not observed when savings were evaluated at the site or EPIA level for the ISO 50001-certified companies. Similar trends for the ISO 50001 companies were observed in the ceramic sector. The number of sites without a monitoring system and including savings data was very low, and for this reason, it was not possible to properly evaluate the effect of the monitoring system on savings.

The services subsectors, retail and banks, showed a similar distribution of EPIAs per site and a ratio for "*general/total*" EPIAs (37% and 31%). In both cases, the certified and monitored sites presented a higher degree of implementation of EPIAs per site compared with the sites without EnMSs or monitoring systems. However, a detailed distribution by EnMS and monitoring was very different. On the one hand, in the retail sector, the number of EPIAs per site was stable (between 1.3 and 1.7), and the *general EPIAs* were concentrated in the ISO and monitored sites. On the other hand, in banks there was a high variability in the number of EPIAs per site (from 1 to 4.1), and it was not possible to identify specific trends due to *general EPIAs*. Additionally, the bank sector is a clear example of the crucial importance of monitoring systems in the implementation of energy efficiency measurements. The savings per site, EPIA, company, and globally were at least sensibly higher in the monitored banks (21.7 toe, 13.4 toe, 86.9 toe, and 0.67%, respectively) compared with the nonmonitored ones.

The use of EnMSs effectively increased energy savings at the corporate level in all the sectors analysed. However, this trend was not fully corroborated at the site or EPIA level. Moreover, it was evident that the presence of a monitoring system was of fundamental importance for the implementation of EPIAs. All four sectors, in fact, had higher "energy savings/company" and "EPIA/site" ratios, where there were an EnMS and a monitoring system. This shows that a correct energy audit must always be accompanied by a specific monitoring plan if it is to be effective and useful to the company decision maker.

The methodology and analysis developed from the four chosen sectors can also be replicated in other sectors, and it would be necessary to implement this analysis also to other productive sectors of the industry or the tertiary sector to effectively evaluate whether the conclusions reached by our analysis can also be extended to other economic sectors.

Author Contributions: Conceptualization, C.H., E.B., C.M., M.S. and C.T.; methodology, C.H., E.B., C.M., M.S. and C.T.; formal analysis, C.H., E.B., C.M., M.S. and C.T.; writing—review and editing, C.H., E.B., C.M., M.S. and C.T.; writing, C.H., E.B., C.M., M.S. and C.T. All authors have read and agreed to the published version of the manuscript.

Funding: This work received funding from the European Union's Horizon 2020 Research and Innovation Programme under Grant Agreement No. 893924 (LEAP4SME Project) and Electrical System Research (PTR 2019–2021) implemented under programme agreements between the Italian Ministry for Economic Development and ENEA, CNR, and RSE S.p.A.

Conflicts of Interest: The authors declare no conflict of interest. The funders had no role in the design of the study; in the collection, analyses, or interpretation of data; in the writing of the manuscript; or in the decision to publish the results.

Nomenclature

EA	energy audit
AFNOR	French Standardization Association
CE	cost effectiveness (EUR/toe saved)
CEI	Italian Electrotechnical Committee
CHP	combined heat and power, cogeneration
CSEA	Environmental Energy Services Fund (in Italian, Cassa per i servizi energetici e ambientali)
CTI	Italian Thermotechnical Committee
EC	European Commission
EE	energy efficiency
EED	European Energy Efficiency Directive
EIB	European Investment Bank
EMS	environmental management systems (e.g., ISO 14001)
ENEA	Italian National Agency for New Technologies, Energy, and Sustainable Economic Development
EnMS	energy management system
EPIA	energy performance improved action
ESCO	energy service company
FIRE	Italian Federation for Energy Efficiency (in Italian, Federazione Italiana per l'uso Razionale dell'Energia)
HVAC	heating, ventilation, and air-conditioning
ISO	International Organization for Standardization
ISO 50001	international standard on energy management systems
LE	large enterprise
LED	light-emitting diode
NACE	Statistical Classification of Economic Activities in the European Community
PBT	simple payback time (y)
RES	renewable energy source
SME	small and medium-sized enterprise
toe	tonne of oil equivalent (=41.868 GJ)

Appendix A. Statistical Analysis

Clean data used for the analysis are presented in this appendix. The main results and the hypothesis derived from this appendix are extensively detailed in the manuscript, and some of the data were not presented in the body of the manuscript to avoid duplicities. Some small variations in data between the appendix and the main sections of the manuscript can be observed due to rounding issues.

In Table A1 is presented the total number of sites (one for each EA) and companies. The final energy consumption and relative distribution are presented. It is possible to observe that the subsectors with very low relative consumption (<2%) were excluded from the analysis, and the sectors with low percentage weight (<10%) were cited in the main text. The implemented EPIA savings are presented as the % with respect to the overall consumptions (as presented in Figures 4, 5, 7, and 8). The importance of general EPIAs in terms of the number of savings and relative weight is highlighted. It is possible to see that the accumulate savings are similar in the sectors that provide these data.

In Table A2 are presented data available about the implemented and planned EPIAs. The "sites with EPIA data" term refers to EAs that have declared the implementation of EPIAs in the last 4 years and EAs that have identified improvement measures. Obviously, the number of "planned" EPIAs is higher due to the intrinsic definition of the EA. One of the aims of the audits is to identify EPIAs. The "EPIAs with savings data" term refers to effective information of energy savings in the EPIAs. There is a high variability in information regarding the effective savings of implemented EPIAs. There is a non-negligible amount of energy audits that specify the details of implemented EPIAs, but without declaring the savings obtained. These EAs vary from 15% to 53% and 11% to 86% for implemented EPIAs and general EPIAs in the different sectors. The quality of these data increases up to 80% in the planned EPIAs. However, these values are not binding estimations. Hence, the analysis of savings was qualitatively carried out in the manuscript.

In Tables A3 and A4 are presented the mean and standard deviation of the main economic indicators (CE and investments by site for implemented EPIAs and CE and PBT for planned EPIAs). It is possible to observe the high standard deviation in all the parameters. These values are reasonable due to the high variability of the EPIAs considered. Overall, EPIAs include measures that vary from the substitution of lighting with led (W scale) to the substitution of furnaces (at the MW scale) and technologies (active vs. passive, process related vs. auxiliary or services related). General EPIAs include capacitation in energy management, implementation of energy management systems, monitoring of energy consumption, extension and improvement of current management and/or monitoring systems, and other actions not strictly related to the production process or technical EE measures. Therefore, it is also strongly heterogeneous. Finally, investment depends on multiple economic (non-energy-related) aspects from the companies (that present a strong variable structure internally to each sector). Therefore, only the mean values of economic indicators are presented, but they are analysed qualitatively.

Table A1. Number of EAs and companies; energy final consumption and savings from EPIAs and general EPIAs.

				Final Energy Consumption		Implemented EPIA Energy Savings (% vs. Consumptions)		General EPIA Energy Savings (% vs. All EPIA Savings)	
22—PLASTICS		**EAs**	**Companies**	**(toe)**	**(%)**	**(toe)**	**(%)**	**(toe)**	**(%)**
	ISO 50001	22	17	17,545	3%	414	2.36%	-	0.00%
	Not ISO 50001	569	509	559,669	97%	4770	0.85%	-	0.00%
	Monitoring System	412	359	473,514	82%	4923	1.04%	185	3.76%
	Not Monitoring System	179	167	103,699	18%	261	0.25%	22	8.28%
	Large Enterprise	104	76	207,955	36%	1272	0.61%	29	2.31%
	SME	487	450	369,259	64%	3912	1.06%	176	4.50%
	Total	**591**	**526**	**577,214**	**100%**	**5184**	**0.90%**	**206**	**3.97%**
23—CERAMICS		**EAs**	**Companies**	**(toe)**	**(%)**	**(toe)**	**(%)**	**(toe)**	**(%)**
	ISO 50001	17	6	175,586	16%	3974	2.26%	-	0.00%
	Not ISO 50001	140	106	938,853	84%	4012	0.43%	-	0.00%
	Monitoring System	133	91	1,047,030	94%	7985	0.76%	-	0.00%
	Not Monitoring System	24	21	67,408	6%	1	0.00%	-	0.00%
	Large Enterprise	69	32	836,010	75%	6424	0.77%	-	0.00%
	SME	88	80	278,429	25%	1562	0.56%	-	0.00%
	Total	**157**	**112**	**1,114,438**	**100%**	**7986**	**0.72%**	**12**	**0.15%**
47—RETAIL		**EAs**	**Companies**	**(toe)**	**(%)**	**(toe)**	**(%)**	**(toe)**	**(%)**
	ISO 50001	105	9	24,376	14%	220	0.90%	164	74.60%
	Not ISO 50001	604	167	146,813	86%	1486	1.01%	43	2.92%
	Monitoring System	458	135	119,165	70%	1362	1.14%	208	15.26%
	Not Monitoring System	251	41	52,024	30%	345	0.66%	-	0.00%
	Large Enterprise	698	162	170,175	99%	1706	1.00%	208	12.18%
	SME	11	14	1014	1%	-	0.00%	-	-
	Total	**709**	**176**	**171,189**	**100%**	**1706**	**1.00%**	**208**	**12.18%**
64—BANKS		**EAs**	**Companies**	**(toe)**	**(%)**	**(toe)**	**(%)**	**(toe)**	**(%)**
	ISO 50001	40	2	17,838	31%	-	0.00%	-	-
	Not ISO 50001	238	51	39,208	69%	364	0.93%	175	47.92%
	Monitoring System	147	13	52,293	92%	348	0.67%	161	46.21%
	Not Monitoring System	131	40	4752	8%	16	0.34%	14	84.41%
	Large Enterprise	275	52	56,970	100%	361	0.63%	174	48.13%
	SME	3	1	76	0%	3	4.53%	1	26.22%
	Total	**278**	**53**	**57,046**	**100%**	**364**	**0.64%**	**175**	**47.92%**

Table A2. Analysis of data available on energy audits: sites, EPIAs, and general EPIAs with information on savings. Implemented and planned EPIAs.

		IMPLEMENTED						PLANNED					
		Sites with EPIA Data		EPIAs with Savings Data		General EPIAs with Savings Data		Sites with EPIA Data		EPIAs with Savings Data		General EPIAs with Savings Data	
		(#)	(%)	(#)	(%)	(#)	(%)	(#)	(%)	(#)	(%)	(#)	(%)
22—PLASTICS	ISO 50001	12	63.6%	49	77.6%	1	33.3%	22	100%	52	91.2%	5	71.4%
	Not ISO 50001	134	39.2%	509	44.8%	16	20.3%	513	90.2%	1426	89.5%	215	71.4%
	Monitoring System	123	47.3%	478	48.5%	17	22.4%	371	90.0%	1018	88.8%	137	69.2%
	Not Monitoring	23	23.5%	80	42.5%	0	0.0%	164	91.6%	460	91.3%	83	75.5%
	Large Cos	64	49.0%	117	68.4%	2	14.3%	94	90.4%	148	58.7%	24	77.4%
	SME	82	38.2%	441	42.2%	15	22.1%	441	90.6%	1330	95.1%	196	70.8%
	Total	**146**	**40.1%**	**558**	**47.7%**	**17**	**20.7%**	**535**	**90.5%**	**1478**	**89.5%**	**220**	**71.4%**
		Sites with EPIA Data		EPIAs with Savings Data		General EPIAs with Savings Data		Sites with EPIA Data		EPIAs with Savings Data		General EPIAs with Savings Data	
		(#)	(%)	(#)	(%)	(#)	(%)	(#)	(%)	(#)	(%)	(#)	(%)
23—CERAMICS	ISO 50001	8	82.4%	37	83.8%	0	0.0%	11	64.7%	14	100%	0	n.a.
	Not ISO 50001	28	54.3%	181	34.3%	4	16.7%	131	93.6%	401	83.5%	32	44.4%
	Monitoring System	35	59.4%	193	47.7%	4	19.0%	119	89.5%	346	83.6%	28	45.2%
	Not Monitoring	1	45.8%	25	4.0%	0	0.0%	23	95.8%	69	86.3%	4	40.0%
	Large Cos	21	71.0%	133	51.1%	2	10.5%	65	94.2%	196	79.4%	14	33.3%
	SME	15	46.6%	85	29.4%	2	20.0%	77	87.5%	219	88.7%	18	60.0%
	Total	**36**	**57.3%**	**218**	**42.7%**	**4**	**13.8%**	**142**	**90.4%**	**415**	**84.0%**	**32**	**44.4%**
		Sites with EPIA Data		EPIAs with Savings Data		General EPIAs with Savings Data		Sites with EPIA Data		EPIAs with Savings Data		General EPIAs with Savings Data	
		(#)	(%)	(#)	(%)	(#)	(%)	(#)	(%)	(#)	(%)	(#)	(%)
47—RETAIL	ISO 50001	43	41.0%	72	100%	58	100%	97	92.4%	339	99.7%	106	100%
	Not ISO 50001	90	14.9%	92	76.0%	6	42.9%	361	59.8%	864	99.3%	86	97.7%
	Monitoring System	110	24.0%	143	87.2%	64	88.9%	331	72.3%	881	99.9%	167	98.8%
	Not Monitoring	23	9.2%	21	72.4%	0	n.a.	127	50.6%	322	98.2%	25	100%
	Large Cos	133	19.1%	164	85.0%	64	88.9%	454	65.0%	1186	99.4%	188	99.5%
	SME	0	n.a	0	n.a	0	n.a.	4	36.4%	17	100%	4	80.0%
	Total	**133**	**18.8%**	**164**	**85.0%**	**64**	**88.9%**	**458**	**64.6%**	**1203**	**99.4%**	**192**	**99.0%**
		Sites with EPIA Data		EPIAS with Savings Data		General EPIAs with Savings Data		Sites with EPIA Data		EPIAS with Savings Data		General EPIAs with Savings Data	
		(#)	(%)	(#)	(%)	(#)	(%)	(#)	(%)	(#)	(%)	(#)	(%)
64—BANKS	ISO 50001	14	35.0%	0	0.0%	1	20.0%	37	92.5%	100	81.3%	21	58.3%
	Not ISO 50001	33	13.9%	37	66.1%	16	76.2%	154	64.7%	482	92.5%	83	89.2%
	Monitoring System	14	9.5%	26	45.6%	9	50.0%	98	66.7%	317	85.4%	45	69.2%
	Not Monitoring	33	25.2%	11	42.3%	8	100%	93	71.0%	265	97.1%	59	92.2%
	Large Cos	45	16.4%	33	41.8%	16	64.0%	188	68.4%	578	90.3%	104	80.6%
	SME	2	66.7%	4	100%	1	100%	3	100%	4	100%	0	n.a.
	Total	**47**	**16.9%**	**37**	**44.6%**	**17**	**65.4%**	**191**	**68.7%**	**582**	**90.4%**	**104**	**80.6%**

Table A3. Analysis of mean and standard deviation of CE and investments for implemented EPIAs.

	IMPLEMENTED EPIAs					
	EPIA CE (EUR/toe)		General EPIA CE (EUR/toe)		Investment per Site (EUR)	
	MEAN	SD	MEAN	SD	MEAN	SD
22—Plastics Total	14,254	24,468	8098	8392	370,991	664,765
23—Ceramics Total	6552	12,747	n.a.	n.a.	482,053	1,099,639
47—Retail Total	8584	6878	5804	4571	81,629	148,132
64—Banks Total	6238	8271	4640	7313	32,690	52,763

Table A4. Analysis of mean and standard deviation of CE and PBT for planned EPIAs.

	PLANNED EPIAs							
	EPIA CE (EUR/toe)		General EPIA CE (EUR/toe)		EPIA PBT (y)		General EPIA PBT (y)	
	MEAN	SD	MEAN	SD	MEAN	SD	MEAN	SD
22—Plastics Total	6028	9953	3277	4641	4.4	4.1	3.2	10.0
23—Ceramics Total	5355	6465	3692	4913	4.2	3.1	2.2	2.1
47—Retail Total	7111	8451	4133	3238	4.0	3.7	2.4	1.9
64—Banks Total	15,201	16,429	7256	4925	8.0	11.1	4.5	4.3

References

1. Directive 2012/27/Eu of the European Parliament and of the Council of 25 October 2012 on Energy Efficiency 2012. Available online: https://eur-lex.europa.eu/legal-content/EN/TXT/HTML/?uri=CELEX:32012L0027&from=EN (accessed on 3 August 2021).
2. Directive (Eu) 2018/2002 of the European Parliament and of the Council of 11 December 2018 Amending Directive 2012/27/Eu on Energy Efficiency 2018. Available online: https://eur-lex.europa.eu/legal-content/EN/TXT/HTML/?uri=CELEX:32018L2002&from=EN (accessed on 3 August 2021).
3. Schleich, J. Do Energy Audits Help Reduce Barriers To Energy Efficiency? An Empirical Analysis for Germany. *Int. J. Energy Technol. Policy* **2004**, *2*, 226–239. [CrossRef]
4. Communication from the Commission Guidelines on State Aid for Environmental Protection and Energy 2014–2020. Available online: https://eur-lex.europa.eu/legal-content/EN/TXT/HTML/?uri=CELEX:52014XC0628(01)&from=ITA (accessed on 3 August 2021).
5. Santino, D.; Biele, E.; Salvio, M. *Guidelines for Energy Audits Under Article 8 of the EED: Italy 's Implementation Practices and Tools*; ENEA: Rome, Italy, 2019. Available online: https://www.efficienzaenergetica.enea.it/component/jdownloads/?task=download.send&id=377&catid=40&Itemid=101. (accessed on 3 August 2021).
6. Schleich, J.; Fleiter, T.; Hirzel, S. Effects of Energy Audits on the Adoption of Energy Efficiency Measures. In Proceedings of the 38th IAEE International Conference, Antalya, Turkey, 25–27 May 2015.
7. Kalantzis, F.; Revoltella, D. *How Energy Audits Promote SMEs' Energy Efficiency Investment*; EIB Working Papers; European Investement Bank: Luxembourg, 2019. Available online: https://op.europa.eu/en/publication-detail/-/publication/0c515b9e-3fcb-11e9-8d04-01aa75ed71a1/language-en (accessed on 3 August 2021).
8. Chiu, T.-Y.; Lo, S.-L.; Tsai, Y.-Y. Establishing an Integration-Energy-Practice Model for Improving Energy Performance Indicators in ISO 50001 Energy Management Systems. *Energies* **2012**, *5*, 5324–5339. [CrossRef]
9. IEA. *Multiple Benefits of Energy Efficiency*; IEA: Paris, France, 2019. Available online: https://www.iea.org/reports/multiple-benefits-of-energy-efficiency (accessed on 3 August 2021).
10. Introna, V.; Cesarotti, V.; Benedetti, M.; Biagiotti, S.; Rotunno, R. Energy Management Maturity Model: An Organizational Tool to Foster the Continuous Reduction of Energy Consumption in Companies. *J. Clean. Prod.* **2014**, *83*, 108–117. [CrossRef]
11. Poveda-Orjuela, P.P.; García-Díaz, J.C.; Pulido-Rojano, A.; Cañón-Zabala, G. ISO 50001: 2018 and Its Application in a Comprehensive Management System with an Energy-Performance Focus. *Energies* **2019**, *12*, 4700. [CrossRef]
12. Poveda-Orjuela, P.P.; García-Díaz, J.C.; Pulido-Rojano, A.; Cañón-Zabala, G. Parameterization, Analysis, and Risk Management in a Comprehensive Management System with Emphasis on Energy and Performance (ISO 50001: 2018). *Energies* **2020**, *13*, 5579. [CrossRef]
13. Solnørdal, M.T.; Thyholdt, S.B. Absorptive Capacity and Energy Efficiency in Manufacturing Firms—An Empirical Analysis in Norway. *Energy Policy* **2019**, *132*, 978–990. [CrossRef]
14. Backlund, S.; Thollander, P.; Palm, J.; Ottosson, M. Extending the Energy Efficiency Gap. *Energy Policy* **2012**, *51*, 392–396. [CrossRef]
15. Cagno, E.; Worrell, E.; Trianni, A.; Pugliese, G. A Novel Approach for Barriers to Industrial Energy Efficiency. *Renew. Sustain. Energy Rev.* **2013**, *19*, 290–308. [CrossRef]
16. Laskurain, I.; Ibarloza, A.; Larrea, A.; Allur, E. Contribution to Energy Management of the Main Standards for Environmental Management Systems: The Case of ISO 14001 and EMAS. *Energies* **2017**, *10*, 1758. [CrossRef]
17. Fiedler, T.; Mircea, P.-M. Energy Management Systems According to the ISO 50001 Standard—Challenges and Benefits. In Proceedings of the 2012 International Conference on Applied and Theoretical Electricity (ICATE), Craiova, Romania, 25–27 October 2012; pp. 1–4.
18. Fuchs, H.; Aghajanzadeh, A.; Therkelsen, P. Identification of Drivers, Benefits, and Challenges of ISO 50001 through Case Study Content Analysis. *Energy Policy* **2020**, *142*, 111443. [CrossRef]
19. Marimon, F.; Casadesús, M. Reasons to Adopt ISO 50001 Energy Management System. *Sustainability* **2017**, *9*, 1740. [CrossRef]

20. António Da Silva Gonçalves, V.; Mil-Homens Dos Santos, F.J. Energy Management System ISO 50001:2011 and Energy Management for Sustainable Development. *Energy Policy* **2019**, *133*, 110868. [CrossRef]
21. Mckane, A.; Therkelsen, P.; Scodel, A.; Rao, P.; Aghajanzadeh, A.; Hirzel, S.; Zhang, R.; Prem, R.; Fossa, A.; Lazarevska, A.M.; et al. Predicting the Quantifiable Impacts of ISO 50001 on Climate Change Mitigation. *Energy Policy* **2017**, *107*, 278–288. [CrossRef]
22. Schleich, J. Barriers To Energy Efficiency: A Comparison Across the German Commercial and Services Sector. *Ecol. Econ.* **2009**, *68*, 2150–2159. [CrossRef]
23. Thollander, P.; Danestig, M.; Rohdin, P. Energy Policies for Increased Industrial Energy Efficiency: Evaluation of a Local Energy Programme for Manufacturing SMEs. *Energy Policy* **2007**, *35*, 5774–5783. [CrossRef]
24. Locmelis, K.; Blumberga, D.; Blumberga, A.; Kubule, A. Benchmarking of Industrial Energy Efficiency. Outcomes of an Energy Audit Policy Program. *Energies* **2020**, *13*, 2210. [CrossRef]
25. Fleiter, T.; Schleich, J.; Ravivanpong, P. Adoption of Energy-Efficiency Measures in SMEs—An Empirical Analysis Based on Energy Audit Data from Germany. *Energy Policy* **2012**, *51*, 863–875. [CrossRef]
26. Sousa Lira, J.M.; Salgado, E.G.; Beijo, L.A. Which Factors Does the Diffusion of ISO 50001 in Different Regions of the World is Influenced? *J. Clean. Prod.* **2019**, *226*, 759–767. [CrossRef]
27. Bonacina, F.; Corsini, A.; De Propris, L.; Marchegiani, A.; Mori, F. Industrial Energy Management Systems in Italy: State of the Art and Perspective. *Energy Procedia* **2015**, *82*, 562–569. [CrossRef]
28. FIRE-CEI-CTI. *Indagine sui Sistemi di Gestione Dell'energia Certificati ISO 50001 in Italia*; FIRE-CEI-CTI: Rome, Italy, 2016. (In Italian). Available online: https://fire-italia.org/indagine-50001-2016 (accessed on 3 August 2021).
29. ACCREDIA. *Le Certificazioni Accreditate per l'Efficienza Energetica*; ACCREDIA—National Accreditation Body: Rome, Italy, 2017. (In Italian). Available online: https://www.accredia.it/app/uploads/2017/07/Quaderno_Osservatorio_ACCREDIA_Le_certificazione_accreditate_per_leffic....pdf (accessed on 3 August 2021).
30. ENEA. *Energy Efficiency Annual Report 2020. Analysis and Results from Energy Efficiency Policies in Italy*; ENEA: Rome, Italy, 2020. (In Italian). Available online: https://www.efficienzaenergetica.enea.it/component/jdownloads/?task=download.send&id=453&catid=40%20&Itemid=101 (accessed on 3 August 2021).
31. Trianni, A.; Cagno, E. Dealing with Barriers to Energy Efficiency and SMEs: Some Empirical Evidences. *Energy* **2012**, *37*, 494–504. [CrossRef]
32. Cantini, A.; Leoni, L.; De Carlo, F.; Salvio, M.; Martini, C.; Martini, F. Technological Energy Efficiency Improvements in Cement Industries. *Sustainability* **2021**, *13*, 3810. [CrossRef]
33. Branchini, L.; Bignozzi, M.C.; Ferrari, B.; Mazzanti, B.; Ottaviano, S.; Salvio, M.; Toro, C.; Martini, F.; Canetti, A. Cogeneration Supporting the Energy Transition in the Italian Ceramic Tile Industry. *Sustainability* **2021**, *13*, 4006. [CrossRef]
34. Timma, L.; Skudritis, R.; Blumberga, D. Benchmarking Analysis of Energy Consumption in Supermarkets. *Energy Procedia* **2016**, *95*, 435–438. [CrossRef]
35. Tallini, A.; Cedola, L. Evaluation Methodology for Energy Efficiency Measures in Industry and Service Sector. *Energy Procedia* **2016**, *101*, 542–549. [CrossRef]
36. Spyropoulos, G.N.; Balaras, C.A. Energy Consumption and the Potential of Energy Savings in Hellenic Office Buildings Used as Bank Branches—A Case Study. *Energy Build.* **2011**, *43*, 770–778. [CrossRef]
37. Aranda, A.; Ferreira, G.; Mainar-Toledo, M.D.; Scarpellini, S.; Llera Sastresa, E. Multiple Regression Models to Predict the Annual Energy Consumption in the Spanish Banking Sector. *Energy Build.* **2012**, *49*, 380–387. [CrossRef]
38. Polzin, F.; Sanders, M. How To Finance the Transition to Low-Carbon Energy in Europe? *Energy Policy* **2020**, *147*, 111863. [CrossRef]
39. Ochieng, E.G.; Jones, N.; Price, A.D.F.; Ruan, X.; Egbu, C.O.; Zuofa, T. Integration of Energy Efficient Technologies in Uk Supermarkets. *Energy Policy* **2014**, *67*, 388–393. [CrossRef]
40. Brown, D.; Sorrell, S.; Kivimaa, P. Worth the Risk? an Evaluation of Alternative Finance Mechanisms for Residential Retrofit. *Energy Policy* **2019**, *128*, 418–430. [CrossRef]

energies

MDPI

Article

Estimation Method of Greenhouse Gas Reduction for Electrical Energy Storage Based on Load-Leveling Application

Dong-Hyun Tae, Hu-Dong Lee, Jian Shen, Byeong-Gill Han and Dae-Seok Rho *

Department of Electrical Engineering, Korea University of Technology & Education (KUT), Cheonan 31253, Korea; tdh2002@koreatech.ac.kr (D.-H.T.); lhd714@koreatech.ac.kr (H.-D.L.); sheengun@koreatech.ac.kr (J.S.); bghan@koreatech.ac.kr (B.-G.H.)

* Correspondence: dsrho@koreatech.ac.kr; Tel.: +82-41-560-1167

Abstract: In recent years, there have been several types of energy storage technologies adopted in many different areas, such as peak shaving, frequency regulation, and renewable stabilization applications. Moreover, technologies of high energy and power density are useful for load leveling, power smoothing for renewable energy systems (RESs), and peak shaving for demand management. Under these circumstances, an estimation technique for assessing environmental issues applied to electrical energy storage (EES) systems is essential in order to promote commercialization of EES systems. Therefore, this paper proposes an estimation method for CO_2 emission in cases where EES systems are introduced and not introduced. It is essential to evaluate environmental issues in EES systems at operation stages of their life cycle and make an effective contribution to environmental improvement and reduce potential adverse environmental impacts. Thus, this paper deals with an evaluation method for CO_2 emission based on an optimal algorithm including a successive approximation method for the best-mix solution of power sources, etc. From the simulation result based on the proposed evaluation algorithm, it is found that the output power of a coal power plant (high CO_2 emission) is replaced by the output powers of the EES systems and the nuclear generator (low CO_2 emission).

Keywords: EES systems; greenhouse gas; load leveling; best-mix solution; optimal operation algorithm

Citation: Tae, D.-H.; Lee, H.-D.; Shen, J.; Han, B.-G.; Rho, D.-S. Estimation Method of Greenhouse Gas Reduction for Electrical Energy Storage Based on Load-Leveling Application. *Energies* **2021**, *14*, 5492. https://doi.org/10.3390/en14175492

Academic Editors: Chiara Martini and Claudia Toro

Received: 28 July 2021
Accepted: 27 August 2021
Published: 2 September 2021

1. Introduction

Recently, the operation of electric power systems has become more difficult because the peak load demand is increasing continuously and also the daily and annual load factors are worsening [1,2]. Furthermore, the global environmental issues need to be considered in electric power systems [3–6]. One countermeasure to overcome these problems is a study of the operation method of electric power systems, including novel energy storage systems such as secondary batteries, superconducting magnets (SMES), and flywheels, which have made astonishing improvements lately [7–9].

In general, the cost of power generation can be reduced if the energy storage system is charged during the off-peak time interval and discharged during the peak time interval [10]. In addition, the benefit of storing electricity is increasing along with the increase in the difference in demand between off-peak and peak time intervals. The result is load leveling by time shifting [11]. Furthermore, the output of power generation can be flatter if the difference in demand between daytime and nightime is reduced using an EES system; as a result, the operation efficiency can be improved and the fuel cost can be reduced [12]. For these reasons, a lot of utilities have built pumped hydro-generators and have started installing large-scaled batteries in substations recently [13].

In contrast, it is expected that an RES such as a PV system or a wind power system will be widely installed and operated in order to overcome global environmental issues [14]. However, operation problems such as output fluctuation or unpredictability may occur if the RES is integrated with the power grid [15]. When the total volume of renewable energies connected to the grid exceeds a certain level, such problems will appear and countermeasures will be needed. To stabilize the fluctuation and to control load management, an EES system is essential for the introduction of large amounts of renewable energy [16–19].

Under these circumstances, an estimation technique for assessing environmental issues applied to EES systems is essential in order to promote commercialization of EES systems. However, when introducing an EES system, many previous studies have been conducted on their merits in terms of economic dispatch, but studies evaluating the environmental merits are insufficient [20,21]. Therefore, this paper presents a concept of an estimation method for CO_2 emission in cases where EES systems are introduced and not introduced. It is based on the idea that can reduce CO_2 emission by existing generator units by operation of EES systems.

It is essential to evaluate environmental issues in EES systems at operation stages of their life cycle and make an effective contribution to environmental improvement and reduce potential adverse environmental impacts. Thus, this paper proposes an evaluation method of CO_2 emission based on an optimal algorithm including a successive approximation method for the best-mix solution of power sources, etc. From the result of calculating the CO_2 emission using the proposed evaluation algorithm of GHG reduction, it is found that the output power of the coal power plant (high CO_2 emission) is replaced by the output powers of the EES systems and the nuclear generator (low CO_2 emission).

2. Formulation of Load-Leveling Application Using EES Systems

2.1. Concepts of Load Leveling

To estimate GHG reduction for EES systems, the proper charging and discharging amounts of EES systems must be obtained in advance, in other words, the composition ratio of generation units at a peak load demand for the two cases where EES systems are introduced and not introduced. This means that the output power of the oil-power plant (high CO_2 emission) is replaced by the output powers of the EES systems and the nuclear power plant (low CO_2 emission). With the allocation of EES systems to distribution systems in Figure 1, the simultaneous load leveling of both the total power system and the distribution systems increases the utilization rates of less expensive generator units, and the benefit of the reduction in the total power operation cost is expected, as shown in Figure 2. In other words, the operation problem of load leveling is to obtain the most appropriate type and number of generators, called an optimal generation mix, in cases where EES systems are introduced and allocated to distribution systems [22–25].

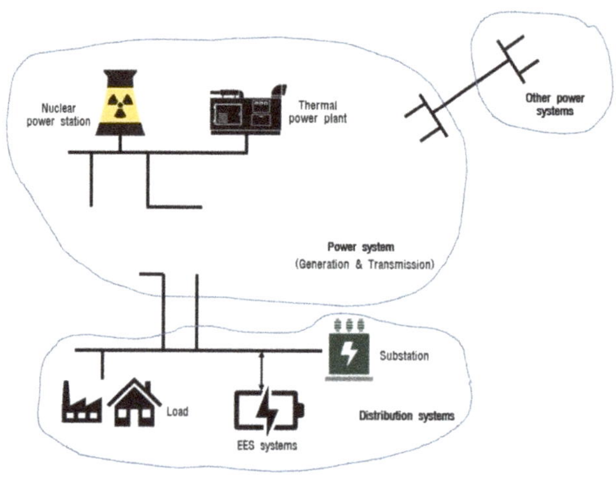

Figure 1. Power system with EES systems.

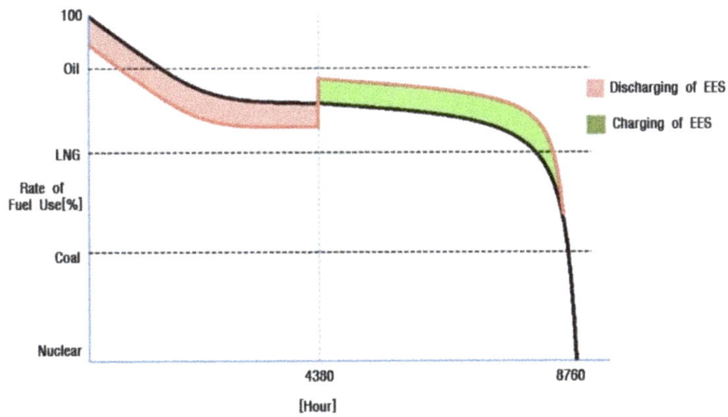

Figure 2. Concept of load duration curve with EES systems.

2.2. Problem Formulation of Load-Leveling Application

As mentioned above, the fundamental problem of load leveling is to decide the most appropriate type and number of generators, called an optimal generation mix, in cases where EES systems are operated in the power (distribution) systems [26]. The optimal generation mix with EES systems is a static problem against the time period and in which the objective is to determine the process in such a manner as to minimize the total cost for load demands provided for a target year [27]. Both both the generation mix (nonlinear integer programming problem) and the operating mode of EES systems (nonlinear programming problem) must be optimized. The problem can be thus formulated and solved as a nonlinear mixed integer programming problem. However, the nonlinear mixed integer programming problem will be complicated, along with the dimension of the problem. The optimal generation mix problem considering EES systems, whose objective is to determine the generation mix that minimizes the total cost for a target year, can be formulated as a nonlinear mixed integer programming problem as follows [28]:

$$MinF_n(x,v) = \sum_{i=1}^{n} [a_i x_i + b_i Q_i(X_{i-1}, X_{i-1} + x_i, v)] + a_s x_s \tag{1}$$

$$\text{Subj. to} \sum_{i=1}^{n} x_i + x_s \geq P_D + P_R \tag{2}$$

$$x_{i\min} \leq x_i \leq x_{i\max},\ i = 1, \cdots, n \tag{3}$$

$$v_{i\min} \leq v_{ik} \leq v_{i\max},\ i = 1, \cdots, n,\ k = 1, \cdots, T \tag{4}$$

$$Q_i(X_{i-1},\ X_{i-1} + x_i, v) = \sum_{k=1}^{K} [z_k \int_{X_{i-1}}^{X_{i-1}+X_i} L_k(u, v_{ik}) du],\ i = 1, \cdots, n \tag{5}$$

$$X_i = \sum_{k=1}^{i} x_k,\ i = 1, \cdots, n,\ X_0 = 0 \tag{6}$$

where F_n is the total cost for the target year; n is the number of generation types; a_i, b_i are the fixed and variable costs, respectively, of generation type I; x_i, x_s are the capacity of generation type i and EES systems, respectively; a_s is the fixed cost of EES systems; v_{ik} is the output power of EES systems at a daily load curve i and time period k; Q_i is the annual energy production for generation type i; X_i is the cumulative capacity up to generation type i; $L_k(u)$ is the time fraction that demand is more than load level u at duration curve k; z_k is the number of days that provide $L_k(u)$; P_D, P_R are the peak demand and spinning reserve, respectively; K is the number of patterns of the daily load duration curve; and T is the number of time intervals for the daily load duration curve.

3. Evaluation Algorithm of GHG Reduction Based on Load Leveling

3.1. Optimal Operation Algorithm for Load Leveling

The problem, as formulated above, which is composed of two kinds of variables, such as generation mix (x) and operating mode of EES systems (v), is a nonlinear mixed integer programming problem. From a theoretical perspective, the problem can be solved by evaluating the objective function for the generation mix (x) under the constraint conditions. However, this method will be complicated with the increase in the system size. Therefore, this paper adapts a successive approximation method considering the parameters of the fixed cost and capacity of EES systems, as shown in Figure 3. The optimization procedure can be illustrated as follows:

<Step1> Assumes system parameters. Put $K_0 = 0$ (fixed cost of EES systems) and $X_0 = 0$ (initial capacity of EES systems).
<Step2> Determines the optimal generation mix for existing generators (x) while fixing the output power of EES systems to zero ($v = 0$). Assume F_0 as the total cost of this solution.
<Step3> Determines the optimal operating mode of EES systems (v) while fixing the generation mix (x). Calculate the optimal generation mix with EES systems, *Fs*.
<Step4> If $F_s \leq F_0$, add the unit size of EES systems ΔX and go to <Step3>. Otherwise, go to the next step.
<Step5> If the introduction capacity of EES systems is zero (X = 0), the algorithm terminates. The generation mix (x) and the capacity and fixed cost of EES systems are the optimal solution. Otherwise, increase the unit fixed cost of EES systems ΔK and go to <Step3>.

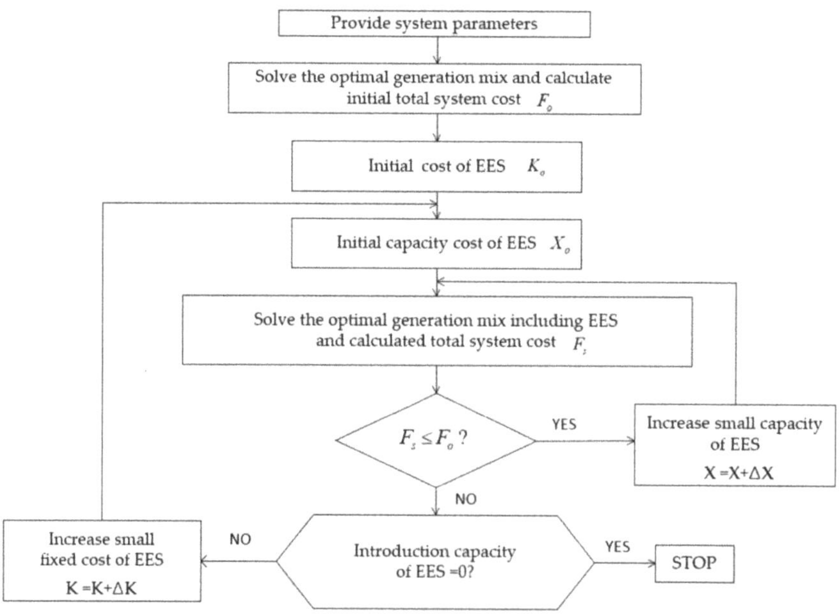

Figure 3. Evaluation algorithm of EES systems.

3.2. Operation Algorithm of EES Systems

This section describes the algorithm to determine the optimal operation of an EES system while fixing the generation mix of existing generators. This paper adopts a gradient method to decide the optimal operation mode of the EES system. Therefore, the economic operating conditions for the EES system are obtained by Equation (7).

$$\eta > \frac{\lambda_{charge}}{\lambda_{discharge}} \tag{7}$$

where η is the round-trip efficiency of the EES system, λ_{charge} is the incremental cost in the charging period, and $\lambda_{discharge}$ shows the incremental cost in the discharging period of the EES system.

The minimization for the objective function F_n, while fixing the generation mix, is obtained by load leveling in order to satisfy the economic operating conditions. The constraint of power (kW) and capacity (kWh) in EES systems must be also satisfied in this procedure. The optimal operation mode of EES systems over the target year is decided, along with all daily duration curves. The procedure is as follows:

<Step1> Decide the lowest and highest load demand periods in the daily load duration curve. Compute the incremental costs λ_{charge} and $\lambda_{discharge}$.
<Step2> If $\eta < \lambda_{charge}/\lambda_{discharge}$, the algorithm terminates. Otherwise, charge a small amount of power ΔP_S in the lowest period, and discharge the power $\eta\,\Delta P_S$ in the highest period.
<Step3> If the maximum storage capacity (kWh) constraint is reached, the algorithm terminates.
<Step4> If the maximum output (kW) constraint for EES systems is reached, eliminate the period from consideration. Go to <Step1>.

In addition, the allocation site of each small amount for EES systems is decided by selecting the lowest $F_n(x)$ for the cases where the above algorithm is applied at all allocation sites that are the distribution substations.

3.3. Estimation Algorithm of GHG Reduction Based on Load Leveling

With the allocation of EES systems to distribution systems, the benefit of the reduction of the total power operation cost is expected because the simultaneous load leveling of both the total power system and the power distribution systems increases the utilization rates of less expensive generator units such as nuclear and coal power plants. In addition, if EES systems are replaced with existing generators in the peak (discharging) time interval, the benefit of CO_2 emission reduction can be expected, where it is ideally assumed that EES systems are charged by the nuclear power unit in the off-peak time and discharged by oil or gas generator units in the peak time. Therefore, the estimation of GHG reduction during a year can be quantified using the formula given below:

$$GHG(y) = \sum_{t=1}^{T}\sum_{i=1}^{n} P_{DE_i}(t) \times \alpha_i - \sum_{t=1}^{T}\sum_{i=1}^{n} P_{CE_i}(t) \times \beta_i - \sum_{t=1}^{T} P_{ESS}(t) \times \eta_{in,out} \times \gamma_{ESS} \tag{8}$$

$$\text{Subj. to } \sum_{i=1}^{n} P_{DE_i} - P_{ESS} = 0 \tag{9}$$

where $GHG(y)$ is the GHG reduction amount a year (kt), $P_{DE_i}(t)$ is the output (kW) of the existing generators in peak (discharging) times, $P_{CE_i}(t)$ is the output (kW) of the existing generators in off-peak (charging) times, $P_{ESS}(t)$ is the charging and discharging output (kW) of the EES systems, α_i is the CO_2 emission coefficient of the generator units in the discharging time, β_i is the CO_2 emission coefficient of the generator units in the charging time, γ_{ESS} is the CO_2 emission coefficient of the EES systems, i is the generator type, t is the time interval, T is the target year, and $\eta_{in,out}$ is the charging and discharging efficiency (loss) of the EES systems.

To find out $GHG(y)$, the proper charging and discharging amounts of EES systems must be obtained in advance, in other words, the composition ratio of generation units at a peak load demand for the two cases where EES systems are introduced and not introduced. This means that the output power of the oil power plant (high CO_2 emission) is replaced by the output powers of the EES systems and the nuclear power plant (low CO_2 emission).

4. Case Studies

4.1. Simulation Conditions

To validate the proposed method, this paper carried out simulations using the model systems and parameters shown in Figure 4 and Table 1. The table is the data of the statistical materials of the Korea Electric Power Cooperation in the fiscal year of 2018. The four load patterns for distribution substations (A, B, C, D) and the peak demand of 10 million kW in Figure 5 were considered. This figure is the typical load pattern in summer, and the load patterns of other seasons were assumed by the same pattern and the size of 70%, 80%, and 90% based on the typical load pattern. In addition, the round-trip efficiencies for EES systems of 70% and 80% were assumed. Furthermore, the CO_2 emission amount of the generator type was assumed, as shown in Table 2, in order to find out GHG reduction based on load leveling in EES systems. This paper found out the optimal generation mix, considering the operation of EES systems under the following assumptions [29–32]:

① The total cost of generators is calculated by the sum of the variable and fixed costs, and the total cost of EES systems is only the fixed cost.
② The maintenance cost of generators is ignored.
③ Unit sizes for the existing generators are previously provided, and unit sizes for new generators are not fixed.

Table 1. Parameters of generation units.

Type	Variable Cost (won/kWh)	Fixed Cost (1000 won/kW)	Rating (MW)	Failure Rate (%)
Nuclear	39.7	2385	-	6.5
Coal	60.9	1399	1000	7.0
LNG	147.2	576	1000	6.0
Oil	184.7	576	-	6.0
EES systems	-	Ca	20 (8 h)	-

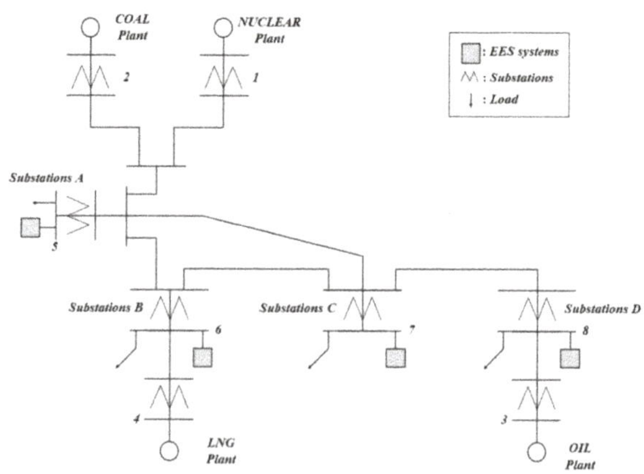

Figure 4. Model power systems.

Table 2. CO_2 emission amount of generator type [33,34]. (Reprint with permission [33,34]; September 2008 and 14 June 2012, ISO and IEC).

Type of Generator	Amount of CO_2 Emission (kT/TWh)
Run-of-the-river hydropower	1
Wind (without back-up production)	9
Solar photovoltaic	13
Hydropower with a reservoir	15
Nuclear	15
Biomass (plantation)	120
Biomass (forestry waste)	120
Natural gas combined cycle	510
Fuel cell (H_2 from CH_4 reforming)	665
Heavy oil steam boiler	780
Diesel	780
Coal steam plant with SO_2 removal	975

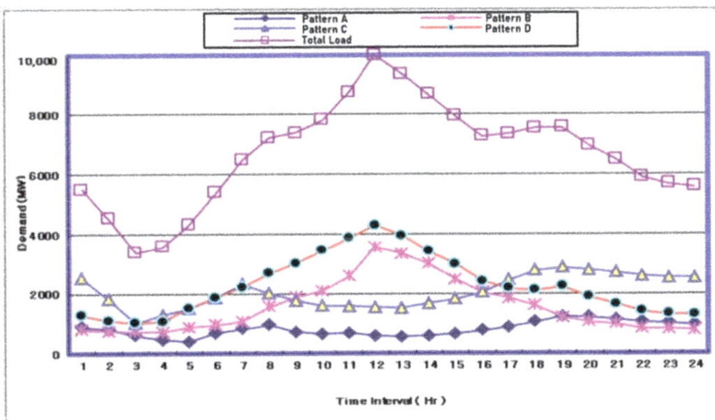

Figure 5. Yearly load patterns of distribution substations.

4.2. Operation Characteristics of Load Leveling

By comparing the total operation cost $F_n(x)$ and $F_0(x)$ for the two cases where EES systems are introduced and not introduced, with the increase in a small unit of the capacity and fixed cost of EES systems, the optimal capacity and fixed cost of EES systems are obtained, as shown in Figure 6. Because of the computation time for parameter analysis, a small unit of the capacity of EES systems is considered as 20 MW (160 MWh, 8 h) and the fixed cost of EES systems is considered as 1000 won. Figure 6 shows that the benefits of the load leveling of EES systems in the distribution substations, which is the fixed cost (C_a), becomes 75,000~94,000 won/kW.

As shown in Figure 6, the marginal and saturated fixed costs are also obtained. The marginal cost, in which the composition ratio of EES systems is zero, represents the economical point for EES systems. In addition, the saturated fixed cost keeps a constant value, although the fixed cost changes, because complete load leveling is accomplished at each fixed cost. Table 3 shows the comparison results for the composition ratios of generation units and the total cost at a fixed cost of 75,000 won for the two cases where EES systems are introduced and not introduced.

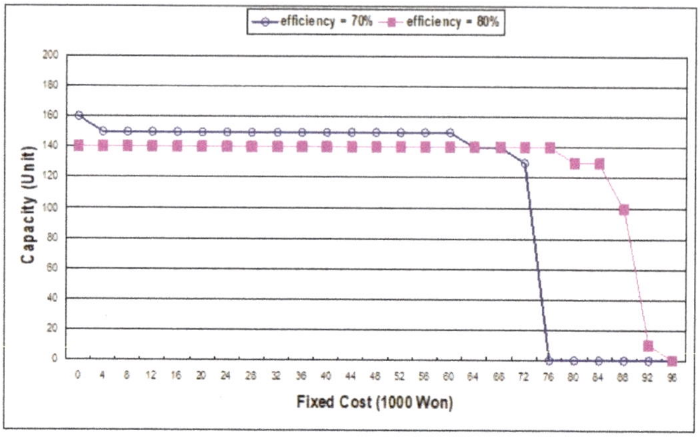

Figure 6. Optimal capacity and fixed cost of EES systems (1 unit: 20 MW, 180 MWh).

Table 3. Composition ratio of generation units (at a fixed cost of 75,000 won).

Type	Output Power without EES Systems (MW)	Output Power with EES Systems (MW)
Nuclear	2899.3	3766.3
Coal	1000.0	1000.0
LNG	1000.0	1000.0
Oil	5100.0	2399.1
EES systems	0.0	2600.0
Total cost	1920.1 million won	1914.1 million won

4.3. Estimation of GHG Reduction Based on Load Leveling

Based on the operation characteristics of EES systems, as shown in Section 4.2, this paper ideally assumes that EES systems are charged by the nuclear power unit in off-peak time and discharged by the oil power plant (heavy oil steam boiler) units in peak time. Here, the total capacity of EES systems is calculated as 2600 MW (8 h, 20,800 MWh) and the efficiency of EES systems is also assumed as 80% for load leveling. Therefore, the amount of GHG reduction during a year can be obtained as 5785.14 kt based on the following procedure:

- $P_{DE_i}(t)/\text{year} = 365$ days $\times$ 20,800 MWh $\times$ 780 kt/MWh $\times 10^{-6} = 5921.8$ kt
- $P_{CE_i}(t)/\text{year} = 365$ days $\times$ 20,800 MWh $\times$ 15 kt/MWh $\times 10^{-6} = 113.88$ kt
- $P_{ESS}(t)\ \eta_{in,out}/\text{year} = 365$ days $\times$ 20,800 MWh $\times (1-0.8) \times$ 15 kt/MWh $\times 10^{-6} =$ 22.78 kt
- $GHG(y) = 5785.14$ kt

However, if the coal power plant is replaced with an oil power plant (heavy oil steam boiler) for discharging EES systems during the peak-time interval, the annual amount of GHG reduction can be calculated as 7265.54 kt based on the following procedure:

- $P_{DE_i}(t)/\text{year} = 365$ days $\times$ 20,800 MWh $\times$ 975 kt/MWh $\times 10^{-6} = 7402.2$ kt
- $P_{CE_i}(t)/\text{year} = 365$ days $\times$ 20,800 MWh $\times$ 15 kt/MWh $\times 10^{-6} = 113.88$ kt
- $P_{ESS}(t)\ \eta_{in,out}/\text{year} = 365$ days $\times$ 20,800 MWh $\times (1-0.8) \times$ 15 kt/MWh $\times 10^{-6} =$ 22.78 kt
- $GHG(y) = 7265.54$ kt

Therefore, it is clear that the amount of GHG reduction with a coal power plant is more effective than with an oil power plant for charging the power of EES systems.

5. Conclusions

This paper presents a concept of an estimation method for CO_2 emission based on an optimal algorithm including a successive approximation method for the best-mix solution of power sources, etc. The main results are summarized as follows.

(1) From the optimal operation algorithm of EES systems, the benefits of load leveling in the distribution substations is calculated as 75,000~94,000 won/kW for 1 year.

(2) The total amount of GHG reduction with an oil power plant is calculated as 5785.14 kt, and that with a coal power plant is obtained as 7265.54 kt. The amount of GHG reduction with a coal power plant is more effective than with an oil power plant for charging the power of EES systems.

(3) It is confirmed that the output of the coal power plant with high CO_2 emission is replaced by the EES systems and the nuclear power plant with low CO_2 emission. Therefore, it is confirmed that EES systems affect the environment at operation stages of their life cycle and contribute to environmental improvement and reduction in potential adverse environmental impacts.

Author Contributions: D.-H.T. and D.-S.R. carried out modeling and simulations and compiled the manuscript. The literature review and simulation analysis were performed by H.-D.L. and B.-G.H., and J.S. collected the data and investigated early works. All authors have read and agreed to the published version of the manuscript.

Funding: This research was funded by the Power Generation and Electricity Delivery Core Technology program of the Korea Institute of Energy Technology Evaluation and Planning (KETEP), with financial resources from the Ministry of Trade, Industry & Energy, Korea.

Institutional Review Board Statement: Not applicable.

Informed Consent Statement: Not applicable.

Data Availability Statement: The raw data analyzed in this article are available online at: https://www.iso.org/standard/41352.html and https://webstore.iec.ch/publication/7520 (accessed on September 2008 and 14 June 2012).

Acknowledgments: This work was supported by the Power Generation and Electricity Delivery Core Technology program of the Korea Institute of Energy Technology Evaluation and Planning (KETEP), with financial resources from the Ministry of Trade, Industry & Energy, Republic of Korea (nos.20191210301940 and 20206910100090).

Conflicts of Interest: The authors declare no conflict of interest.

References

1. IEA. *Impact of Smart Grid Technologies on Peak Load to 2050*; IEA: Paris, France, 2011.
2. Ministry of Trade, Industry and Energy. *The Fifth National Basic Plan for New and Renewable Energies (2014–2035)*; Ministry of Trade, Industry and Energy: Seoul, Korea, 2020.
3. Alsharif, M.H.; Yahya, K.; Geem, Z.W. Strategic Market Growth and Policy Recommendations for Sustainable Solar Energy Deployment in South Korea. *J. Electr. Eng. Technol.* **2020**, *15*, 803–815. [CrossRef]
4. Dincer, I. Renewable energy and sustainable development: A crucial review. *Renew. Sustain. Energy Rev.* **2000**, *4*, 157–175. [CrossRef]
5. Elnozahy, A.; Yousef, A.M.; Ghoneim, S.S.; Abdelwahab SA, M.; Mohamed, M.; Abo-Elyousr, F.K. Optimal Economic and Environmental Indices for Hybrid PV/Wind-Based Battery Storage System. *J. Electr. Eng. Technol.* **2021**, 1–16.
6. Bahramara, S.; Parsa Moghaddam, M.; Haghifam, M.R. Optimal planning of hybrid renewable energy systems using HOMER: A review. *Renew. Sustain. Energy Rev.* **2016**, *62*, 609–620. [CrossRef]
7. Tixador, P. Superconducting Magnetic Energy Storage: Status and Perspective; Grenoble INP/Institut Néel–G2Elab, B.P. 166, 38 042 Grenoble Cedex 09, France. 2008. Available online: https://snf.ieeecsc.org/sites/ieeecsc.org/files/CR5_Final3_012008.pdf (accessed on 20 August 2021).

8. Mishra, R.; Saxena, R. Comprehensive review of control schemes for battery and super-capacitor energy storage system. In Proceedings of the 2017 7th International Conference on Power Systems (ICPS), Pune, India, 21–23 December 2017; pp. 702–707.
9. Jin, C.; Jiang, X.; Zhong, G.; Li, X. Research on coordinated control strategy of flywheel energy storage array for island microgrid. In Proceedings of the 2017 IEEE Conference on Energy Internet and Energy System Integration (EI2), Beijing, China, 27–28 November 2017; pp. 1–6. [CrossRef]
10. Hao, C.; Yanbing, J.; Jin, Z.; Yanfang, Z.; Gang, L.; Dong, X. Energy Storage Frequency Regulation Energy Management Strategy Based on K-Means Analysis. In Proceedings of the 2019 IEEE 3rd International Conference on Green Energy and Applications (ICGEA), Taiyuan, China, 16–18 March 2019; pp. 163–166. [CrossRef]
11. Park, M.; Jin, Y.; Lee, W.; Won, D. Optimal Operation of Aggregated Industrial Loads Coupled with Energy Storage System. In Proceedings of the 2019 IEEE Milan PowerTech, Milan, Italy, 23–27 June 2019; pp. 1–6. [CrossRef]
12. Morrissey, K.; Kahrobaee, S.; Ioan, A. Optimal Energy Storage Schedules for Load Leveling and Ramp Rate Control in Distribution Systems. In Proceedings of the 2020 IEEE Conference on Technologies for Sustainability (SusTech), Santa Ana, CA, USA, 23–25 April 2020; pp. 1–4. [CrossRef]
13. Pienaar, S.B.; Kusakana, K.; Manditereza, P.T. Usage of Battery Energy Storage Systems to Defer Substation Upgrades. In Proceedings of the 2018 Open Innovations Conference (OI), Johannesburg, South Africa, 3–5 October 2018; pp. 151–156. [CrossRef]
14. Jagtap, K.K.; Patil, G.; Katti, P.K.; Kulkarni, S.B. Techno-economic modeling of Wind-Solar PV and Wind-Solar PV-Biomass hybrid energy system. In Proceedings of the 2016 IEEE International Conference on Power Electronics, Drives and Energy Systems (PEDES), Trivandrum, India, 14–17 December 2016; pp. 1–6. [CrossRef]
15. Manabe, Y.; Hara, R.; Kita, H.; Takitani, K.; Tanabe, T.; Ishikawa, S.; Oomura, T. Cooperation of energy storage systems and biogas generator for stabilization of renewable energy power plants. In Proceedings of the Power Systems Computation Conference, Lyngby, Denmark, 6–9 October 2013; pp. 1–7. [CrossRef]
16. IEC White Paper "Electrical Energy Storage", 3 rue de Varembé PO Box 131 1211 Geneva, 20 Switzerland. 2011. Available online: https://www.academia.edu/31825759/Electrical_Energy_Storage (accessed on 20 August 2021).
17. Divya, K.C.; Ostergaard, J. Battery energy storage technology for power systems-An overview. *Electr. Power Syst. Res.* **2009**, *79*, 511–520. [CrossRef]
18. Vasilije, P.L. Optimal Operation Policy for Energy Stoage. *IEEE Trans.* **1982**, *PAS-101*, 3295–3302. [CrossRef]
19. Jim, E.; Garth, C. *Energy Storage for the Electricity Grid: Benefits and Market Potential Assessment Guide*; Sandia Report; Sandia National Laboratories: Albuquerque, NM, USA; Livermore, CA, USA, 2010.
20. Ye, J.; Xue, J.; Sang, B.; Lu, D.; Liu, H. Economic value and government compensation calculative method of energy storage system. In Proceedings of the 2016 IEEE 8th International Power Electronics and Motion Control Conference (IPEMC-ECCE Asia), Hefei, China, 22–26 May 2016; pp. 536–540. [CrossRef]
21. Choi, S.; Min, S. Optimal Scheduling and Operation of the ESS for Prosumer Market Environment in Grid-Connected Industrial Complex. *IEEE Trans. Ind. Appl.* **2018**, *54*, 1949–1957. [CrossRef]
22. Daeseok, R. Estimation Method of Green House Gas Reduction for EES systems Based on Use-case. In Proceedings of the 146, Dublin 2017 Symposium, CIGRE, Dublin, Ireland, 29 May–2 June 2017; Volume 6.
23. Daeseok, R. Estimation Technique of Green House Gas Reduction for Load Leveling Application in EES systems. In Proceedings of the Annual Conference, IEEJ, Kyushu, Japan, 20–24 October 2018; Volume 3.
24. Saboori, H.; Abdi, H. Application of a grid scale energy storage system to reduce distribution network losses. In Proceedings of the 18th Electric Power Distribution Conference, Kermanshah, Iran, 30 April–1 May 2013; pp. 1–5. [CrossRef]
25. Oudalov, A.; Cherkaoui, R.; Beguin, A. Sizing and Optimal Operation of Battery Energy Storage System for Peak Shaving Application. In Proceedings of the 2007 IEEE Lausanne Power Tech, Lausanne, Switzerland, 1–5 July 2007; pp. 621–662. Available online: https://ieeexplore.ieee.org/document/4538388 (accessed on 20 August 2021).
26. Park, J.; Wu, L.; Choi, J.; Cha, J.; El-Keib, A.A.; Watada, J. Fuzzy theory-based best generation mix considering renewable energy generators. In Proceedings of the 2009 IEEE International Conference on Fuzzy Systems, Jeju, Korea, 20–24 August 2009; pp. 1462–1467. [CrossRef]
27. Gkaidatzis, P.A.; Bouhouras, A.S.; Sgouras, K.I.; Doukas, D.I.; Labridis, D.P. Optimal distributed generation placement problem for renewable and DG units: An innovative approach. In Proceedings of the Mediterranean Conference on Power Generation, Transmission, Distribution and Energy Conversion (MedPower 2016), Belgrade, Serbia, 6–9 November 2016; pp. 1–7. [CrossRef]
28. Viswanath, A.; Goel, L.; Peng, W. Mixed integer programming formulation techniques and applications to Unit Commitment problem. In Proceedings of the 2012 10th International Power & Energy Conference (IPEC), Ho Chi Minh City, Vietnam, 12–14 December 2012; pp. 25–30. [CrossRef]
29. Khan, M.Y.; Ali, M.; Qaisar, S.; Naeem, M.; Chrysostomou, C.; Iqbal, M. Placement Optimization for Renewable Energy Sources: Ontology, Tools, and Wake Models. *IEEE Access* **2020**, *8*, 72781–72800. [CrossRef]
30. Putranto, L.M.; Yasirroni, M. Impact of Sizing and Placement on Energy Storage System in Generation Scheduling Considering Transmission Losses. In Proceedings of the 2019 International Conference on Technologies and Policies in Electric Power & Energy, Yogyakarta, Indonesia, 21–22 October 2019; pp. 1–6. [CrossRef]
31. Lee, H.D.; Tae, D.H.; Kim, J.M. Evaluation Method for Hosting Capacity of PV System in Distribution System with Micro Hydropower Generator. *J. Electr. Eng. Technol.* **2020**, *15*, 2489–2499. [CrossRef]

32. Choi, S.; Kang, M.; Lee, H.; Nam, Y.; Rho, D. A Stable Operation Strategy in Micro-grid Systems without Diesel Generators. *J. Electr. Eng. Technol.* **2018**, *13*, 114–123.
33. ISO Guide 64:2008. Guide for Addressing Environmental Issues in Product Standards, September. 2008. Available online: https://www.iso.org/standard/41352.html (accessed on 20 August 2021).
34. IEC Guide 109:2012. Environmental Aspects—Inclusion in Electrotechnical Product Standards. 14 June 2012. Available online: https://webstore.iec.ch/publication/7520 (accessed on 20 August 2021).

energies

Article

The IRC-PD Tool: A Code to Design Steam and Organic Waste Heat Recovery Units

Youcef Redjeb [1], Khatima Kaabeche-Djerafi [1], Anna Stoppato [2] and Alberto Benato [2,*]

[1] LGPDDPS Laboratory, Department of Process Engineering, National Polytechnic School of Constantine (ENPC), BP 75, Ali Mendjeli, Constantine 25000, Algeria; redjeb.youcef@gmail.com (Y.R.); kaabeche_khatima@yahoo.fr (K.K.-D.)

[2] Department of Industrial Engineering, University of Padova, Via Venezia 1, 35131 Padova, Italy; anna.stoppato@unipd.it

* Correspondence: alberto.benato@unipd.it; Tel.: +39-049-827-6752

Abstract: The Algerian economy and electricity generation sector are strongly dependent on fossil fuels. Over 93% of Algerian exports are hydrocarbons, and approximately 90% of the generated electricity comes from natural gas power plants. However, Algeria is also a country with huge potential in terms of both renewable energy sources and industrial processes waste heat recovery. For these reasons, the government launched an ambitious program to foster renewable energy sources and industrial energy efficiency. In this context, steam and organic Rankine cycles could play a crucial role; however, there is a need for reliable and time-efficient optimization tools that take into account technical, economic, environmental, and safety aspects. For this purpose, the authors built a mathematical tool able to optimize both steam and organic Rankine units. The tool, called Improved Rankine Cycle Plant Designer, was developed in MATLAB environment, uses the Genetic Algorithm toolbox, acquires the fluids thermophysical properties from CoolProp and REFPROP databases, while the safety information is derived from the ASHRAE database. The tool, designed to support the development of both RES and industrial processes waste heat recovery, could perform single or multi-objective optimizations of the steam Rankine cycle layout and of a multiple set of organic Rankine cycle configurations, including the ones which adopt a water or an oil thermal loop. In the case of the ORC unit, the working fluid is selected among more than 120 pure fluids and their mixtures. The turbines' design parameters and the adoption of a water- or an air-cooled condenser are also optimization results. To facilitate the plant layout and working fluid selection, the economic analysis is performed to better evaluate the plant economic feasibility after the thermodynamic optimization of the cycle. Considering the willingness of moving from a fossil to a RES-based economy, there is a need for adopting plants using low environmental impact working fluids. However, because ORC fluids are subjected to environmental and safety issues, as well as phase out, the code also computes the Total Equivalent Warming Impact, provides safety information using the ASHRAE database, and displays an alert if the organic substance is phased out or is going to be banned. To show the tool's potentialities and improve the knowledge on waste heat recovery in bio-gas plants, the authors selected an in-operation facility in which the waste heat is released by a 1 MW_{el} internal combustion engine as the test case. The optimization outcomes reveal that the technical, economic, environmental, and safety performance can be achieved adopting the organic Rankine cycle recuperative configuration. The unit, which adopts Benzene as working fluid, needs to be decoupled from the heat source by means of an oil thermal loop. This optimized solution guarantees to boost the electricity production of the bio-gas facility up to 15%.

Keywords: optimization; organic rankine cycle; steam rankine cycle; energy analysis; economic analysis; environmental analysis

Citation: Redjeb, Y.; Kaabeche-Djerafi, K.; Stoppato, A.; Benato, A. The IRC-PD Tool: A Code to Design Steam and Organic Waste Heat Recovery Units. *Energies* **2021**, *14*, 5611. https://doi.org/10.3390/en14185611

Academic Editors: Claudia Toro, Chiara Martini and T. M. Indra Mahlia

Received: 11 July 2021
Accepted: 1 September 2021
Published: 7 September 2021

1. Introduction

The availability of cheap energy is a key factor in the socio-economic development of the humankind. However, the ever-growing population and the subsequent continuous rise of the energy demand is boosting the human-caused carbon dioxide (CO_2) emissions at an yearly rate ranging from 0.5% to 2%, an unsustainable trend of growth considering that it is well-known that CO_2 and other greenhouse gases (GHGs) emissions are one of the driver of the global climate change [1]. Thus, to fight GHGs effects and meet the ambitious target of limiting global warming to "well below" 2 °C above pre-industrial levels [1], there is an urgent need to shift from a fossil- to a renewable-based energy generation system, as well as supporting the energy efficiency implementation, at least at the industrial level. In fact, the efficient use of energy in the industrial sector is a cornerstone because, based on the International Energy Agency (IEA) estimations, it is the largest producer of CO_2, with over 21.4 Gton [2]. This quantity is not equally shared by neither the industries nor the countries. The power industry (25%), the iron and steel production (7.2%), the cement manufacturing (5%), and the chemicals and petrochemicals (3.6%) are responsible for approximately 40% of the global CO_2 emissions, while Japan (32%), China (28%), US (15%), and India (7%) contribute to over 80% of the total [2]. As known, the CO_2 emissions are the result of the heavy usage of combustion processes adopting oil, natural gas and coal as fuel [3], a fact that highlights the still poor penetration in the industrial sector of low-carbon and renewable-based generation options. However, CO_2 and GHGs are not the unique emissions of the industrial sector because, due to the lack of waste heat recovery units (WHRUs), the sector largely also contributes to thermal pollution. Therefore, to substitute conventional fuels, secure the production, and abate the pollutants, there is a need to simultaneously support the spread of renewable energy source (RES) plants and waste heat recovery units.

In this context, the European Union (EU) is the leading region because (i) it is reaching its GHG emissions reduction targets set for 2020, (ii) it has planned to further cut emissions by at least 55% by 2030, and (iii) it aims to become the world's first climate-neutral continent by 2050 [4,5]. However, despite the EU ambitious goals, global warming and environmental pollution need to be counteracted at a worldwide level, particularly in countries extremely dependent on fossil resources, as well as rich of RES and processes with high energy recovery potential.

This is the case of Algeria, the largest African country in surface area, the gate of the African continent, and a strategic hub for shipping raw materials to EU. Currently, the country's economy is extremely dependent on fossil fuels because oil and natural gas constitute the 93.6% of its export. In addition, the electricity generation sector is strictly bounded to fossil resources; 90% of the electricity is generated in thermal power plants fed by natural gas. In addition, the Algerian soil hosts several energy intensive and large pollutants emitter industries, such as the cement manufacturing.

However, regardless the current structure of the power industries, the country can easily meet its national energy demand by exploiting the considerable RES and waste heat recovery potential [6]. In fact, looking the country's map, it is clear that the major contribution to electricity production can be supplied by solar photovoltaic (PV) installations because the Saharan region, which covers approximately the 86% of the Algerian soil, can meet the national demand, as well as part of the EU one, thanks to an average energy and sunshine duration of 2650 kWh m^{-2} per year and 3500 h per year, respectively [7]. However, to move toward a renewable and sustainable energy generation mix able to (i) cut 193 million tons of CO_2 by 2030 [6], (ii) shift Algeria from the third most significant emitters of CO_2 among the African countries to the forefront of the climate-neutral ones [8], and (iii) free up energy resources for export [8], it is also mandatory to exploit wastes, such as biomass fermentable substances and the waste heat released by industrial processes, RES plants, etc.

To this end, it is important to highlight that biomass fermentable resources can play a crucial role not only in the Algerian energy mix but also in the country's waste management. In 2014, the household and similar wastes sent to landfill reached 14 million tons, and, among them, 8.7 million tons could be used for energy purposes [9]. Therefore, thanks to the anaerobic digestion process, the 8.7 million tons of fermentable biomass can produce 974 million cubic meter of bio-gas, a volume that could generate at least 1685 GWh of electricity, which, in turn, could cover the annual demand of approximately one million of Algerian inhabitants [9]. In this context, it is clear that, for recovering energy from both industrial processes (see, e.g., Reference [10–14]) and RES (see, e.g., Reference [15–19]), the major obstacle is the availability of flexible and time-efficient tools able to properly select the waste heat recovery unit type and design it (plant scheme, as well as devices characteristics, such as turbine type (axial or radial), heat exchanger dimensions, etc.) without requiring modifications, regardless of the heat source, in terms of type, mass flow, and temperature.

To this end, the authors developed the "Improved Rankine Cycle Plant Designer" (IRC-PD), an optimization tool able to design and select the most suitable WHRU between the steam and the organic Rankine cycles (SRC and ORC). In this manner, the tool can provide the WHRU design for each heat source type (liquid or gas) and temperature range (from high to low temperature). The SRC layout implemented in the code is a base one, given the WHRU aim of simplicity the cost-effectiveness, while the ORC configuration is selected among a multiple set of layouts, including the ones which adopt a water or an oil thermal loop, and the working fluid is selected among more than 120 pure fluids and their mixtures. For both SRC and ORC, the optimizer selects the most suitable turbine configuration (single or multi-stage and axial or radial) and condenser type (water- or air-cooled), as well as provides a preliminary design of the entire set of devices that make up the WHRU. Then, according to the optimization goal, which can be single or multiple (e.g., maximum design power), the IRC-PD tool ranks the solutions, and, for each of them, it performs an economic and a safety analysis. Finally, the tool provides a series of maps in which the most promising WHRU design characteristics are classified based on plant safety. To the authors' knowledge, the aforementioned features are points of novelty because in literature no one has integrated into an unique optimization tool (i) the selection and design of both SRC and ORC units considering different plant layouts (including thermal loop) and pure working fluids and their mixtures, (ii) the preliminary design of the expander in both axial and radial configurations, (iii) the condenser type selection, and (iv) the economic and safety analyses.

To test the code ability of selecting the most performing WHRU, the authors choose an internal combustion engine (ICE) fed by bio-gas as a test case. This choice is driven by (i) the Algerian's researchers need to evaluate the waste heat recovery potential in the bio-gas sector and (ii) the Italian's researchers need to improve the knowledge in the ICE's waste heat recovery potential, considering SRC and different ORC configurations. As said, the bio-gas production can be a way for Algeria to both produce electricity near the users and properly manage the fermentable wastes, while, for the Italian bio-gas market, this analysis can help in the further development of a strategic renewable resource, especially after the cessation of feed-in tariffs. In this regard, it is important to highlight that WHRUs in Italian bio-gas plants are rare; therefore, for Algerian and Italian bio-gas ICEs, this study can (i) point out the energetic and environmental benefits, (ii) push the legislator to promote the use of such technologies, (iii) help in the development of new production changes, and (iv) exchange knowledge between two countries with a strategic role in the Mediterranean Sea.

The rest of the work is organized as follows. In Section 2, the IRC-PD tool is presented in detail, while its validation is summarized in Section 3. Section 4 describes the selected case study and the tool settings, while the outcomes of the optimization process are presented and discussed in Section 5. Finally, the concluding remarks are given in Section 6.

2. The Improved Rankine Cycle Plant Designer—IRC-PD

The Improved Rankine Cycle Plant Designer is the result of a joint research between the University of Padova (Italy) and the National Polytechnic School of Constantine (Algeria). The IRC-PD tool, which was developed in MATLAB environment [20], is an "in-house" optimization code able to design two kinds of waste heat recovery unit: the one adopting the steam Rankine cycle and the WHRU operating with the organic Rankine cycle. The code is an updated version of the ORC-PD tool developed by the University of Padova research group starting from the year 2016 [17,21].

The IRC-PD adopts the genetic algorithm (GA) tool available into the MATLAB Global Optimization Toolbox [22] to perform single or multi-objective optimization, while the thermodynamic properties of the fluids are acquired from REFPROP [23] and CoolProp [24] databases. The code is also linked with the ASHRAE 34-2019 database [25] in order to provide fluids safety information based on toxicity and flammability data.

The steam Rankine cycle configuration implemented into the IRC-PD tool is shown in Figure 1. The plant layout is the most simple in order to keep the WHRU cost, weight, and occupied volume as low as possible [26,27]. As a need arising from the fact that the heat source is a waste flux, hence, the SRC design needs to be guided by layout simplicity and, subsequently, cost-effectiveness. Note that, in order to also make available WHRU based on SRC in applications characterized by particular needs (e.g., a gaseous heat source and an area for the WHRU installation far away from the source outlet), the IRC-PD tool can design both the SRC and the thermal loop (using water or oil) in order to decouple the heat source from the WHRU.

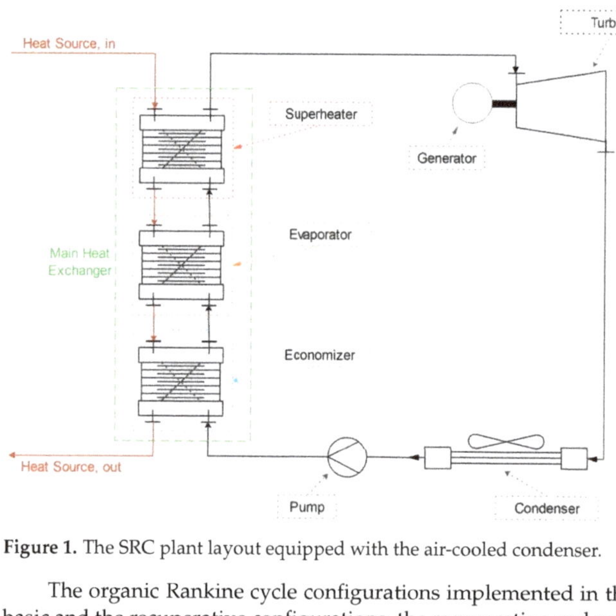

Figure 1. The SRC plant layout equipped with the air-cooled condenser.

The organic Rankine cycle configurations implemented in the IRC-PD tool are the basic and the recuperative configurations, the regenerative and recuperative architecture presented by Branchini et al. [28], the dual pressure and dual fluid layout developed by Shokati et al. [29], and the dual-stage scheme proposed by Meinel et al. [30]. In the aforementioned plant layouts, there is a direct heat exchange between the heat source fluid and the ORC medium, but this is a dangerous way of transferring the heat in case of flammable organic fluids. Therefore, to avoid the risk of flammability that can be occurred in the case of direct contact between the heat source medium and the organic fluid, the IRC-PD tool automatically couples the aforementioned configurations with a thermal loop, which, in turn, adopts water or diathermic oil as working fluid. Specifically, based on plant manufacturers' experience, the water is used for low temperature heat sources because, usually, it reaches a maximum temperature of 160 °C. The experience also suggests to use the diathermic oil (in the present study Therminol 66 or Therminol VP-1) for medium to high temperature heat source because it can operate at a maximum temperature of 360 °C (Therminol 66) or 400 °C (Therminol VP-1).

Note that the adoption of both water and oil thermal loop can be forced in the IRC-PD tool by the user in the case of specific needs arising from the under-investigation test case. As for the thermal loop, the user can also exclude one or more ORC configurations; as an example, the authors can explore only the basic and the recuperative ORC layouts because the others are much too complex or difficult to be managed for the selected case.

Figure 2 depicts the scheme of a recuperative ORC unit equipped with the thermal loop. The condenser is depicted as a water-cooled one, but, as for SRC unit, the user can select the type between the latter and the air-cooled condenser depicted in Figure 1.

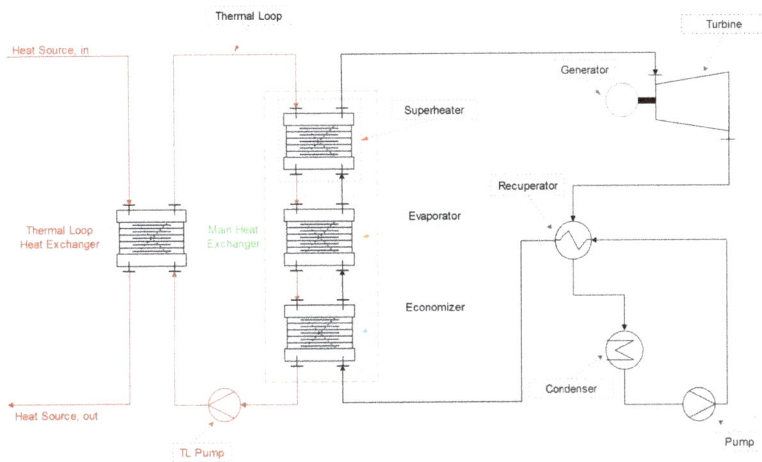

Figure 2. The layout of the recuperative ORC unit equipped with the thermal loop and the water-cooled condenser.

All the proposed ORC layouts can be designed as sub- or trans-critical cycles. The selection of the most appropriate layout and the determination of the necessity of superheating the working fluids are results of the optimization process.

Thanks to the selected way of implementing the IRC-PD tool, it is possible with a unique code to optimize and design ORC units working with (i) low (geothermal, solar, etc.), as well as medium and high, temperature heat sources (exhaust gases released by, e.g., gas turbines, internal combustion engines, industrial processes, etc.); and (ii) heat sources characterized by different nature (e.g., gaseous medium, such as exhaust gases of a boiler or liquid substances, such as the geothermal water).

As extensively discussed in literature, the working fluid of an ORC unit operating with a low or ultra-low temperature heat source is different from the one which needs to be adopted in an ORC linked to a high temperature heat source. To this end, more than 120 pure fluids (HydroCarbons, HydroFluoroCarbons, PerFluoroCarbons, Siloxanes, etc.) and their mixtures are included into the code, thanks to the direct link of the IRC-PD tool with both REFPROP [23] and CoolProp [24] databases. The collection of two databases ensures covering the fluid lack, which can be observed adopting a single source of data.

Regarding the possibility of selecting a mixture as ORC working fluid instead of pure one (see, e.g., Reference [31–33]), it is important to highlight that the IRC-PD tool can (i) use the pre-defined mixture implemented on CoolProp database and (ii) set up the mixture starting from both REFPROP and CoolProp available fluids. To do that avoiding unfeasible composition and time-consuming computations, the method proposed by Venkatarathnam and Timmerhaus [34] is implemented to select the mixture components. Despite the fact that the method was developed for cryogenic refrigerants, Chys et al. [35] suggested its adoption for ORC applications, as well.

The availability in a unique code of two cycle types (SRC and ORC), different configurations for both cycles (including the thermal loop), and, for the ORC unit, a large set of possible working fluid candidates (including pure fluids and mixtures) are hallmarks of the IRC-PD tool. Despite that fact, these features do not guarantee to cover the entire user's needs arising from the fact that the tool has to be able to optimize the WHRU independently to the heat source type and thermophysical properties. To this end, the tool was built in such a way that the user can run both single- and multi-objective optimization using several objective functions (e.g., the maximization of the net electric power, the thermal efficiency, the net present value, etc. or the minimization of the simple pay back time, the exergy losses, etc.) requiring only a few parameters as input:

- Heat source:
  - Medium;
  - Mass flow rate;
  - Inlet temperature;
  - Inlet pressure.
- Pump:
  - Isentropic efficiency;
  - Mechanical efficiency.
- Electric motor efficiency;
- Electric generator efficiency;
- Expander mechanical efficiency.

The variables that can be optimized by the IRC-PD tool, for each working fluid, are:

- the heat source outlet temperature, $T_{\text{hot,out}}$;
- the evaporation pressure of the working fluid, p_{ev};
- the turbine inlet temperature, TIT;
- the ORC medium concentration if the fluid is a mixture, X_1;
- the minimum temperature difference in the main heat exchanger, $\Delta T_{pp,MHE}$;
- the minimum temperature difference in the recuperator if it is present into the cycle, $\Delta T_{pp,rec}$;
- the recuperator efficiency if the components is included into the cycle, E;
- the Minimum temperature difference in the condenser, $\Delta T_{pp,cond}$; and
- the condensation temperature, T_{cond}.

In the case where the thermal loop is adopted, there is also a need to fix the oil or water pump isentropic and mechanical efficiency, while the code also provides, as optimization variables, the oil or water mass flow rate and the returning temperature of such intermediate fluid.

To show how the IRC-PD tool performs the thermodynamic, economic, environmental, and safety analyses, see Sections 2.1–2.3.

2.1. The Thermodynamic Analysis

For the computation of the thermodynamic points and other SRC and ORC parameters, the equations presented in Appendix A are implemented into the code. Appendix A also lists the fluids available into the code (see Table A1).

Compared to previous code [17,21] and with the aim of making the code able to design both SRC and ORC, several innovative features have been implemented.

Firstly, the main heat exchanger, the recuperator (if present), the condenser, and the thermal loop heat exchanger are discretized into "n" elements, and, for each of them, the tool calculates the thermodynamic states of the fluids in exchange in input as in output. Then, the code checks the pinch point and verifies the constraints violation in each element. This process is necessary as a means to better match the hot fluid profile with the cold one, since the pinch point position is not predefined, a feature which guarantees to design sub- and trans-critical cycle without modifying the code. Obviously, a high number of discretized elements is required, as a means to identify the exact position of the pinch point and avoid its violation. The non-predefined pinch point position and the non-fixed pinch point value in heat exchangers are innovative features implemented in the IRC-PD tool. The tool, in case of mixtures, employs the method suggested by Bell and Ghaly [36] to correct the heat transfer coefficient obtained by Shah [37]. This approach is also used to precisely design both the water- and the air-cooled condenser. The possibility of selecting a particular type of condenser based on water availability in the WHRU installation site or the user needs is an important point of novelty that, to the authors' best knowledge, is not present in the literature in this kind of optimization tools.

Another novelty introduced into the IRC-PD tool is its ability of providing, as an optimization variable, the efficiency of the single or multi-stage turbine, as well as its preliminary design. In particular, the overall efficiency of the single-stage or the multi-stage steam turbine for the SRC cycle is computed using the correlations and the correction factors presented by Bahadori and Vuthaluru [38], where, in the case of multi-stage steam turbine, the number of stages is computed fixing the maximum specific enthalpy drop to each stage equal to 150 kJ kg^{-1}.

The isentropic efficiency of the single-stage turbine used for the ORC cycle is estimated by employing the method proposed by Macchi and Perdichizzi [39] for axial flow turbine and the one presented by Perdichizzi and Lozza [40] for radial flow machine.

Whilst, in the case of a multi-stage turbine, the isentropic efficiency is determined by employing the method developed by Astolfi et al. [41], in this method, two limits are added to the tool to be able to compute the number of stages, symbolized in the volume ratio and the specific enthalpy drop. The volume ratio in each stage is fixed equal to 4, whereas the specific enthalpy drop is calculated based on the load coefficient (k_{is}) and the mean peripheral speed. The coefficients are, respectively, set equal to 2 and 255 m s^{-1}, as suggested by Astolfi et al. [41] and Martelli et al. [42]. Therefore, the specific enthalpy drop is computed as:

$$\Delta h_{is,Stage} = \frac{u_m^2 k_{is}}{2} \tag{1}$$

In case the mentioned limits are outstripped, the expansion process is divided into the minimum number of stages that fulfills the previously described conditions. Then, if the number of stages is higher than 3, the value is reset to 3 stages, and the new specific enthalpy drop and the new volume ratio are computed again under the new condition. This change was introduced since some studies mentioned that a number of stages that exceeds 3 does not provide a significant improvement in the turbine's efficiency but increases the complexity and the costs [43].

So, once the specific enthalpy drop and the volume ratio are determined, the tool computes the size parameter and the specific speed of each stage. Then, it estimates the isentropic efficiency for the turbine, whatever its arrangement, axial or radial, by employing the methods presented by Macchi and Perdichizzi [39] and Perdichizzi and Lozza [40], respectively.

Given the code's ability to design both SRC and ORC, there is a need to identify the fluid type in order to properly evaluate the need or not of superheating the fluid itself. To this end, the IRC-PD tool classifies the fluids according to their vapor saturation line into three categories: dry, isentropic, or wet, by employing the method proposed by Liu et al. [44]. The method consists of the derivation of the specific entropy by the temperature:

$$\xi = \frac{ds}{dT_H} \tag{2}$$

The fluid is dry when $\xi > 0$, while it is isentropic and wet in the case of $\xi \approx 0$ and $\xi < 0$, respectively.

For the prediction of the slope, Liu et al. [44] simplified the equation using the relations of the ideal gases:

$$\xi = \frac{c_p}{T_H} = \frac{\frac{n\ T_{rH}}{1-dT_{rH}} + 1}{T_H^2} \Delta h_H \tag{3}$$

where T_H is the standard boiling point, Δh_H is the evaporation specific enthalpy change, n is a coefficient generally varying in the range between 0.375 and 0.380, and c_p is the specific heat, while T_{rH} is defined as:

$$T_{rH} = \frac{T_H}{T_c} \tag{4}$$

where T_c is the critical temperature, and T_H is, again, the standard boiling point.

For the sake of clarity, Figure 3 depicts the vapor saturation line of the three different types of fluids.

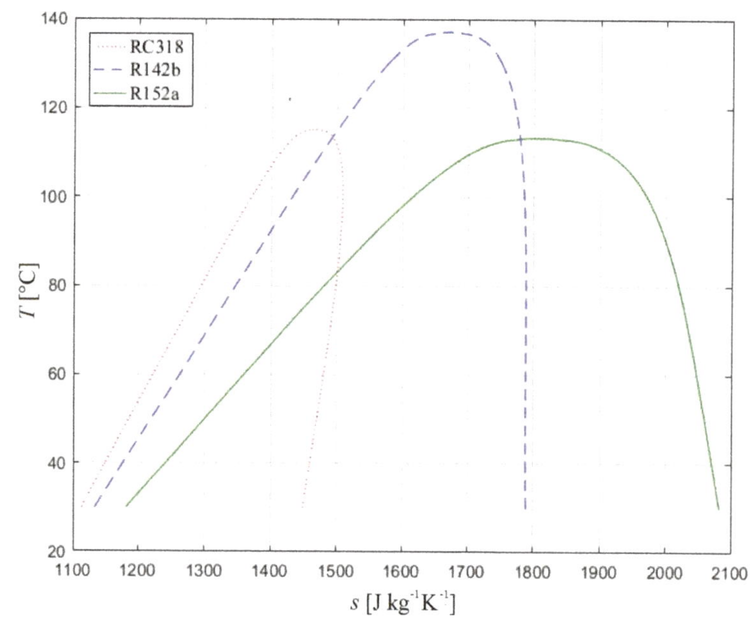

Figure 3. Vapor saturation line of a dry (RC318), an isentropic (R142b), and a wet fluid (R152a).

During the design of both SRC and ORC, there is also a need for checking the vapor quality at the end of the expansion process to avoid the turbine's blade erosion, an aspect which is directly linked to the need for superheating the fluid. To manage this issue, the IRC-PD tool adopts the approach proposed by Wang et al. [45], which introduces the concept of the turning point for the isentropic and dry fluids. This point corresponds with the maximum value of the fluid evaporation temperature that can safely enter the turbine without the need for superheating, since all values below this limit are acceptable, as opposed to higher values, for which superheating is mandatory. However, Wang et al. [45] do not consider the behavior of some fluids, which may show a minimum turning point, as well, where their saturation line takes a reverse path in some points, e.g., Benzene. Therefore, to better describe the behavior of these particular fluids and avoid the turbine's blade erosion caused by a reverse path of the saturation line, the authors introduce the concept of the minimum turning point and combined it with the maximum one defined by Wang et al. [45]. This constitutes an important point of novelty that no one previously adopted into an ORC optimization tool. For the sake of clarity, the two points are depicted in Figure 4.

The check is performed employing Equation (5), where the temperature T is equal to the turning point T_{tn} in the saturated vapor (sv) line:

$$\xi = \left(\frac{ds}{dT}\right)_{sv,T=T_{tn}} = 0 \tag{5}$$

In the event that the evaporation temperature is higher than the maximum or below the minimum turning point, the tool performs another check as a way to guarantee that there is no liquid formation during the expansion process. In case where the liquid is present, the tool runs the simulation again at another evaporation temperature.

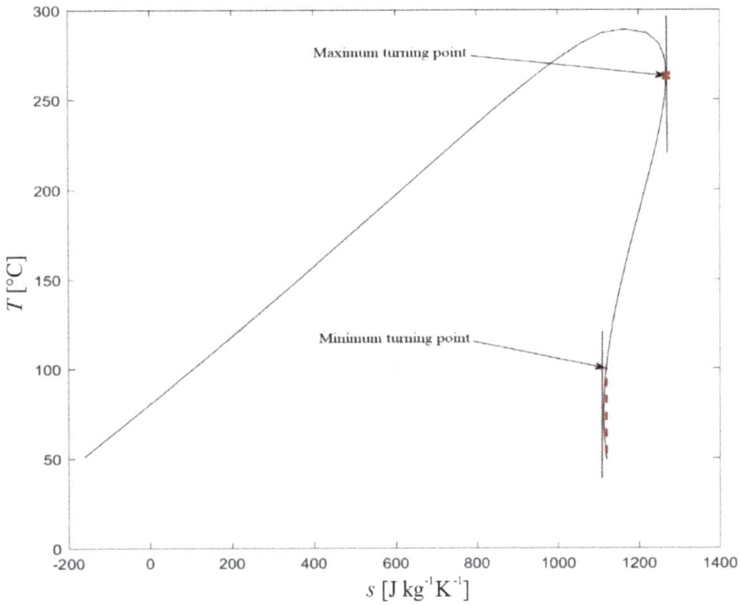

Figure 4. Turning points for the Benzene.

The aforementioned distinctive traits of the IRC-PD tool are coupled with several checks that guarantee to avoid pinch point violations in the heat exchangers, the presence of liquid at the turbine inlet, a low value of steam quality at the turbine outlet, and, if the recuperator is part of the layout, guarantee that the evaporation process does not take place in this device. The IRC-PD tool also provides a warning if the organic fluid has been banned or if it is phasing out. In addition, in the tool, there is another set of checks and warnings devoted to the detection of possible numerical issues that can occur during the evaluation of the fluid thermodynamic properties, especially during the acquisition of values from REFPROP and CoolProp databases.

In the case of a single objective optimization aiming to maximize, e.g., the net output power or the cycle efficiency, at the end of the optimization process, the user can decide whether or not to perform the exergetic, economic, and environmental analyses. For a detailed description of the exergetic analysis, the reader can refer to Pezzuolo et al. [21], while the economic and environmental analyses descriptions are given in Sections 2.2 and 2.3, respectively. The latter is another distinctive trait of the IRC-PD tool because, to the authors' best knowledge, this kind of analysis has not been included in previously presented optimization codes.

In the case of a multi-objective optimization, the code handles the thermodynamic, exergetic, economic, and environmental analysis based on the selected optimization goals.

To the authors' knowledge, the implemented features and the adopted structure of the IRC-PD tool guarantee higher optimization flexibility and lower computation time compared to other optimization tools, such as the ORC-PD one (the latter has also been developed by the University of Padova research group [17,21]).

2.2. Economic Analysis

To estimate the purchase costs of the different devices and carry out a preliminary economic analysis, several correlations and equations are included in the IRC-PD tool in order to cover the equipment size range from small to large. The selection of the equations is performed by the code, depending on: (i) the equipment size, (ii) the case study, and (iii) the desired application.

The economic evaluation is computed adopting the technique of preliminary cost estimation for chemical plants, named Module Costing Method (MCT). This technique guarantees to estimate the purchase costs considering both direct and indirect charges, according to basic estimated costs.

The Main Heat Exchanger (MHE), the recuperator, the water-cooled condenser, and the heat exchanger in the thermal loop can be of different sizes (as said, ranging from small to large), depending on the case study. Thus, the purchase costs of these heat exchangers are estimated by the equation presented in Reference [46] for small-scale WHRUs (heat transfer area lower than 80 m^2), while, in the case of heat transfer area ranging between 80 and 4000 m^2, the code adopts the equation proposed by Smith [47].

The purchase costs of the single-stage turbine, of the pumps adopted in the SRC, the ORC, and the thermal loop, as well as the one of the air cooled condenser, are also computed based on the equations proposed by Turton et al. [46], while the approach suggested by Smith [47] is employed for the fans, including their electric motors. The cost of the electric generator is predicted using the equation developed by Toffolo et al. [48].

For the computation of the purchase cost of the ORC multi-stage turbine, the equation proposed by Astolfi et al. [41] is employed, while the equation suggested by Manesh et al. [49] is used in the case of a steam multi-stage turbine.

In a nutshell, the general form of the purchase cost equation for heat exchangers with an area ranging from 80 m^2 to 4000 m^2, the fan, including its electric motors, and the electric generator is:

$$C_p^0 = C_B \cdot \left(\frac{N}{Q_B}\right)^M \tag{6}$$

where C_p^0 is the purchased equipment cost, C_B is the base cost of the equipment, M is a constant peculiar to each device, and N is the capacity or the size parameter of the equipment, while Q_B is a coefficient.

In contrast, the equation used for other devices, including the heat exchangers with an area lower than 80 m^2, can be expressed as:

$$log_{10}C_P^0 = K_1 + K_2 \cdot log_{10}N + K_3 \cdot (log_{10}N)^2 \tag{7}$$

where N is the capacity or the size parameter of the equipment, and K_1, K_2, and K_3 are correction factors which depend on each piece of equipment.

To compute the bare module cost, C_{BM}, several correction factors, F_{BM}, are needed. Their values are selected based on the system's pressure, the material selected to build the device, and, in some cases, the device's operating temperature. So, the bare module cost general equation for the different devices can be written as:

$$C_{BM} = C_P^0 \cdot F_{BM} \tag{8}$$

where, for the heat exchangers and the fan, including its electric motor, F_{BM} is given as:

$$F_{BM} = F_M \cdot F_P \cdot F_T \tag{9}$$

where F_M is the material correction factor, F_P is the pressure correction factor, and F_T is the temperature correction factor. These correction factors can be determined using the indexes listed in Reference [47].

The remaining devices' bare module correction factors F_{BM} are computed as suggested by Turton et al. [46]:

$$F_{BM} = (B_1 + B_2 \cdot F_M \cdot F_P) \tag{10}$$

where the pressure correction factor, F_P, can be computed as:

$$F_P = 10^{(C_1 + C_2 \cdot log_{10} p + C_3 \cdot (log_{10} p)^2)} \tag{11}$$

C_1, C_2, and C_3 are correction factors peculiar to each device. Conversely, for the electric motor and the electric generator, the F_{BM} is directly given in Reference [46].

The purchase cost equation of the multi-stage expander adopted in the ORC cycle is derived from Reference [41], where the cost is a function of the number of stages n, and the last stage size parameter SP_{LS}:

$$C_{Exp} = C_0 \left(\frac{n}{n_0}\right)^{0.5} \left(\frac{SP_{LS}}{SP_0}\right)^{1.1} \tag{12}$$

where C_0 = 1230 k€, n_0 = 2, and SP_0 = 0.18 m.

In contrast, the cost of the steam turbine, C_{ST}, is calculated employing the steam mass flow rate ($\dot{m}$) and the turbine's inlet pressure (p_{ev}), as suggested by Manesh et al. [49]. The purchase cost is expressed in M$, and it is computed as:

$$C_{ST} = 3.165 + 0.1048 \cdot \dot{m} + 0.01636 \cdot p_{ev} \tag{13}$$

In addition, the chemical engineering plant cost index (CEPCI 2019: 607.5) is adopted to calculate the "new purchase costs" for the different devices. Therefore, the new bare module cost of each equipment is computed in accordance with the equation proposed by Zhang et al. [50].

$$C_{BM_{New}} = C_{BM_{ref}} \frac{CEPCI_{new}}{CEPCI_{ref}} \tag{14}$$

Then, the total cost of the cycle, C_{BM_T}, is calculated as:

$$C_{BM_T} = \sum_{1}^{n} C_{BM_i} \tag{15}$$

where C_{BM_i} represents the investment of each device.

In accordance with Pezzuolo et al. [21], the cost of the site, C_{site}, is determined by multiplying the total cycle cost by 1.4:

$$C_{site} = 1.4\ C_{BM_T} \tag{16}$$

while the operation and maintenance cost, $C_{O\&M}$, is calculated by multiplying the site cost by 0.02:

$$C_{O\&M} = 0.02\ C_{site} \tag{17}$$

The annual benefits deriving from the electricity sale (also named cash flow), CF, can be estimated as:

$$CF = (1 - t_{corp})(S_{annual} - C_{O\&M} - C_{fuel}) \tag{18}$$

where

$$H_{annual} = f \cdot 365 \cdot 24 \tag{19}$$

$$E_{el} = H_{annual}\ P_{el} \tag{20}$$

$$S_{annual} = E_{el}\, s_E \tag{21}$$

t_{corp} is the corporate tax rate, S_{annual} is the annual income from the sale of electricity, f is the operational factor, s_E is the price of electricity, and E_{el} is the annual production of electricity.

The net present value, *NPV,* therefore, is computed as:

$$NPV = CF \cdot RF - C_{site} \tag{22}$$

where the capital recovery factor, *RF*, is given as:

$$RF = \sum_{1}^{n} \frac{1}{(1+i)^n} \tag{23}$$

where i is the annual interest rate and n is the expected life. Thus, the *NPV* can be rewritten as:

$$NPV = \sum_{1}^{n} \frac{CF}{(1+i)^n} - C_{site} \tag{24}$$

The profitability index, *IP*, is calculated by dividing the net present value by the total costs, including the site one:

$$IP = \frac{NPV}{C_{site}} \tag{25}$$

while the Levelized Cost of Energy, *LCOE*, is defined as the cost associated with each unit of produced electrical energy, and it is calculated as:

$$LCOE = \frac{C_{site} + C_{O\&M} \cdot RF}{E_{annual} \cdot RF} \tag{26}$$

Finally, the simple payback, *SPB*, which is the ratio between the total costs and the annual benefits, can be expressed as:

$$SPB = \frac{C_{site}}{CF} \tag{27}$$

More details about parameters and assumption adopted in the economic analysis can be found in Appendix B.

2.3. Environmental and Safety Analysis

The environmental impact and the safety conditions are considered as a key point during the selection of suitable fluids. In fact, the fluids should have low toxicity and flammability to avoid the need for site protection measures.

So, as a way to assess the safety of the fluids, the concept introduced by ASHRAE standard 34 [51] is employed, where the tool can classify the working fluid based on its safety level in accordance with the classes defined in Table 1.

Table 1. Classification of fluids according to ASHRAE 34 [51].

	Low Toxicity	High Toxicity
High flammability	A3	B3
Lower flammability	A2 A2L	B2 B2L
No flame propagation	A1	B1

Then, from an environmental point of view, the value of Ozone Depletion Potential (ODP) and Global Warming Potential (GWP) need to be in the controlled zone as specified by the international treaties and regulations, such as the Montreal protocol for ODP [52] and the EU Directive 2006/40 for GWP [53]. For this purpose, the Total Equivalent Warming Impact (*TEWI*) method is adopted and implemented in the tool for environmental analysis.

The *TEWI* general formula for the estimation of the environmental impacts is defined in accordance with Gullo et al. [54]:

$$TEWI = TEWI_{direct} + TEWI_{indirect} \tag{28}$$

where $TEWI_{direct}$ represents the direct emissions of greenhouse gases, including the leakages of refrigerant into the atmosphere [55]. The latter can be calculated employing the *GWP* of the fluid:

$$TEWI_{direct} = GWP \cdot L \cdot n + GWP \cdot M \cdot (1 - \alpha) \tag{29}$$

where L, M, α, and n are the fluid leakage (kg year^{-1}), the refrigerant charge (kg), the recycling factor (%), and the system lifetime (year).

The $TEWI_{indirect}$ refers to the CO_2 emissions due to the generation of the consumed electricity [55,56], and it is computed as:

$$TEWI_{indirect} = Ea \cdot \beta \cdot n \tag{30}$$

where Ea, β, and n are the consumed energy (kWh), the carbon dioxide emission factor (kg$CO_{2_{eq}}$ kWh^{-1}), and the system lifespan (year).

To perform the environmental and safety analysis, the IRC-PD tool is linked to a separated database that contains the different environmental and safety parameters for the different fluids, as well as information regarding the organic fluid phase out. These parameters are derived from ASHRAE 34 or predicted from material safety data sheets of the fluids which are available in the literature (see, e.g., Reference [57]). In this way, it is possible to rank the ORC optimize configurations based on their safety and environmental friendliness.

Note that, for the refrigerant charge, the method suggested by Collings et al. [58] is implemented in the tool to estimate its value, where it depends on the energy transferred through a heat exchanger and the power extracted through the turbine.

3. The Tool Validation Procedure

The validation process of a new tool is a fundamental step, especially in the case of highly complex tools, such as the IRC-PD one. To this purpose, the code needs to be tested using different plant configurations and heat sources as suggested by, e.g., Pezzuolo et al. [21], where several cases are taken from the literature and use as reference for the comparison with the developed tool.

For testing the ORC basic configuration, the IRC-PD tool was set up with the specifications presented by Vaja and Gambarotta [59]. Three fluids have been screened, the first one without superheating, and the rest with the superheating. For the first one, Benzene, the results obtained by the IRC-PD tool exhibit a deviation of about 3% in terms of net power output, while the maximum deviation is around 3.8% if the turbine inlet volumetric flow is considered. The other parameters show a deviation lower than 1%.

The results obtained with the IRC-PD tool for the fluids R11 and R134a, and the same conditions reported by Vaja and Gambarotta [59], denote a deviation of about 2.5% in the net power output for R11, and lower than 1.7% for R134a. For the mass flow rate of the working fluid, the adoption of R11 showed a deviation less than 2.6%, while a deviation less than 1.5% is observed in the case of R134a. The other parameters, as previously, denote a deviation lower than 1%. A clear overview of the obtained results are given in Table 2. The small deviations in the results can be mainly attributable to a not precise estimation of the heat source mass flow rate. In fact, this value is not listed in ref. [59]; thus, the authors estimate it.

To validate the recuperative configuration, the authors selected the case study presented by Chys et al. [35], in which five pure fluids are considered as possible working medium candidates. The comparison between the reference case and the IRC-PD tool results are given in Table 3. The deviations analysis clearly pointed out that the discrepancy are in the 0.34–1.25% for the generated power, while, for the mass flow rate and the efficiency, the maximum deviation reaches 1.79% and 1.06%, respectively. As in the case of the basic configuration, the recuperative one can also be considered validated, given that the IRC-PD tool results are in line with the one reported in the selected reference.

The IRC-PD validation also included the configurations adopting the thermal oil loop as a medium to transfer the heat from the source to the ORC. In this case, the reference work is the one presented by Liu et al. [60], where the adopted oil is the so-called Dowtherm Q. As for the other validation scenarios, the results revealed that the highest deviation is observed in the working fluid's mass flow rates and the cycle efficiency (1.6% and 1.5%, respectively), while the other parameters present a deviations below 1%, as listed in Table 4.

Table 2. Comparisons between the results obtained by the IRC-PD tool and the ones listed in ref. [59].

Fluid	P_{ORC} [kW]	η_{ORC} [-]	p_{cond} [kPa]	p_{ev} [kPa]	T_{ev} [K]	$\dot{m}_{wf}$ [kg s^{-1}]	$\dot{V}_3$ [m^3 s^{-1}]	$\dot{V}_4/\dot{V}_3$ [-]	
Benzene	349.3	0.198	19.6	2000	494.5	2.737	0.052	107	Ref.
	338.0	0.198	19.7	2000	494.56	2.647	0.05	107.7	IRC-PD
R11	290.3	0.166	147.9	3835.9	461	7.487	0.03	32	Ref.
	283.0	0.166	148	3836	461.56	7.293	0.03	32.3	IRC-PD
R134a	147.5	0.0852	883.3	3723.4	369.9	8.9667	0.041	5	Ref.
	145.0	0.0852	883	3723	369.94	8.833	0.041	5.1	IRC-PD

Table 3. Comparisons between the results obtained by the IRC-PD tool and the ones listed in ref. [35].

Fluid	p_{ev} [bar]	p_{cond} [bar]	p_{ratio} [-]	$\dot{m}_{wf}$ [kg s^{-1}]	P_P [kW]	P_{Gen} [kW]	η_{ORC} [%]	
Cyclopentane	21.2	1.85	11.5	3.84	−13.3	247.5	13.02	Ref.
	21.71	1.85	11.73	3.82	−13.6	248.6	13	IRC-PD
Toluene	4.6	0.26	17.6	3.76	−2.5	239.1	13.15	Ref.
	4.71	0.26	18.11	3.76	−2.5	242.1	13.3	IRC-PD
Cyclohexane	9.5	0.7	13.6	3.92	−5.9	244.7	13.28	Ref.
	9.56	0.7	13.65	3.93	−5.9	246	13.3	IRC-PD
OMTS	2.1	0.05	38.6	7.1	−2.3	242.5	13.35	Ref.
	2.07	0.05	39.05	7.1	−2.3	243.9	13.4	IRC-PD
HMDS	7.4	0.35	21	6.37	−7.8	249.3	13.42	Ref.
	7.45	0.35	21.28	6.37	−7.9	250.3	13.4	IRC-PD

Table 4. Comparisons between the results obtained by the IRC-PD tool and the ones listed in ref. [60]. In the thermal loop flows the Dowtherm Q oil.

Fluid	P_{ORC} [kW]	η_{ORC} [%]	T_{cond} [°C]	$\dot{m}_{wf}$ [kg s^{-1}]	p_{ev} [kPa]	
R245FA	87.2	13.3	35	2.6	2000	Ref.
	87.4	13.5	35	2.56	2000	IRC-PD

With the aim of exploring the cycle's performance improvements and the computational cost of adopting a detailed method to predict the turbine's performance and its design, the authors constrained the IRC-PD tool as the ORC-PD one and considered the case presented by Benato and Macor [17]. The simulation outcomes are given in Table 5.

Table 5. Comparison between the results obtained with the IRC-PD tool using the same settings of ref. [17] but with the updated turbine's features.

Fluid	P_{ORC} [kW]	*Stages* [-]	*Model*
Toluene	137.8	1-R	Ref.
	169.9	3-R	IRC-PD
Benzene	134.1	1-R	Ref.
	161.6	3-R	IRC-PD
Acetone	123.6	1-R	Ref.
	156.5	3-R	IRC-PD

The results show that the adoption of a multi-stage turbine instead of a single-stage one guarantees higher isentropic efficiency of the ORC turbine, thus contributing to higher net output power. In fact, the cycle working with Toluene and adopting a multi-stage turbine generates a net output power 23% higher than the same cycle mounting a single-stage turbine [17]. Similarly, adopting a multi-stage turbine instead of a single one (as given in Ref. [17]) guarantees reaching a net power output 15% and 14% higher in the case of Benzene and Acetone, as well. In terms of computational efforts, the introduced features do not drastically affect the speed. In fact, the computational time only increases 2%. These outcomes clearly show the importance of adopting a detailed model for the turbine at the optimization stage, as well, because this is the only way that it is possible to properly predict the generated electricity and the cycle performance, as well as provide a preliminary design of the machine.

Finally, the steam cycle configuration was tested replicating the plant setup presented by Nord et al. [26]. As previously, the simulation outcomes derived with the IRC-PD tool are perfectly in line with the one reported in the reference because the highest observed deviation is smaller than 2%.

In a nutshell, considering the obtained findings and the large variety of performed tests, it is possible to claim that the code is able to replicate the results reported in the references; thus, it can considered validated.

4. Case Study and Optimization Settings

To test the code ability of selecting the most performing WHRU, the authors chose an internal combustion engine fed by bio-gas as the test case. This choice was driven by the need for the Algerian's research group to evaluate the benefits of adding a WHR to upcoming bio-gas installations, while the Italian's researchers want to evaluate the ICE's recovery potential considering both SRC and ORC configurations.

The selected ICE is an Italian power system installed on a bio-gas plant located in Northern Italy, as well as the power unit that will be installed on future Algerian bio-gas plants. The bio-gas ICE is a GE power unit [61], and the nameplate data is listed in Table 6.

The plant adopts a standard and well-established layout made up of a reception tank, two primary fermenters and two secondary fermenters, a gas holder, an overflow tank, and the bio-gas engine. The ICE waste heat recovery can be done only in the exhaust gases stage because of the heat of the cooling water and lube oil already recovered and used to maintain the digesters at a temperature of 42–44 °C.

Table 6. Technical data of the bio-gas ICE at design point conditions as given by the manufacturer [61].

Parameter	Value
Number of cylinders	20
Bore [mm]	135
Stroke [mm]	170
Displacement/cylinder [liters]	2.433
Rotational speed [rpm]	1500
Mean speed of the piston [m s^{-1}]	8.5
Electrical power [kW]	999
Thermal power [kW]	2459
Electrical efficiency [%]	40.58
Mass flow rate of the flue gases [kg h^{-1}]	5312
Temperature of the flue gases [°C]	457

The data used to design the WHRUs are not the nameplate ones but the results of an experimental campaign that demonstrated a large mismatch between the measured and the nameplate exhaust gases mass flow rate and temperature [17]. In particular, the measured mass flow rate and temperature of the flue gases are 6477 kg h^{-1} and 503 °C, while the exhaust gas compositions on mole fractions are: CO_2 (6%), N_2 (74%), O_2 (14%), H_2O (5%), and Argon (1%). More details about the experimental campaign can be found in ref. [17].

The use of these data guarantees selection and design of the WHRU based on real data, as well as comparison of the obtained findings with the ones presented by Benato and Macor [17], where a less advanced optimization tool was adopted.

Note that, despite the fact that the literature presents a large number of innovative technologies to recover the ICE's waste heat (see, e.g., Kalina cycle [62] and super-critical CO_2 cycle [63]), acceptable performance improvements and techno-economic feasibility are today reachable only with WHRUs based on the steam and the organic Rankine cycle.

As an example, Yu et al. [64] proposed to recover the exhaust gases heat content of a heavy-duty diesel engine by means of a cascaded dual-loop WHRU composed by an SRC and an ORC. The results revealed that the 101.5 kW of waste heat can be generated up to 12.7 kW, ensuring a 5.6% power increment of the system. Similarly, Liu et al. [65] analyzed the possibility of recovering the waste heat of a 14-cylinders marine engine using a WHRU which combines an SRC and the dual pressure ORC. They also compared the performance of this configuration with the one reachable with a WHRu composed by an SRC or a dual pressure ORC. The results outline that combining the SRC and the dual pressure ORC guarantees a fuel saving of 9355 tons per year and an improvement of the system's efficiency of 4.42%. Contrary, the SRC and the ORC alone guarantee a higher simplicity but lower thermal efficiency improvements, 2.68% and 3.42%, respectively. Andreasen et al. [66] studied how to improve the performance of a 23-MW two-stroke MAN diesel engine working at a load variable between 25 and 100% of the design power. They proposed to adopt a dual pressure steam Rankine cycle or an ORC. The results of the simulations indicate that the SRC unit is able to improve the power of 18%, while the ORC unit adopting MM produces 33% more power.

In contrast to previous studies, Yang et al. [67], Wang et al. [68], and Song and Gu [69] proposed to recover the diesel engine waste heat by means of a dual loop ORC, while Shu et al. [70] evaluated the system performance improvements reachable with a recuperative ORC layout using a mixture as working fluid. In particular, the use of a mixture of Benzene and R11 can increase the system thermal efficiency up to 16.7%.

As said, several works available in literature study the diesel ICE's waste heat recovery using ORC, especially for marine applications (see, e.g., References [59,71–75]), while only a few are focused on engines fed by bio-gas.

Schulz et al. [76] and Kane et al. [77] were among the first to suggest the use of the ORC technology to improve the agricultural bio-gas plants performance and to apply the ORC to these engines. In particular, Kane et al. [77] investigated the use of an ORC unit for waste heat recovery from the cooling jacket of a 200 kW_e bio-gas ICE. They found that the ORC can reach an efficiency up to 7%.

In addition, Meinel et al. [30] explored the benefits of adding an ORC to a bio-gas ICE. However, in their investigations, an innovative two-stage ORC configuration was proposed and the best working fluid selected among 4 media. The ORC heat source is the constituted by the exhaust gases at 490 °C and 1 bar. The outcomes of the study indicate that the two-stage layout without the recuperator boosts the performance of wet and isentropic fluids, while the one with the recuperator is more appropriate for dry fluids. In terms of performance, the two-stage non-recuperative configuration operating with isentropic fluids improves the thermodynamic efficiency up to 2.25% compared to conventional ORC layout, while the recuperative one operated with dry fluids reaches efficiency improvement up to 2.68% compared to standard ORC. Contrary to Meinel et al. [30], Dumont et al. [78] and Koç et al. [79] proposed to improve the bio-gas ICE performance adopting standard configurations: a non-recuperative ORC layout and a regenerative one, respectively. Dumont et al. [78] performed a thermo-economic optimization with the aim of defining the architecture, the working fluid, and the plant components, while Koç et al. [79], after a parametric optimization, carried out an exergy analysis. In Reference [78], the results indicate that R1233zd(E), R245fa, and Ethanol guarantee higher electricity production compared to R134a and R1234yf, but they require higher investments, while, in Reference [79], the authors observed that the higher exergy destruction is in the evaporator. However, the overall thermal and exergetic efficiency are 19.17% and 32.41% for the sub-critical recuperative ORC layout, while they become 18.50% and 31.67% for the super-critical unit. This is not a marginal increment, considering that the sub- and super-critical non-recuperative ORC configuration can reach thermal and exergetic efficiency equal to 15.51% and 27.20% and 15.93% and 27.76%, respectively.

Finally, Saravia et al. [80] and Uusitalo et al. [81] proposed to retrofit the bio-gas ICE with an ORC to reduce engine's fuel consumption and GHG emissions, respectively. In the first case, an ORC was added to a 6 MW_{el} in-operation ICE fed with landfill bio-gas; a system improvement that guarantees higher overall power production by recovering 5–10% of the fuel energy content. In the second case, using the LCA approach, Uusitalo et al. [81] evaluated the environmental benefits in terms of GHG emissions reduction introduced with the ORC unit. They observed that adding the ORC guarantees a GHG emissions reduction in the range 280–820 ton of $CO_{2,eq}$, depending on the type of substituted electricity, while the impact of the ORC and its working fluid is only the 0.1% of the total bio-gas ICE power plant GHG emissions.

Starting from the literature analysis, it is clear that it is not convenient to study all the possible configurations of the ORC cycle because some of them are characterized by high complexity and costs. In addition, the use of mixtures as working fluids is not convenient (also see Benato and Macor [17]), but the authors set the IRC-PD tool free to explore all the possible configurations and the use of pure, as well as mixtures, as working fluid to be sure that the code excludes these not-performing and less cost-effective configurations and fluids. For the entire set of WHRUs, the adoption of the water and the oil thermal loop is explored, as well as configurations in which the ICE's exhaust and the working fluid directly exchange the heat. Again, considering the motivation of the work and, in particular, a need to maximize the waste heat recovery, the authors performed a single-objective optimization aiming to maximize the net output power of the WHRU, followed by an economic and environmental analysis. For the sake of clarity, the optimization steps are summarized in Figure A2 (Appendix A), and ORC of the upper and lower bound of the optimized variables is listed in Table 7.

Table 7. Upper and lower bound used in the optimization of both SRC and ORC cycles.

Parameter	**LB**	**UB**
Heat source outlet temperature, $T_{\text{hot,out}}$ [°C]	90	$T_{\text{hot,in}}$
Evaporation pressure of the steam for the SRC, p_{ev} [bar]	15	40
Evaporation pressure of the organic medium for the ORC, p_{ev} [bar]	p_{cond}	p_{max}
First mixture component concentration, X_1 [-]	0	1
Turbine Inlet Temperature, *TIT* [°C]	TIT_{min}	TIT_{max}
Condensation temperature, T_{cond} [°C]	30	90
Recuperator efficiency, *E* [-]	0	0.8
Minimum temperature difference in the MHE, $\Delta T_{\text{pp,MHE}}$ [°C]	25	100
Minimum temperature difference in the recuperator, $\Delta T_{\text{pp,rec}}$ [°C]	20	100
Minimum temperature difference in the condenser, $\Delta T_{\text{pp,cond}}$ [°C]	10	100
In case of adopting the thermal loop	**LB**	**UB**
Minimum temperature difference in the TL HX, $\Delta T_{\text{pp,TLHX}}$ [°C]	10	100
Thermal loop "oil" outlet temperature from the cycle $\Delta T_{\text{oil,out}}$ [°C]	90	$T_{\text{oil,in}}$

The minimum and the maximum evaporation pressure for the SRC cycle are fixed equal to 15 bar and 40 bar in accordance with the specifications provided by Nord et al. [26].

The maximum pressure, p_{max}, of the ORC cycle is assumed to be the minimum between the critical pressure, p_{crit}, and 35 bar.

$$p_{max} = min(p_{crit}, 35\ bar) \tag{31}$$

The selection of this value guarantees reasonable pumping conditions as given by Marcuccilli and Zouaghi [82], as well as reduces the material expenses and improves the plant safety, as underlined by Javanshir et al. [83]. Therefore, the adoption of a pressure lower than 35 bar avoids the need for very expensive pipes, heat exchangers, etc., as well as control and management systems.

On the other side, from the thermophysical point of view, the fluid must have adequate chemical stability in the desired temperature ranges and should have good compatibility with the material in contact with, as the organic fluids prove chemical deterioration and decomposition at high temperature, as pointed out in References [35,84]. To this end, the maximum temperature of the cycle, TIT_{max}, is selected as the minimum between:

$$TIT_{max} = T_{hot,in} - \Delta T_{pp,MHE} \quad \text{and} \tag{32}$$

$$TIT_{max} = T_{decomposition} \tag{33}$$

where $T_{hot,in}$ is the temperature of the fluid entering the ORC main heat exchanger (MHE), and $\Delta T_{pp,MHE}$ is the minimum temperature difference in the MHE, while $T_{decomposition}$ is the organic fluid decomposition temperature.

When a thermal loop is adopted, the user can select the oil type among Therminol VP-1, Therminol 66, and Dowtherm Q. In the analyzed case, the authors' choice fell on Therminol VP-1 due to its high safety level and stability, no toxicity, and availability at an acceptable price. The oil inlet temperature ($T_{oil,in}$) is assumed as follows:

$$T_{oil,in} = min(T_{hot,in} - \Delta T_{pp,TLHX}, T_{max,bulk} - 40\ °\text{C}) \tag{34}$$

where $T_{max,bulk}$ represents the maximum operating temperature of the thermal oil without risk of thermal degradation, while $T_{hot,in}$ is, in this case, the temperature of the hot source at the inlet section of the thermal loop heat exchanger. In the case of Therminol VP-1, T_{Bulk} is equal to 400 °C.

In regard to the setting of the water loop inlet and outlet temperatures, the authors fix them equal to 160 °C and 140 °C, respectively, a choice driven by the manufacturers' experience which prescribe to limit the water pressure in the thermal loop and, consequently, the cost of the devices that made up the loop itself.

The different parameters assumed for the genetic algorithm set up are:

- Population size: 500;
- Generation size: 350;
- Crossover Fraction: 0.8;
- Migration Fraction: 0.2.

These assumed values are checked in term of computational time and results accuracy, and they confirmed that adopting higher values for both population and generation (e.g., 700 instead of 500 in population and 500 instead of 350 in generation) do not provide more accurate results but only increases the computational time of up to 30%. For the same reasons, the number of elements in which the heat exchangers have been discretized is assumed equal to 50.

To perform the economic and environmental analysis, the authors assumed a WHRU lifespan equal to 15 years, while the value of the fluid leakage (L_{rate}) in the ORC and of the recycling factor (α) is considered equal to 0.02 and 0.8, as prescribed by Gerber and Maréchal [85].

5. Results and Discussion

In this work, the authors perform an optimization aimed at finding the most appropriate technology between SRC and ORC cycles for the selected case study, besides the determination of the most suitable plant configuration and working fluid that guarantees the maximization of the net output power. Concerning the economic analysis of the optimized solutions, it is important to point out that it is difficult to perform it in terms of net present value or simple pay back due to the need for estimating the electricity selling price, a difficult task considering the uncertainties linked to support schemes establish by Governments. So, the authors do not perform an economic analysis but only compute the investment costs of the WHRU.

The tool is set in such a way that, at the end of the optimization process, it provides a set of plots where, for each safety category, the 3 best working fluids that guarantee the maximization of the net output power are shown versus the cost of the WHR unit.

The picture is given for the case with no thermal loop (see Figure 5), as well as for the water and oil thermal loop (see Figures 6 and 7, respectively).

For the sake of clarity, the authors list the results obtained for the SRC when the heat is exchanged directly between the ICE's exhaust and the cycle (see Table A4), as well as in the case of adopting a water and an oil thermal loop (see Tables A5 and A6, respectively). Similarly, Table A7 lists the optimization findings in the case of the ORC without thermal loop, while Tables A8 and A9 report the results of the optimizations when a water and an oil loop is adopted. For compactness, only the first five fluids are listed for each plant configuration. The cost analysis of the different ORC arrangements is given in Tables A10–A12.

Analyzing the results, it is clear that neither the mixtures nor the regenerative and recuperative, the dual pressure, the dual fluids, and the dual stage are appropriate architectures for the analyzed test case. Additionally, the SRC also does not guarantee acceptable performance compared to ORCs employing pure fluids and a recuperative configuration.

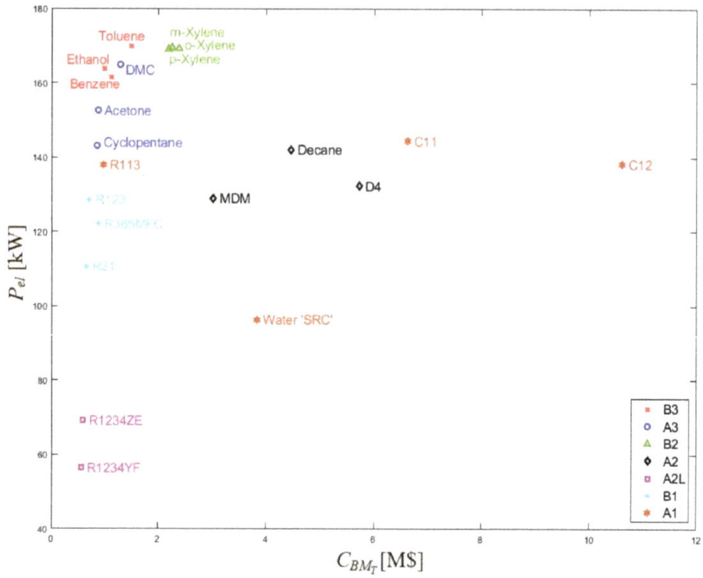

Figure 5. Net output power versus plant cost of the best 3 fluids of each safety category in the case of no thermal loop.

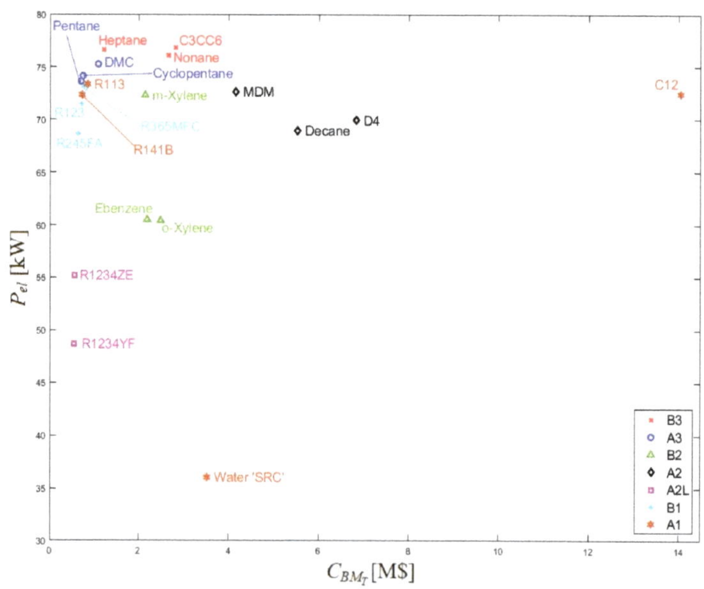

Figure 6. Net output power versus plant cost of the best 3 fluids of each safety category in the case of water thermal loop.

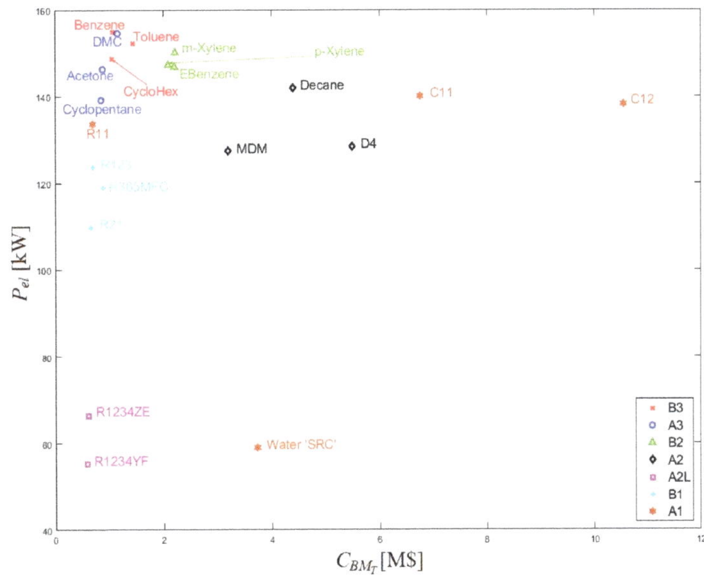

Figure 7. Net output power versus plant cost of the best 3 fluids of each safety category in the case of oil thermal loop.

The SRC which guarantees the highest performance (96.37 kW) exchanges directly with the ICE exhaust gases and adopts a multi-stage steam turbine with 7 stages and an isentropic efficiency of 52.8%. The latter is an estimated value perfectly in line with the one expected by the steam turbine manufacturer that collaborates with the authors and the available in literature (see, e.g., Reference [86]). In contrast, the cycle has an optimal evaporation pressure of 19.22 bar, a condensation pressure of 0.15 bar, and a thermal efficiency of 13.96% as listed in Table A4. In addition, the plant is characterized by unfeasible cost per kW_{el} (41 k$ per kW_{el}) because the plant cost reaches the 3.83 M$. It is also important to note that the use of a water or an oil thermal loop drastically reduces the waste heat recovery and increases the plant costs compared to a direct exchange layout (see Tables A5 and A6). In addition, because water is a non-flammable and non-toxic fluid, there is no safety reason that justifies the adoption of a thermal loop. Finally, it is important to remark that the authors expected that the steam cycle was not a feasible solution for the selected test case due to the small amount of available heat.

Focusing on the ORC solutions, it is clear that the best performance in terms of net power maximization is guaranteed by a direct exchange between the ICE exhaust and the ORC working fluid. In particular, as shown in Figure 5, the highest performance is reached with Toluene (169.89 kW), followed by M-xylene and O-xylene with 169.50 kW and 169.27 kW (see Table A7 for more details). However, Toluene is a fluid belonging to category B3 (high toxicity and flammability); thus, for safety reasons, it is not convenient to build a plant in which a high toxic and flammable fluid exchange the heat directly with ICE exhaust. Therefore, the use of such fluid is excluded.

However, the adoption of M-xylene and O-xylene is also not applicable due to the really low condensation pressure, as highlighted by Figure 8. Additionally, these ORC units are characterized by costs ranging from 1.5 to 2.2 M$; thus, approximately 9 k$ per kW_{el}, an unsustainable investment for a bio-gas owner considering that the investment for a bio-gas system is, excluding the ORC, approximately 4.5–5 M$.

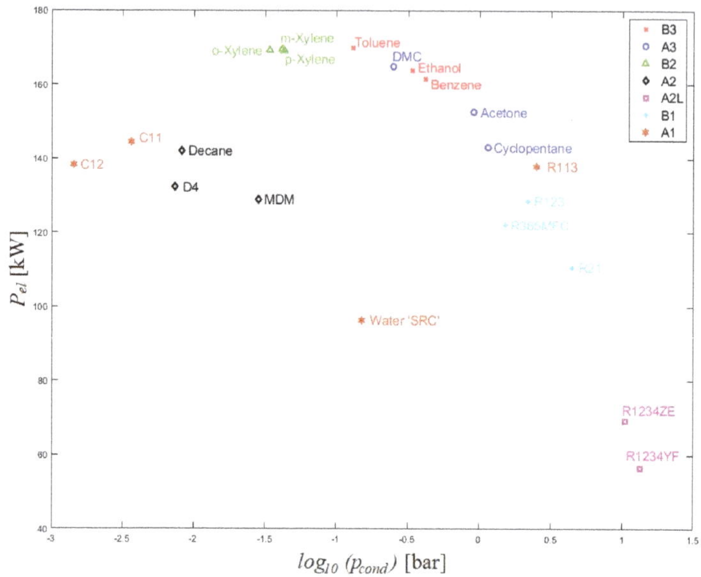

Figure 8. Net output power versus condensation pressure (in logarithmic scale) of the best 3 fluids of each safety category in the case of no thermal loop.

Comparing Figure 5 with Figure 6, it is clear that the adoption of a water loop halved the power producible by the ORC. Therefore, the insertion of a water loop must be avoided. Contrary, the insertion of an oil loop (see Figure 7) adopting Therminol VP-1 guarantees, in the best case (Benzene), a reduction of the ORC net output power of 10% compared to the case without thermal loop (detailed values are given in Tables A8 and A9). Benzene guarantees a net power output only 5% higher (161.60 kW) if directly coupled with ICE's exhaust compared to using an oil loop, while the use of DMC, Toluene, Cyclohexane, and M-xylene ensures higher net power output: 6%, 12%, 2%, and 14%. Conversely, the adoption of a water thermal loop provokes a 50% reduction of the power producible by Benzene compared to the cycle adopting Therminol VP-1. Thus, for this application, the use of a thermal oil loop seems the preferable choice. For the sake of clarity, in Table 8, the best and the worst 5 fluids are listed, along with cycle and turbine characteristics.

Table 8. The 5 best and worst fluids and in the case of an ORC adopting an oil thermal loop.

Fluid	P_{el} [kW]	p_{ev} [bar]	p_{cond} [bar]	E [%]	η_{ORC} [%]	$\eta_{is,T}$ [%]	*Stages* [-]	Type [-]	Safety [-]
Benzene	154.92	32.25	0.42	53	20.45	0.855	3-R	Dry	B3
DMC	154.62	34.94	0.26	44	20.46	0.852	3-R	Dry	A3
Toluene	152.21	15.68	0.15	77	21.68	0.864	3-R	Dry	B3
M-xylene	150.17	7.58	0.05	68	21.33	0.868	3-R	Dry	B2
Cyclohexane	148.75	34.03	0.42	78	21.91	0.852	3-R	Dry	B3
...	...	...	...	...	...	...	...	...	...
...	...	...	...	...	...	...	...	...	...
CycloPropane	67.590	34.98	13.74	53	9.01	0.852	2-R	Wet	A3
R1234ze	66.435	34.48	10.45	47	8.43	0.859	1-R	Dry	A2L
R236fa	65.057	23.51	6.13	30	8.16	0.867	2-R	Dry	A1
R134a	64.861	34.62	13.80	73	8.69	0.862	1-R	Wet	A1
R1234yf	55.347	33.15	13.29	50	7.08	0.862	1-R	Dry	A2L

In fact, the results reveal that the cycle can achieve the maximum net power output using Benzene as working fluid, followed by Dimethylcarbonate (DMC), a medium that guarantees a reduction of 0.1% in terms of net output power. Conversely, Toluene, M-xylene, and Cyclohexane guarantee a net output power only 2%, 3%, and 4% lower than the one given with Benzene (see Figure 7).

M-xylene and Toluene are not preferable from a technical point of view, and they can be excluded from the list since their condensation pressures are very low compared to other fluids which show reasonable values (see Figure 9 and Table 8). In fact, as depicted in Figure 10, such low condensation pressure leads to higher purchase cost of the ORC unit, besides to more complexity in the plant. In return, the optimization results exhibit that the use of these promising fluids requires a thermal loop as a means to avoid a direct contact between the heat source and the working medium since all of them are flammable. Then, the recuperative configuration linked with a thermal loop and using Benzene as working fluid can be considered the most promising option, despite the fact that its cycle efficiency (20.45%) is lower than the one reachable with Cyclohexane (21.91%) or Toluene (21.68%). Regardless of the fluid, the expander is a 3-stage radial turbine which exhibits an isentropic efficiency higher than 85.2%. Specifically, the higher value is registered with M-xylene (86.4%), while the lower with DMC (85.2%); the turbine using Benzene reaches an isentropic efficiency of 85.5%.

To sum up, the thermodynamic optimization of both SRC and ORC technologies reveals that the use of ORC is more favorable for this application since it guarantees a net power output that can reach 160% of the one generated by SRC. Thus, the authors suggest to adopt the ORC technology for ICE waste heat recovery using an oil thermal loop to guarantee the system safety.

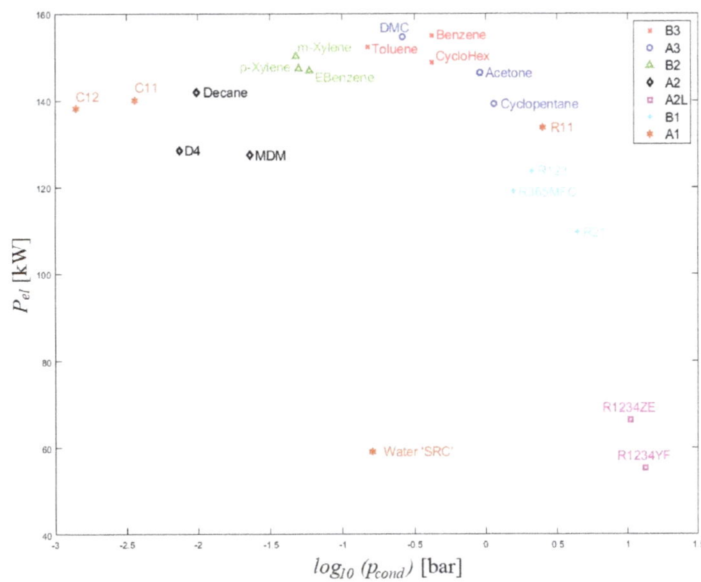

Figure 9. Net output power versus condensation pressure (in logarithmic scale) of the best 3 fluids of each safety category in the case of oil thermal loop.

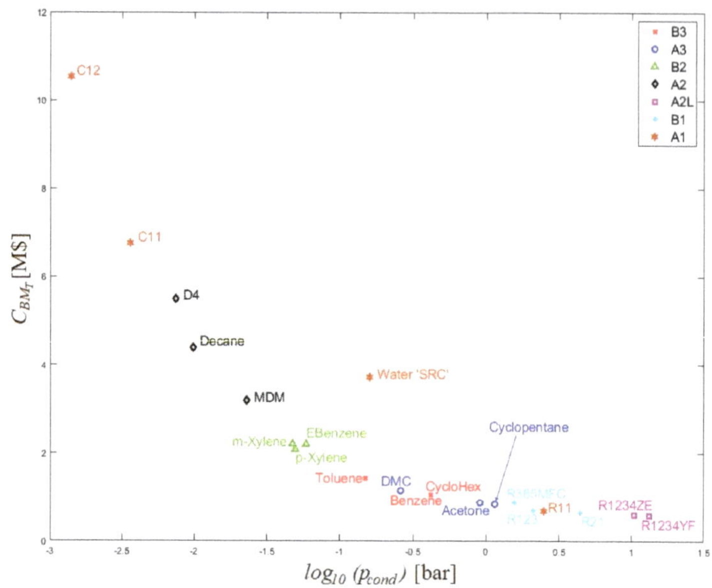

Figure 10. Plant cost versus condensation pressure (in logarithmic scale) of the best 3 fluids of each safety category in the case of oil thermal loop.

The analysis of the turbine's parameters (see Table A12) reveals that, among the most promising fluids, Cyclohexane and Benzene show the smaller value of the volumetric flow rate, resulting in a small last stage size parameter (0.0691 m and 0.0690 m, respectively) and, then, leading to a cheaper price of the expander. The latter results in cheaper total price of the ORC unit. These plants are followed by DMC and Toluene, and, finally, the M-xylene. The value of the volumetric flow of M-xylene is too large compared with the one of the other fluids, which results in a large last stage size parameter (0.1825 m), a condition that leads to a higher purchase cost of ORC unit. These results coincide with the ones reported by Astolfi et al. [41], as they mentioned that the use of high critical temperature fluids is associated with very low condensation pressure, which results in a high specific volume at turbine exhaust and, consequently, leads to high costs of the expander, a fact encountered in the case of M-xylene and Toluene.

Therefore, the economic analysis and the determination of the investment costs of the expander reveal that the cycle using Benzene or Cyclohexane as working fluids guarantees the best economic results, as they show the cheapest price for the expander device, which leads to the lowest purchase cost of the ORC unit. Contrary, Toluene and M-xylene are not preferable from an economic point of view since they are characterized by large volumetric flow and large last stage size parameter. So, overall, the higher the turbine cost is, the higher the ORC unit investment cost turns out.

However, it is also interesting to point out that the ORC purchasing costs drop from 8.86 k$ per kW_{el} (no thermal loop and Toluene as working fluid) to 6.8 k$ per kW_{el} (oil thermal loop and Benzene), a more reasonable price for this kind of unit. Therefore, as put forward by the thermodynamic analysis, the costs analysis confirms that Benzene is a good choice for this application.

Since this study examined all the fluids listed in Table A1 and classified them according to their categories, there is a need to note that numerous of them may be nominated as promising fluids for this application, but they are in fact ineligible to be suitable, since many of them have been banned from the application according to many international regulations or because of their environmental impacts. As an example, R123 and R11 in categories B1 and A1, respectively, are candidates to be suitable for use from a thermodynamic point of view in their categories, but they are not adoptable since they are banned from the application by international regulations. Therefore, additional checks and an environmental analysis are required to ensure that the chosen fluids are not banned and are not the source of environment impact.

Thus, the TEWI method was employed to carry out the environmental analysis, and as a means to determine the most environmentally friendly fluids among the promising ones (see Table 9).

Table 9. *TEWI* value for the promising fluids.

Fluid	Toluene	Benzene	Cyclohexane	DMC
GWP	3.3 [24]	<2.6 [87]	Very low [88]	3.2 [89]
M_{charge} [kg kW^{-1}]	1.909	2.024	1.890	2.023
TEWI [ton $CO_{2,eq}$]	0.480	0.407	Very low	0.500
% of DMC	96.00	81.40	Very low	100

Among the 5 best fluids listed in Table 8, the highest value of *TEWI* is associated with Toluene and DMC, with their GWP values being the highest ones. Benzene ranks third, while the TEWI value of Cyclohexane and M-xylene cannot be estimated due to the lack of numerical values for their GWP. In particular, for Cyclohexane, Li et al. [88] claim that the GWP value of this fluid is "very low". In addition, given the indirect *TEWI* linked to the CO_2 emissions caused by the generation of the consumed electricity, in the analyzed case, it is not computable. Therefore, with the direct TEWI linked to the GWP, the *TEWI* value is directly linked to the GWP of the fluid.

The comparison of *TEWI* values is limited to 3 fluids, and the results show that Benzene is the most environmentally friendly, given its total lifetime CO_2 emissions equal to 0.407 ton. Therefore, based on the thermodynamic, economic, and environmental results, to recover the bio-gas ICE waste heat, the best option is to use an ORC unit equipped with a thermal oil loop and using Benzene as working fluid. Thus, considering the Algerian and Italian situation in bio-gas sector, the recuperative ORC configuration using Benzene can improve the electricity production up to 15%, a not-negligible electricity improvement that abate the ICE's emissions, as well as the thermal pollution.

For the sake of clarity, the T-s diagram and the T-Q diagram of the Main Heat Exchanger for this cycle are depicted in Figure 11, while Table 10 lists the cycle thermodynamic points.

Table 10. Calculated points of Benzene for the ORC cycle.

Point	T [°C]	s [J kg^{-1} K^{-1}]	p [bar]	h [kJ kg^{-1}]
1	55.23	−140.59	32.25	−44.140
2	79.35	−9.10	32.25	0.617
3	255.81	905.32	32.25	403.828
4	258.85	1284.61	32.25	604.479
5	118.62	1354.01	0.42	448.443
6	85.86	1234.78	0.42	403.686
7	53.85	1119.08	0.42	363.996
8	53.85	−143.46	0.42	−48.854

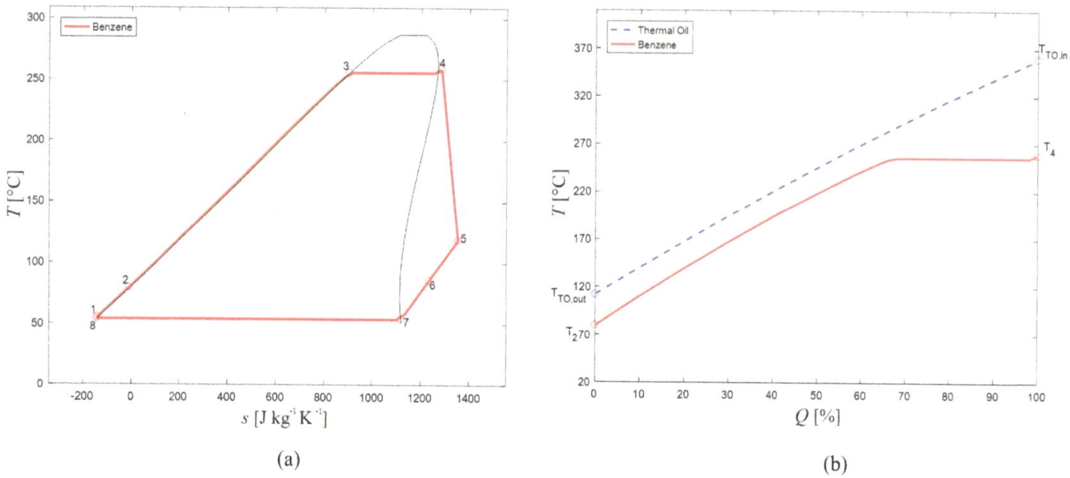

Figure 11. T-s (**a**) and T-Q (**b**) diagrams of the ORC employing Benzene as the working fluid.

6. Conclusions

Steam and organic Rankine cycles are viable solutions for waste heat recovery from both fossil and renewable-based plants, as well as industrial processes. However, there is a need for reliable and time-efficient optimization tools that take into account technical, economic, environmental, and safety aspects.

To this end, the authors of the present work developed a versatile tool named Improved Rankine Cycle Plant Designer (IRC-PD), characterized by a wide variety of options that made it adaptable for different cases and give the ability to design and optimize both SRC and ORC units. In addition, as a way to examine it, a real case study of a bio-gas engine was used as the test case. The engine's nameplate power is 1 MW_{el}, while the design of the waste heat recovery unit is based on real measurements in terms of composition, temperature, and mass flow rate of the engine's exhaust gases.

Results exhibit that the ORC technology is more appropriate for the examined case compared to SRC technology, mainly because it ensures higher net power output and better economic results. The analysis of the various fluids (more than 120 fluids) for the ORC unit show that Benzene is the most promising fluid from a thermodynamic, as well as an economic, point of view. In particular, the latter ensures the best option for this case by employing it in a recuperative organic Rankine cycle unit which does not recover directly the waste heat source but using an oil loop, where Therminol VP-1 is adopted as thermal medium. The ORC unit, equipped with a 3-stage radial turbine, characterized by an isentropic efficiency of 85.5%, is able to generate 154.92 kW_{el}, a solution that can boost the electricity production of the plant up to 15%.

On the other hand, environmental analysis cannot be considered a major criterion for the selection of the most suitable fluid for this application, since the GWP values of all the promising fluids are very low or approximately zero, as well as the calculated values of *TEWI*, which are directly related to the fluid GWP.

Therefore, the IRC PD tool is an excellent choice for assessing waste heat recovery for different applications, given the multiple options available on it, the small number of required input from the user, and the ability to evaluate and study the possibility of heat recovery regardless of the field of application. These features make the tool able to design waste heat recovery units based on the ORC and SRC technology with nameplate power ranging from kW to MW.

Author Contributions: Conceptualization, Y.R., K.K.-D., A.S. and A.B.; Data curation, Y.R., K.K.-D., A.S. and A.B.; Methodology, Y.R., K.K.-D., A.S. and A.B.; Validation, Y.R., K.K.-D., A.S. and A.B.; Writing—original draft, A.B.; Writing—review & editing, Y.R., K.K.-D., A.S. and A.B. All authors have read and agreed to the published version of the manuscript.

Funding: This research received no external funding.

Data Availability Statement: Not applicable.

Conflicts of Interest: The authors declare no conflict of interest.

Abbreviations

CO_2	carbon dioxide
GHG	greenhouse gases
IEA	International Energy Agency
WHRU	plants and waste heat recovery unit
RES	renewable energy source
EU	European Union
PV	PhotoVoltaic
IRC-PD	Improved Rankine Cycle Plant Designer
SRC	Steam Rankine Cycle
ORC	Organic Rankine Cycle
ICE	Internal Combustion Engine
GA	Genetic Algorithm
CEPCI	Chemical engineering plant cost index
GWP	Global Warming Potential
ODP	Ozone Depletion Potential
TEWI	Total equivalent warming impact
HC	HydroCarbons
HFC	HydroFluoroCarbons
HCFC	HydroChloroFluoroCarbons
CFC	ChloroFluoroCarbons
PFC	PerFluoroCarbons
HFO	HydroFluoroOlefins
MHE	Main Heat Exchanger
TL	Thermal Loop
HX	Heat Exchange
A	Axial
R	Radial
Symbols	
T	temperature (K or °C)
T_H	standard boiling point temperature (K or °C)
T_c	critical temperature (K or °C)
T_{tn}	turning point temperature (K or °C)
P	power (W)
p	pressure (bar)
h	specific enthalpy (kJ kg^{-1})
E	recuperator efficiency (-)
s	specific entropy (J kg^{-1} K^{-1})
C_{BM}	bare module cost ($)
C_P^0	purchased equipment cost base conditions ($)
A	heat transfer surface (m^2)
C_{site}	cost of the site ($)
$C_{O\&M}$	operation and maintenance cost ($)
i	interest rate (%)
f	plant availability factor (-)
t_{corp}	corporate tax rate (%)
RF	capital recovery factor ($)
k_{is}	load coefficient (-)

M	Refrigerant charge (kg)
$\dot{m}$	mass flow rate (kg s^{-1})
n	system operating lifetime (year)
NPV	Net Present Value ($)
SP	Size Parameter (m)
CF	Cash Flow ($)
IP	Profitability Index (-)
SPB	Simple PayBack (year)
$LCOE$	Levelized Cost Of Energy ($ kW^{-1})
L	annual leakage (kg $year^{-1}$)
s_E	electricity sell price ($)
E_{el}	annual electricity production (kWh)
H_{annual}	annual operating hour (hour)
S_{annual}	annual incomes ($)
$TEWI$	Total equivalent warming impact (ton $CO_{2,eq}$)
u_m	mean peripheral speed (m s^{-1})
$\dot{V}$	volumetric flow rate (m^3 s^{-1})
Subscripts	
$cond$	
el	electrical
Exp	expander
in	inlet
hot	hot source
is	isentropic
mec	mechanical
max	maximum
min	minimum
out	outlet
ev	evaporation
P	Pump
pp	Pinch Point
ST	steam turbine
sv	saturated vapor
tn	turning point
T	turbine
TO	thermal oil
t	total
wf	working fluid
Greek symbols	
α	recycling factor (-)
β	carbon dioxide emission factor ($kg_{CO_{2,eq}}$ kWh^{-1})
Δ	difference
η	efficiency (%)
ξ	Slope (-)

Appendix A. Thermodynamic Points Computations

The equations implemented into the IRC-PD tool for the recuperative configuration are listed in the following considering the thermodynamic points shown in Figure A1.

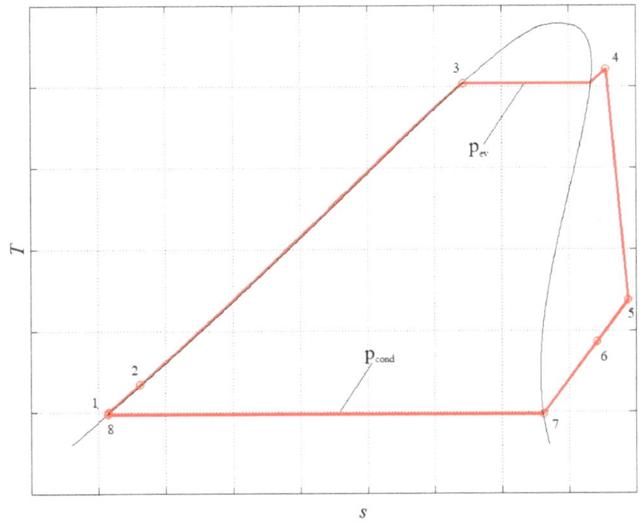

Figure A1. T-s diagram of the recuperative cycle implemented in the code. The thermodynamic state points are included in the graph to better identify their position.

$$\eta_{is,P} = \frac{(h_{is,P} - h_8)}{(h_1 - h_8)} \tag{A1}$$

$$E = \frac{(h_5 - h_6)}{(h_5 - h_7)} \tag{A2}$$

$$T_1 = f(h_1,\ p_{ev}) \tag{A3}$$

$$h_{is,P} = f(p_{ev},\ s_8) \tag{A4}$$

$$h_{is,T} = f(p_{cond},\ s_4) \tag{A5}$$

$$h_3 = f(p_{ev},\ x = 0) \tag{A6}$$

$$h_7 = f(p_{cond},\ x = 1) \tag{A7}$$

$$h_8 = f(p_{cond},\ x = 0) \tag{A8}$$

$$s_8 = f(h_8,\ p_{cond}) \tag{A9}$$

$$h_4 = f(T_4,\ p_{ev}) \tag{A10}$$

$$s_4 = f(T_4,\ p_{ev}) \tag{A11}$$

$$h_6 = f(T_6,\ p_{cond}) \tag{A12}$$

$$h_2 = h_5 - h_6 + h_1; \tag{A13}$$

$$P_T = \dot{m}_{Cycle}\ (h_4 - h_5)\ \eta_{mec,T}\ \eta_{el,gen} \tag{A14}$$

$$P_P = \dot{m}_{Cycle}\ \frac{h_1 - h_8}{\eta_{mec,P}\ \eta_{el,mot}} \tag{A15}$$

$$\eta_{th,Cycle} = \frac{P_{el}}{\dot{m}_{hot}\ (h_{hot,in} - h_{hot,out})} \tag{A16}$$

The net output power is computed according to the selected configuration. For the configuration without thermal loop, it is computed as:

$$P_{el} = (P_T - P_P) \tag{A17}$$

while, for the configuration including a thermal loop, it is given as:

$$P_{el} = \left(P_T - P_P - P_{P_{TL}}\right) \tag{A18}$$

For the configuration including a thermal loop and air-cooled condenser, the net output power is computed as:

$$P_{el} = \left(P_T - P_P - P_{P_{TL}} - P_{fans}\right) \tag{A19}$$

while, for the configuration with an air-cooled condenser and without a thermal loop, it is derived from:

$$P_{el} = \left(P_T - P_P - P_{fans}\right) \tag{A20}$$

The list of the available working fluids is presented in Table A1.

Table A1. HCs, HFCs, Siloxanes, PCFs, HCFCs, CFCs, and other candidates implemented in the code.

HCs		HFCs	Siloxanes	PCFs	HCFCs	CFCs	Cryogens
N-octane	N-decane	R32	D4	C4F10	R123	R11	Neon
1-butene	Neopentane	R125	D5	C5F12	R124	R113	Nitrogen
Acetone	N-heptane	R134A	D6	R116	R141B	R114	Ohydrogen
Benzene	N-hexane	R13I1	MD2M	R1216	R142B	R115	Oxygen
C1CC6	N-nonane	R143A	MD3M	R14	R21	R12	Phydrogen
C2BUTENE	N-Undecane	R152A	MD4M	R161	R22	R13	Xenon
C3CC6	O-xylene	R227EA	MDM	R218			Argon
Cyclohexane	Pentane	R23	MM				CO
Cyclopentane	Propane	R236EA					Deuterium
Cyclopropane	Propene	R236FA					Fluorine
E-Benzene	Propyne	RC318					Helium
Ethylene	P-xylene	R245CA					Hydrogen
Isobutane	R365MFC	R245FA					Krypton
Isobutene	SES36	R40					
N-dodecane	Toluene	R404A					
Isohexane	trans-Butene	R407C					
Isopentane	Iso-octane	R41					
M-xylene		R410A					
N-butane		R507A					
HFOs	**Ethers**	**FAMEs**	**Inorganics**	**Alcohols & Esters**	**Others**	**ICF**	
R1233zd(E)	DEE	Mlinolea	Ammonia	Ethanol	d2o	ThermVP1	
R1234YF	DME	Mlinolen	CO_2	Methanol	Novec649	DowQ	
R1234ZDE	RE143A	Moleate	Water	DMC	Propylene	Therm66	
R1234ZE	RE245cb2	Mpalmita			H2S		
R1234ze(Z)	RE245fa2	Mstearat			HCL		
	RE347MCC				SO2		

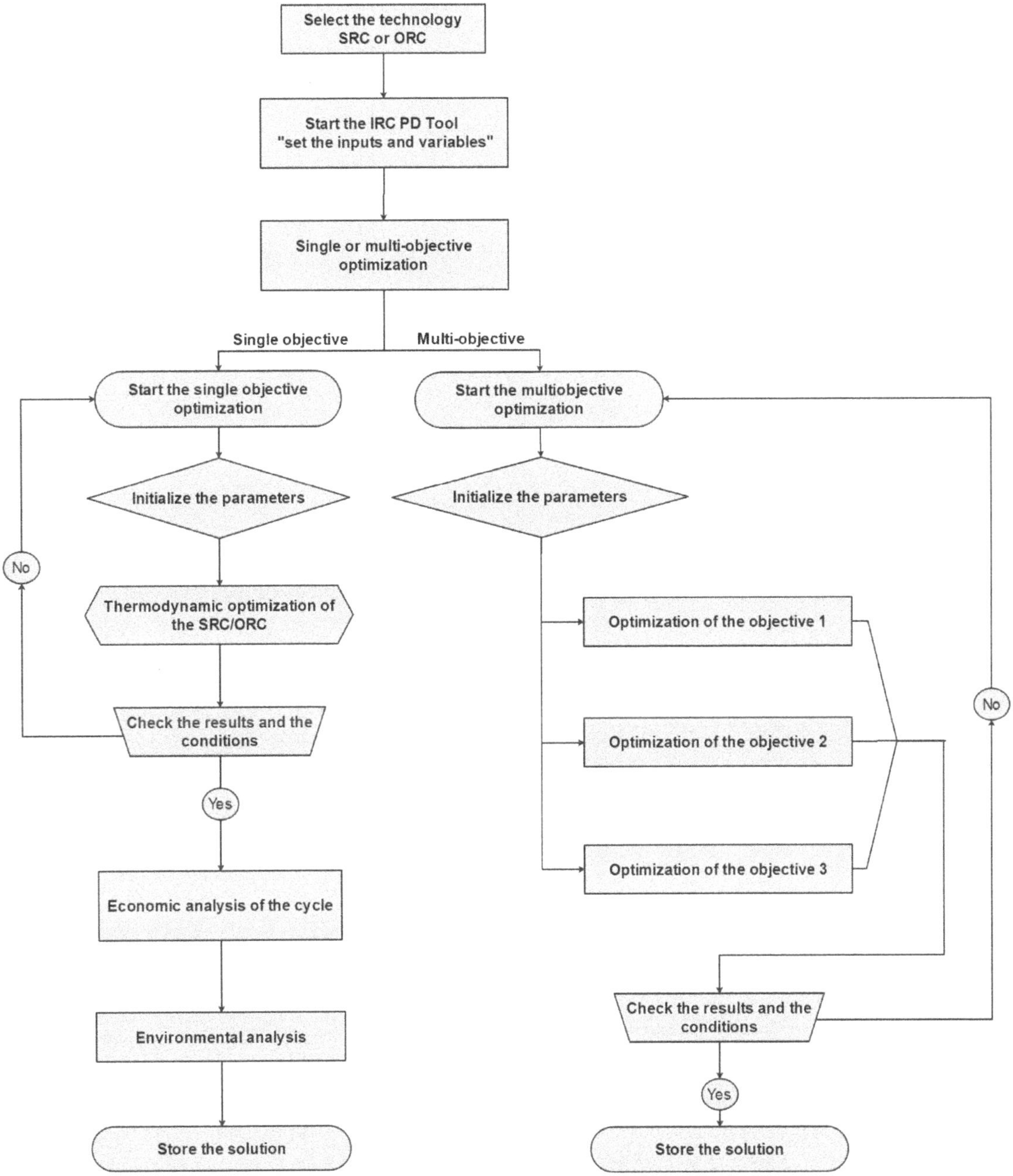

Figure A2. Steps involved during the optimization using the IRC-PD Tool.

Appendix B. Parameters Adopted in the Economic Analysis

In order to perform the economic analysis, the logarithmic mean temperature approach is used to compute the heat exchanger area and the adopted formula is given in Equation (A21) [90]. The exchange heat is given multiplying the overall heat transfer coefficient U, by the area of exchange A, and the logarithmic mean temperature ΔT_{lm}.

$$Q = U \cdot A \cdot \Delta T_{lm} \tag{A21}$$

Thus, for the computation of the area of exchange for each heat exchanger, the overall heat transfer coefficient, U, is selected according to Dimian and Bildea [91] and given as follows:

$$U_{cond,A} = 300\ \mathrm{Wm^{-2}K^{-1}} \tag{A22}$$

$$U_{cond,B} = 700\ \mathrm{Wm^{-2}K^{-1}} \tag{A23}$$

$$U_{rec} = 300\ \mathrm{Wm^{-2}K^{-1}} \tag{A24}$$

$$U_{eco,TO} = 400\ \mathrm{Wm^{-2}K^{-1}} \tag{A25}$$

$$U_{ev,TO} = 700\ \mathrm{Wm^{-2}K^{-1}} \tag{A26}$$

$$U_{sup,TO} = 300\ \mathrm{Wm^{-2}K^{-1}} \tag{A27}$$

$$U_{eco} = 50\ \mathrm{Wm^{-2}K^{-1}} \tag{A28}$$

$$U_{ev} = 70\ \mathrm{Wm^{-2}K^{-1}} \tag{A29}$$

$$U_{sup} = 30\ \mathrm{Wm^{-2}K^{-1}} \tag{A30}$$

$$U_{TL,HE} = 200\ \mathrm{Wm^{-2}K^{-1}} \tag{A31}$$

The coefficients used for preforming the economic analysis of each device are listed in Tables A2 and A3.

Table A2. Parameters and range of applications used in the equations.

Component	N	Size Range	K_1 K_2 K_3	C_1 C_2 C_3	B_1 B_2	F_M	F_{BM}	Ref.
Shell and tube heat exchanger	A [m^2]	10–1000	4.3247 −0.3030 0.1634	0.03881 −0.11272 0.08183	1.63 1.66	1	-	[46]
Pump	P [kW]	1–300	3.3892 0.0536 0.1538	−0.3935 0.3957 −0.00226	1.89 1.35	1.575	-	[46]
Pump electrical motor	P [kW]	-	2.4604 1.4191 −0.1798	-	-	-	1.5	[46]

Table A3. Parameters and range of applications used in the equations.

Component	N	Size Range	C_B	Q_B	M	F_M	F_p	F_T	F_{BM}	Ref.
Shell and tube heat exchanger	A [m^2]	80–4000	32,800	80	0.68	1	1.5	1.6	-	[47]
Electric generator	P [kW]	-	1,850,000	11,800	0.94	-	-	-	1.5	[48]

Appendix C. Optimized Variables for the Different SRC and ORC Configurations

The thermodynamic optimization of SRC in terms of net power output is carried out adopting the basic configuration of SRC and a multi-stage steam turbine in which a specific enthalpy drop of 150 kJ kg^{-1} and a fixed rotational speed equal to 6000 rpm [92] are adopted. In the case of a steam cycle coupled with a water loop, the optimization process is unfeasible if the lower bound of the evaporation pressure is set equal to 15 bar. So, to show the code ability of providing a solution (unfeasible from the technical point of view), the lower bound of the evaporation pressure is set equal to 1 bar.

Table A4. Results of the thermodynamic optimization and economic analysis of the SRC without thermal loop.

P_{el} [kW]	$T_{hot,out}$ [°C]	TIT [°C]	p_{ev} [bar]	p_{cond} [bar]	η_{SRC} [%]	$\eta_{is,ST}$ [%]	$\Delta T_{pp,MHE}$ [-]	$\Delta T_{pp,cond}$ [°C]	$Stages$ [-]	C_{BM_t} [M$]	C_{ST} [M$]
96.38	156.9	477.9	19.22	0.15	13.96	0.528	25.0	10.1	7	3.8302	3.502

Table A5. Results of the thermodynamic optimization and economic analysis of the SRC adopting a water thermal loop.

P_{el} [kW]	$T_{hot,out}$ [°C]	$T_{water,out}$ [°C]	TIT [°C]	p_{ev} [bar]	p_{cond} [bar]	η_{SRC} [%]	$\eta_{is,ST}$ [%]	$\Delta T_{pp,MHE}$ [-]	$\Delta T_{pp,cond}$ [°C]	$Stages$ [-]	C_{BM_t} [M$]	C_{ST} [M$]
36.06	150.9	140	134.9	1.82	0.16	4.92	0.397	25.1	10.8	3	3.517	3.225

Table A6. Results of the thermodynamic optimization and economic analysis of the SRC adopting an oil thermal loop.

P_{el} [kW]	$T_{hot,out}$ [°C]	$T_{oil,out}$ [°C]	TIT [°C]	p_{ev} [bar]	p_{cond} [bar]	η_{SRC} [%]	$\eta_{is,ST}$ [%]	$\Delta T_{pp,MHE}$ [-]	$\Delta T_{pp,cond}$ [°C]	$Stages$ [-]	C_{BM_t} [M$]	C_{ST} [M$]
59.10	190.9	180.9	333.9	15.00	0.16	9.45	0.403	25.4	10.6	6	3.725	3.433

Table A7. Most promising ORC fluids in the case of direct exchange between ICE's exhaust and the cycle.

Fluid	P_{el} [kW]	$T_{hot,out}$ [°C]	TIT [°C]	p_{ev} [bar]	p_{cond} [bar]	E [%]	η_{ORC} [%]	$\eta_{is,ST}$ [-]	$\Delta T_{pp,MHE}$ [°C]	$\Delta T_{pp,cond}$ [°C]	$Stages$ [-]
Toluene	169.89	133.9	329.9	33.5	0.13	51	23.13	0.847	25.6	10.3	3-R
M-xylene	169.50	113.9	306.9	20.98	0.04	35	21.94	0.844	26.1	10.1	3-R
O-xylene	169.27	135.9	332.9	23.02	0.03	47	23.17	0.838	25.4	10.1	3-R
P-xylene	169.08	129.9	330.9	25.34	0.04	44	22.78	0.839	25.2	10.1	3-R
E-Benzene	167.37	147.9	317.9	23.85	0.05	62	23.65	0.843	25.3	10.5	3-R

Table A8. Most promising fluids in the case of adopting an ORC and a water loop.

Fluid	P_{el} [kW]	$T_{hot,out}$ [°C]	$T_{water,out}$ [°C]	TIT [°C]	p_{ev} [bar]	p_{cond} [bar]	E [%]	η_{ORC} [%]	$\eta_{is,ST}$ [-]	$\Delta T_{pp,MHE}$ [°C]	$\Delta T_{pp,cond}$ [°C]	$Stages$ [-]
C3CC6	76.84	150.9	140	131.9	0.34	0.03	49	11.00	0.903	26.7	10.2	2-A
Heptane	76.64	150.9	140	133.9	1.84	0.21	53	10.97	0.891	25.5	10.2	2-A
Nonane	76.13	151.9	140	124.9	0.41	0.03	45	10.90	0.902	26.4	10.2	2-A
Octane	75.53	151.9	140	128.9	0.82	0.08	50	10.81	0.901	27.4	10.2	2-A
DMC	75.28	150.9	140	131.9	2.35	0.27	17	10.77	0.887	26.2	10.4	2-A

Table A9. Most promising fluids in the case of adopting an ORC and an oil loop.

Fluid	P_{el} [kW]	$T_{hot,out}$ [°C]	$T_{oil,out}$ [°C]	TIT [°C]	p_{ev} [bar]	p_{cond} [bar]	E [%]	η_{ORC} [%]	$\eta_{is,ST}$ [-]	$\Delta T_{pp,MHE}$ [°C]	$\Delta T_{pp,cond}$ [°C]	$Stages$ [-]
Benzene	154.92	121.9	111.9	258.9	32.25	0.42	53	20.45	0.855	26.4	10.8	3-R
DMC	154.62	122.9	112.9	266.9	34.94	0.26	44	20.46	0.852	25.4	10.6	3-R
Toluene	152.21	150.9	140.9	250.9	15.68	0.15	77	21.68	0.864	25.8	11.4	3-R
M-xylene	150.17	149.9	139.9	241.9	7.58	0.05	68	21.33	0.868	25.3	10.2	3-R
Cyclohex	148.75	162.9	152.9	275.9	34.03	0.42	78	21.91	0.852	26.0	11.2	3-R

Table A10. Economic analysis results of the ORC fluids in the case of a direct exchange between the ICE exhaust and the working fluid.

Fluid	C_{BM_t} [M$]	C_{Exp} [M$]	$\frac{C_{Exp}}{C_{BM_t}}$ [%]	SP_{LS} [m]	$\dot{V}_5$ [m^3 s^{-1}]
Toluene	1.505	1.023	67.98	0.1081	3.45
E-Benzene	2.062	1.607	77.94	0.1630	7.66
P-xylene	2.197	1.716	78.10	0.1729	8.88
M-xylene	2.257	1.774	78.60	0.1783	9.32
O-xylene	2.383	1.904	79.87	0.1901	10.95

Table A11. Economic analysis results of the ORC fluids in the case of using water as heat transfer medium in the thermal loop.

Fluid	C_{BM_t} [M$]	C_{Exp} [M$]	$\frac{C_{Exp}}{C_{BM_t}}$ [%]	SP_{LS} [m]	$\dot{V}_5$ [m^3 s^{-1}]
DMC	1.100	0.707	64.31	0.0929	1.68
Heptane	1.225	0.853	69.61	0.1101	2.25
Octane	1.791	1.406	78.49	0.1735	5.49
Nonane	2.670	2.279	85.35	0.2692	13.24
C3CC6	2.825	2.444	86.50	0.2869	15.00

Table A12. Economic analysis results of the ORC fluids in the case of using oil as heat transfer medium in the thermal loop.

Fluid	C_{BM_t} [M$]	C_{Exp} [M$]	$\frac{C_{Exp}}{C_{BM_t}}$ [%]	SP_{LS} [m]	$\dot{V}_5$ [m^3 s^{-1}]
Benzene	1.067	0.625	58.54	0.0690	1.23
DMC	1.159	0.752	62.86	0.0817	1.74
Toluene	1.440	1.031	71.64	0.1089	3.00
M-xylene	2.218	1.820	82.06	0.1825	8.31
Cyclohex	1.058	0.625	59.07	0.0691	1.23

References

1. Horowitz, C.A. Paris agreement. *Int. Leg. Mater.* **2016**, *55*, 740–755. [CrossRef]
2. International Energy Agency. IEA statistics, 2019. In *CO_2 Emissions from Fuel Combustion: Highlights*; International Energy Agency: Paris, France, 2019.
3. Meinshausen, M.; Meinshausen, N.; Hare, W.; Raper, S.C.; Frieler, K.; Knutti, R.; Frame, D.J.; Allen, M.R. Greenhouse-gas emission targets for limiting global warming to 2 C. *Nature* **2009**, *458*, 1158–1162. [CrossRef] [PubMed]
4. Sikora, A. European Green Deal–legal and financial challenges of the climate change. In *ERA Forum*; Springer: Berlin/Heidelberg, Germany, 2021; Volume 21, pp. 681–697.

5. Skjærseth, J.B. Towards a European Green Deal: The evolution of EU climate and energy policy mixes. *Int. Environ. Agreem. Politics Law Econ.* **2021**, *21*, 25–41. [CrossRef]
6. Algerian Ministry of Energy. Energies Nouvelles, Renouvelables et Maitrise de l'Energie; Algiers, Algeria. 2021. Available online: https://www.energy.gov.dz/?rubrique=energies-nouvellesrenouvelables-et-maitrise-de-lrenergie (accessed on 4 September 2021).
7. Bouzid, Z.; Ghellai, N. Overview of solar potential, state of the art and future of photovoltaic installations in Algeria. *Int. J. Renew. Energy Res. IJRER* **2015**, *5*, 427–434.
8. Bouraiou, A.; Necaibia, A.; Boutasseta, N.; Mekhilef, S.; Dabou, R.; Ziane, A.; Sahouane, N.; Attoui, I.; Mostefaoui, M.; Touaba, O. Status of renewable energy potential and utilization in Algeria. *J. Clean. Prod.* **2020**, *246*, 119011. [CrossRef]
9. Akbi, A.; Saber, M.; Aziza, M.; Yassaa, N. An overview of sustainable bioenergy potential in Algeria. *Renew. Sustain. Energy Rev.* **2017**, *72*, 240–245. [CrossRef]
10. Campana, F.; Bianchi, M.; Branchini, L.; De Pascale, A.; Peretto, A.; Baresi, M.; Fermi, A.; Rossetti, N.; Vescovo, R. ORC waste heat recovery in European energy intensive industries: Energy and GHG savings. *Energy Convers. Manag.* **2013**, *76*, 244–252. [CrossRef]
11. Benato, A. Improving the efficiency of a cataphoresis oven with a cogenerative organic Rankine cycle unit. *Therm. Sci. Eng. Prog.* **2018**, *5*, 182–194. [CrossRef]
12. Papapetrou, M.; Kosmadakis, G.; Cipollina, A.; La Commare, U.; Micale, G. Industrial waste heat: Estimation of the technically available resource in the EU per industrial sector, temperature level and country. *Appl. Therm. Eng.* **2018**, *138*, 207–216. [CrossRef]
13. Cavazzini, G.; Bari, S.; McGrail, P.; Benedetti, V.; Pavesi, G.; Ardizzon, G. Contribution of Metal-Organic-Heat Carrier nanoparticles in a R245fa low-grade heat recovery Organic Rankine Cycle. *Energy Convers. Manag.* **2019**, *199*, 111960. [CrossRef]
14. Fierro, J.J.; Escudero-Atehortua, A.; Nieto-Londoño, C.; Giraldo, M.; Jouhara, H.; Wrobel, L.C. Evaluation of waste heat recovery technologies for the cement industry. *Int. J. Thermofluids* **2020**, *7*, 100040. [CrossRef]
15. Prando, D.; Renzi, M.; Gasparella, A.; Baratieri, M. Monitoring of the energy performance of a district heating CHP plant based on biomass boiler and ORC generator. *Appl. Therm. Eng.* **2015**, *79*, 98–107. [CrossRef]
16. Pezzuolo, A.; Benato, A.; Stoppato, A.; Mirandola, A. Fluid selection and plant configuration of an ORC-biomass fed system generating heat and/or power. *Energy Procedia* **2016**, *101*, 822–829. [CrossRef]
17. Benato, A.; Macor, A. Biogas engine waste heat recovery using organic Rankine cycle. *Energies* **2017**, *10*, 327. [CrossRef]
18. DeLovato, N.; Sundarnath, K.; Cvijovic, L.; Kota, K.; Kuravi, S. A review of heat recovery applications for solar and geothermal power plants. *Renew. Sustain. Energy Rev.* **2019**, *114*, 109329. [CrossRef]
19. Freeman, J.; Hellgardt, K.; Markides, C.N. Working fluid selection and electrical performance optimisation of a domestic solar-ORC combined heat and power system for year-round operation in the UK. *Appl. Energy* **2017**, *186*, 291–303. [CrossRef]
20. The Mathworks. *Guide, MATLAB User's*; The Mathworks: Natick, MA, USA, 2021.
21. Pezzuolo, A.; Benato, A.; Stoppato, A.; Mirandola, A. The ORC-PD: A versatile tool for fluid selection and Organic Rankine Cycle unit design. *Energy* **2016**, *102*, 605–620. [CrossRef]
22. The Mathworks. Global Optimization Toolbox. 2021. Available online: https://it.mathworks.com/help/gads/?s_tid=srchbrcm (accessed on 4 September 2021).
23. NIST. NIST Reference Fluid Thermodynamic and Transport Properties Database (REFPROP): Version 9.1. 2010. Available online: http://www.nist.gov/srd/nist23.cfm (accessed on 4 September 2021).
24. Bell, I.H.; Wronski, J.; Quoilin, S.; Lemort, V. Pure and Pseudo-pure Fluid Thermophysical Property Evaluation and the Open-Source Thermophysical Property Library CoolProp. *Ind. Eng. Chem. Res.* **2014**, *53*, 2498–2508. [CrossRef]
25. ASHRAE. ANSI/ASHRAE 34-2019. Designation and Safety Classification of Refrigerants. 2021. Available online: https://www.ashrae.org/technical-resources/standards-and-guidelines/ashrae-refrigerant-designations (accessed on 4 September 2021).
26. Nord, L.O.; Martelli, E.; Bolland, O. Weight and power optimization of steam bottoming cycle for offshore oil and gas installations. *Energy* **2014**, *76*, 891–898. [CrossRef]
27. Pierobon, L.; Benato, A.; Scolari, E.; Haglind, F.; Stoppato, A. Waste heat recovery technologies for offshore platforms. *Appl. Energy* **2014**, *136*, 228–241. [CrossRef]
28. Branchini, L.; De Pascale, A.; Peretto, A. Systematic comparison of ORC configurations by means of comprehensive performance indexes. *Appl. Therm. Eng.* **2013**, *61*, 129–140. [CrossRef]
29. Shokati, N.; Ranjbar, F.; Yari, M. Exergoeconomic analysis and optimization of basic, dual-pressure and dual-fluid ORCs and Kalina geothermal power plants: A comparative study. *Renew. Energy* **2015**, *83*, 527–542. [CrossRef]
30. Meinel, D.; Wieland, C.; Spliethoff, H. Effect and comparison of different working fluids on a two-stage organic rankine cycle (ORC) concept. *Appl. Therm. Eng.* **2014**, *63*, 246–253. [CrossRef]
31. Weith, T.; Heberle, F.; Preißinger, M.; Brüggemann, D. Performance of siloxane mixtures in a high-temperature Organic Rankine Cycle considering the heat transfer characteristics during evaporation. *Energies* **2014**, *7*, 5548–5565. [CrossRef]
32. Oyewunmi, O.A.; Taleb, A.I.; Haslam, A.J.; Markides, C.N. An assessment of working-fluid mixtures using SAFT-VR Mie for use in organic Rankine cycle systems for waste-heat recovery. *Comput. Therm. Sci. Int. J.* **2014**, *6*, 266. [CrossRef]
33. Heberle, F.; Brüggemann, D. Thermo-economic analysis of zeotropic mixtures and pure working fluids in organic Rankine cycles for waste heat recovery. *Energies* **2016**, *9*, 226. [CrossRef]
34. Venkatarathnam, G.; Timmerhaus, K.D. *Cryogenic Mixed Refrigerant Processes*; Springer: Berlin/Heidelberg, Germany, 2008.

35. Chys, M.; van den Broek, M.; Vanslambrouck, B.; De Paepe, M. Potential of zeotropic mixtures as working fluids in organic Rankine cycles. *Energy* **2012**, *44*, 623–632. [CrossRef]
36. Bell, K.; Ghaly, M. An Approximate Generalized Design Method for Multicomponent-Partial Condenser. *AIChEn Symp. Ser.* **1972**, *69*, 72–79.
37. Shah, M.M. Chart correlation for saturated boiling heat transfer: Equations and further study. *ASHRAE Trans.* **1982**, *88*, 185–196.
38. Bahadori, A.; Vuthaluru, H.B. Estimation of performance of steam turbines using a simple predictive tool. *Appl. Therm. Eng.* **2010**, *30*, 1832–1838. [CrossRef]
39. Macchi, E.; Perdichizzi, A. Efficiency prediction for axial-flow turbines operating with nonconventional fluids. *J. Eng. Gas Turbines Power* **1981**, *103*, 718–724. [CrossRef]
40. Perdichizzi, A.; Lozza, G. Design criteria and efficiency prediction for radial inflow turbines. In Proceedings of the ASME 1987 International Gas Turbine Conference and Exhibition, Anaheim, CA, USA, 31 May–4 June 1987; American Society of Mechanical Engineers: New York, NY, USA, 1987.
41. Astolfi, M.; Romano, M.C.; Bombarda, P.; Macchi, E. Binary ORC (Organic Rankine Cycles) power plants for the exploitation of medium-low temperature geothermal sources—Part B: Techno-economic optimization. *Energy* **2014**, *66*, 435–446. [CrossRef]
42. Martelli, E.; Capra, F.; Consonni, S. Numerical optimization of combined heat and power organic Rankine cycles—Part A: Design optimization. *Energy* **2015**, *90*, 310–328. [CrossRef]
43. Macchi, E.; Astolfi, M. *Organic Rankine Cycle (ORC) Power Systems: Technologies and Applications*; Woodhead Publishing: Shaxton, UK, 2016.
44. Liu, B.T.; Chien, K.H.; Wang, C.C. Effect of working fluids on organic Rankine cycle for waste heat recovery. *Energy* **2004**, *29*, 1207–1217. [CrossRef]
45. Wang, X.; Levy, E.K.; Pan, C.; Romero, C.E.; Banerjee, A.; Rubio-Maya, C.; Pan, L. Working fluid selection for organic Rankine cycle power generation using hot produced supercritical CO_2 from a geothermal reservoir. *Appl. Therm. Eng.* **2019**, *149*, 1287–1304. [CrossRef]
46. Turton, R.; Bailie, R.C.; Whiting, W.B.; Shaeiwitz, J.A. *Analysis, Synthesis and Design of Chemical Processes*; Pearson Education: London, UK, 2018.
47. Smith, R. *Chemical Process: Design and Integration*; John Wiley & Sons: Hoboken, NJ, USA, 2016.
48. Toffolo, A.; Lazzaretto, A.; Manente, G.; Paci, M. A multi-criteria approach for the optimal selection of working fluid and design parameters in Organic Rankine Cycle systems. *Appl. Energy* **2014**, *121*, 219–232. [CrossRef]
49. Manesh, M.K.; Navid, P.; Marigorta, A.B.; Amidpour, M.; Hamedi, M. New procedure for optimal design and evaluation of cogeneration system based on advanced exergoeconomic and exergoenvironmental analyses. *Energy* **2013**, *59*, 314–333. [CrossRef]
50. Zhang, X.; Cao, M.; Yang, X.; Guo, H.; Wang, J. Economic analysis of organic Rankine cycle using R123 and R245fa as working fluids and a demonstration project report. *Appl. Sci.* **2019**, *9*, 288. [CrossRef]
51. ANSI/ASHRAE. *ANSI/ASHRAE Standard 34-2010, Designation and Safety Classification of Refrigerants*; ANSI/ASHRAE: Peachtree Corners, GA, USA, 2010.
52. United Nations Environment Programme. Ozone Secretariat. *Handbook for the Montreal Protocol on Substances That Deplete the Ozone Layer*; UNEP/Earthprint: Kenya, Nairobi, 2006.
53. European Union. Directive 2006/40/EC of the European Parliament and of the Council of 17 May 2006 relating to emissions from air-conditioning systems in motor vehicles and Amending Council Directive 70/156/EEC. *Off. J. Eur. Union L* **2006**, *161*, 12–18.
54. Gullo, P.; Tsamos, K.M.; Hafner, A.; Banasiak, K.; Yunting, T.G.; Tassou, S.A. Crossing CO_2 equator with the aid of multi-ejector concept: A comprehensive energy and environmental comparative study. *Energy* **2018**, *164*, 236–263. [CrossRef]
55. Ye, Z.; Yang, J.; Shi, J.; Chen, J. Thermo-economic and environmental analysis of various low-GWP refrigerants in Organic Rankine cycle system. *Energy* **2020**, *199*, 117344. [CrossRef]
56. Bellos, E.; Tzivanidis, C. A theoretical comparative study of CO_2 cascade refrigeration systems. *Appl. Sci.* **2019**, *9*, 790. [CrossRef]
57. Material Safety Data Sheets. Available online: https://www.nationalref.com/technical/msds/ (accessed on 4 September 2021).
58. Collings, P.; Yu, Z.; Wang, E. A dynamic organic Rankine cycle using a zeotropic mixture as the working fluid with composition tuning to match changing ambient conditions. *Appl. Energy* **2016**, *171*, 581–591. [CrossRef]
59. Vaja, I.; Gambarotta, A. Internal combustion engine (ICE) bottoming with organic Rankine cycles (ORCs). *Energy* **2010**, *35*, 1084–1093. [CrossRef]
60. Liu, P.; Shu, G.; Tian, H.; Wang, X. Engine load effects on the energy and exergy performance of a medium cycle/organic Rankine cycle for exhaust waste heat recovery. *Entropy* **2018**, *20*, 137. [CrossRef] [PubMed]
61. GE. GE Power Generation. 2016. Available online: https://powergen.gepower.com/applications/bio-gas.html (accessed on 4 September 2021).
62. Bombarda, P.; Invernizzi, C.M.; Pietra, C. Heat recovery from Diesel engines: A thermodynamic comparison between Kalina and ORC cycles. *Appl. Therm. Eng.* **2010**, *30*, 212–219. [CrossRef]
63. Pan, M.; Zhu, Y.; Bian, X.; Liang, Y.; Lu, F.; Ban, Z. Theoretical analysis and comparison on supercritical CO_2 based combined cycles for waste heat recovery of engine. *Energy Convers. Manag.* **2020**, *219*, 113049. [CrossRef]
64. Yu, G.; Shu, G.; Tian, H.; Huo, Y.; Zhu, W. Experimental investigations on a cascaded steam-/organic-Rankine-cycle (RC/ORC) system for waste heat recovery (WHR) from diesel engine. *Energy Convers. Manag.* **2016**, *129*, 43–51. [CrossRef]

65. Liu, X.; Nguyen, M.Q.; Chu, J.; Lan, T.; He, M. A novel waste heat recovery system combing steam Rankine cycle and organic Rankine cycle for marine engine. *J. Clean. Prod.* **2020**, *265*, 121502. [CrossRef]
66. Andreasen, J.G.; Meroni, A.; Haglind, F. A comparison of organic and steam Rankine cycle power systems for waste heat recovery on large ships. *Energies* **2017**, *10*, 547. [CrossRef]
67. Yang, F.; Dong, X.; Zhang, H.; Wang, Z.; Yang, K.; Zhang, J.; Wang, E.; Liu, H.; Zhao, G. Performance analysis of waste heat recovery with a dual loop organic Rankine cycle (ORC) system for diesel engine under various operating conditions. *Energy Convers. Manag.* **2014**, *80*, 243–255. [CrossRef]
68. Wang, E.; Zhang, H.; Fan, B.; Ouyang, M.; Yang, F.; Yang, K.; Wang, Z.; Zhang, J.; Yang, F. Parametric analysis of a dual-loop ORC system for waste heat recovery of a diesel engine. *Appl. Therm. Eng.* **2014**, *67*, 168–178. [CrossRef]
69. Song, J.; Gu, C.W. Parametric analysis of a dual loop Organic Rankine Cycle (ORC) system for engine waste heat recovery. *Energy Convers. Manag.* **2015**, *105*, 995–1005. [CrossRef]
70. Shu, G.; Gao, Y.; Tian, H.; Wei, H.; Liang, X. Study of mixtures based on hydrocarbons used in ORC (Organic Rankine Cycle) for engine waste heat recovery. *Energy* **2014**, *74*, 428–438. [CrossRef]
71. Wang, E.; Zhang, H.; Fan, B.; Ouyang, M.; Zhao, Y.; Mu, Q. Study of working fluid selection of organic Rankine cycle (ORC) for engine waste heat recovery. *Energy* **2011**, *36*, 3406–3418. [CrossRef]
72. Tian, H.; Shu, G.; Wei, H.; Liang, X.; Liu, L. Fluids and parameters optimization for the organic Rankine cycles (ORCs) used in exhaust heat recovery of Internal Combustion Engine (ICE). *Energy* **2012**, *47*, 125–136. [CrossRef]
73. Peris, B.; Navarro-Esbrí, J.; Molés, F. Bottoming organic Rankine cycle configurations to increase Internal Combustion Engines power output from cooling water waste heat recovery. *Appl. Therm. Eng.* **2013**, *61*, 364–371. [CrossRef]
74. Grljušić, M.; Medica, V.; Radica, G. Calculation of efficiencies of a ship power plant operating with waste heat recovery through combined heat and power production. *Energies* **2015**, *8*, 4273–4299. [CrossRef]
75. Valencia Ochoa, G.; Piero Rojas, J.; Duarte Forero, J. Advance exergo-economic analysis of a waste heat recovery system using ORC for a bottoming natural gas engine. *Energies* **2020**, *13*, 267. [CrossRef]
76. Schulz, W.; Heitmann, S.; Hartmann, D.; Manske, S.; Peters-Erjawetz, S.; Risse, S. *Utilization of Heat Excess from Agricultural Biogas Plants*; Bremer Energie Institut, Universität Bremen, Institut für Umweltverfahrenstechnik: Bremen, Germany, 2007.
77. Kane, E.H.M.; Favrat, D.; Gay, B.; Andres, O. *Scroll Expander Organic Rankine Cycle (ORC) Efficiency Boost of Biogas Engines*; Technical Report; Università degli Studi Di Padova: Padova, Italy, 2007.
78. Dumont, O.; Dickes, R.; De Rosa, M.; Douglas, R.; Lemort, V. Technical and economic optimization of subcritical, wet expansion and transcritical Organic Rankine Cycle (ORC) systems coupled with a bio-gas power plant. *Energy Convers. Manag.* **2018**, *157*, 294–306. [CrossRef]
79. Koç, Y.; Yağlı, H.; Koç, A. Exergy analysis and performance improvement of a subcritical/supercritical organic rankine cycle (ORC) for exhaust gas waste heat recovery in a bio-gas fuelled combined heat and power (CHP) engine through the use of regeneration. *Energies* **2019**, *12*, 575. [CrossRef]
80. Saravia, J.R.; Galaz, J.R.V.; Villena, A.C.; Ortiz, J.N. Technical and economical evaluation of landfill-bio-gas fired combined cycle plants. *Distrib. Gener. Altern. Energy J.* **2012**, *27*, 7–25.
81. Uusitalo, A.; Uusitalo, V.; Grönman, A.; Luoranen, M.; Jaatinen-Värri, A. Greenhouse gas reduction potential by producing electricity from bio-gas engine waste heat using organic Rankine cycle. *J. Clean. Prod.* **2016**, *127*, 399–405. [CrossRef]
82. Marcuccilli, F.; Zouaghi, S. Radial inflow turbines for Kalina and organic Rankine cycles. *System* **2007**, *2*, 1.
83. Javanshir, A.; Sarunac, N.; Razzaghpanah, Z. Thermodynamic analysis of ORC and its application for waste heat recovery. *Sustainability* **2017**, *9*, 1974. [CrossRef]
84. Delgado-Torres, A.M.; García-Rodríguez, L. Preliminary assessment of solar organic Rankine cycles for driving a desalination system. *Desalination* **2007**, *216*, 252–275. [CrossRef]
85. Gerber, L.; Maréchal, F. Environomic optimal configurations of geothermal energy conversion systems: Application to the future construction of Enhanced Geothermal Systems in Switzerland. *Energy* **2012**, *45*, 908–923. [CrossRef]
86. Darrow, K.; Tidball, R.; Wang, J.; Hampson, A. *Catalog of CHP Technologies*; US Environmental Protection Agency Combined Heat and Power Partnership: Washington, DC, USA, 2015.
87. Pierobon, L.; Nguyen, T.V.; Mazzucco, A.; Larsen, U.; Haglind, F. Part-load performance of awet indirectly fired gas turbine integrated with an organic rankine cycle turbogenerator. *Energies* **2014**, *7*, 8294–8316. [CrossRef]
88. Li, J.; Li, P.; Pei, G.; Alvi, J.Z.; Ji, J. Analysis of a novel solar electricity generation system using cascade Rankine cycle and steam screw expander. *Appl. Energy* **2016**, *165*, 627–638. [CrossRef]
89. Garcia-Herrero, I.; Cuéllar-Franca, R.M.; Enriquez-Gutierrez, V.M.; Alvarez-Guerra, M.; Irabien, A.; Azapagic, A. Environmental assessment of dimethyl carbonate production: Comparison of a novel electrosynthesis route utilizing CO_2 with a commercial oxidative carbonylation process. *ACS Sustain. Chem. Eng.* **2016**, *4*, 2088–2097. [CrossRef]
90. Sinnott, R. *Chemical Engineering Design: Chemical Engineering Volume 6*; Elsevier: Amsterdam, The Netherlands, 2005.
91. Dimian, A.C.; Bildea, C.S. *Chemical Process Design: Computer-Aided Case Studies*; John Wiley & Sons: Hoboken, NJ, USA, 2008.
92. GPSA. *GPSA Engineering Data Book*; Gas Processors Association: Tulsa, OK, USA, 2004.

energies

Article

Mission-Oriented Policies and Technological Sovereignty: The Case of Climate Mitigation Technologies

Serenella Caravella [1], Valeria Costantini [1,2,*] and Francesco Crespi [1,2]

[1] Department of Economics, Roma Tre University, 00145 Rome, Italy; serenella.caravella@uniroma3.it (S.C.); francesco.crespi@uniroma3.it (F.C.)
[2] SEEDS Interuniversity Research Centre, Department of Economics and Management, University of Ferrara, 44121 Ferrara, Italy
* Correspondence: valeria.costantini@uniroma3.it

Abstract: The rapid decarbonization of the global economy represents the main challenge for the next decades to combat climate change. The European Union (EU) is leading the negotiation process under the Paris Agreement and recently approved an ambitious unilateral mitigation strategy known as the European Green Deal (EGD). In this paper, we present a novel approach based on the analysis of patent data related to climate change and mitigation technologies (CCMTs) with the aim of describing the evolutionary pattern of the EU in green technology. Based on our analysis, two of our main results deserve attention. First, at the global level, the pace of generation of new green technologies as measured by patent data is slowing down in recent years. This trend, if not inverted, casts some doubts on the economic sustainability of the ambitious environmental targets set by the EC. Second, the current EU technological positioning with respect to green areas appears to be problematic in terms of technological sovereignty, with serious risks of potential technological dependences from other countries. Given the radical technological shift required for the implementation of a full decarbonization pattern, the EU must realize a mission-oriented technology policy with additional and directed investments to ensure technological independence, together with a low-carbon and energy secure economy.

Keywords: clean energy technologies; European Green Deal; fit for 55; patent family; rarity index; revealed technological advantages; technological sovereignty

Citation: Caravella, S.; Costantini, V.; Crespi, F. Mission-Oriented Policies and Technological Sovereignty: The Case of Climate Mitigation Technologies. *Energies* **2021**, *14*, 6854. https://doi.org/10.3390/en14206854

Academic Editors: Claudia Toro and Chiara Martini

Received: 19 September 2021
Accepted: 15 October 2021
Published: 19 October 2021

1. Introduction

The European Union (EU) is a key player in international climate negotiations given the ambitious mitigation targets declared under the Paris Agreement (PA) framework. Such a challenging decarbonization pattern is the only road to a more sustainable development of human activities as emphasized by the main available assessments based on future projections of greenhouse gas (GHG) emissions related to anthropic activities [1,2].

Given the attitude of the EU to design complex strategies devoted to the mitigation process by using different regulatory and voluntary measures, the final evolution of the institutional setting is based on the carbon-neutral policy package formed by the blueprint for this transformational change represented by the European Green Deal (EGD) and the latest practical planning known as the Fit for 55 package. The EGD represents an ambitious long-term strategy with the primary objective to ensure the complete decarbonization of the EU by involving all policy levels as well as the civil society and the private economic sectors [3]. With the Fit for 55 package, the EU shapes an additional set of inter-connected proposals in order to find practical solutions to ensure a fair, competitive, and green transition by 2030 and beyond by exploiting potential benefits arising from the specific measures designed until now [4]. The involvement of a high number of stakeholders is a source of positive spillovers emerging from the fruitful interaction of different agents and the co-evolution of knowledge and skills [5]. At the same time, relevant transaction costs might slow down the technological shift required for the sustainable transition if adequate complementary policies and instruments fail to provide the institutional and economic environment required to support the weaker players such as energy-intensive industries or sectors with a low speed in adapting to new technological solutions [6].

Given that the rationale underlying the EGD is based on long-term economic growth, the targets highlighted in the roadmap to make Europe the first climate-neutral continent by 2050 must rely on a set of measures to assist the economic and industrial transformation [7]. If we look at the difference between the EU carbon emissions projected under a baseline scenario [8] and the full decarbonization pattern compatible with the target declared by the EU, the emissions gap is so steep that the cost of achieving the final 2050 CO_2 level might be prohibitive without a massive investment in the technological transition [9].

According to Tagliapietra et al. [9], among the priorities the EU strategy must address in the near future to be on track with the transition process, a key role is played by the resources invested in the development of new technologies and, more importantly, in the achievement of a leading competitive position in the international market.

The design of an ambitious policy strategy is a necessary condition to push the system toward a radical shift in the technological content of the economic system. Nonetheless, such a process is formed by different stages in the innovation process, from the invention of new technologies, through to the development of commercial devices and the diffusion among final consumers. All these stages are associated with different policy instruments that mainly impact one dimension of the transition process. For instance, instruments designed to influence the final demand of those technologies aim at respecting the standards, also known as demand-pull policies, which are typically associated with the deployment of existing technologies or to the development of incremental innovation [10,11]. In contrast, the direct financial support to research and development (R&D) activities carried out by the public and private sectors is more suitable for tracing a radical shift in the technological trajectory, allowing the system to escape from technological lock-in [12]. The recent development of the EU energy strategy seems to go in this direction thanks to the introduction of the Innovation Fund (IF), designed to support the development of low-carbon technologies with a budget of around EUR 10 billion over the period 2020–2030. Nonetheless, according to Bassi et al. [7], the resources to be directed toward the development and adoption of clean energy technologies should amount at 10 times the value of the IF. The estimates are based on a dynamic computable general equilibrium model that allows including the contribution of green technologies in achieving the decarbonization target at the lowest possible cost. In other words, they quantify the financial resources to be directed to the technology-based radical shift of the EU economy in order to transform the cost of cutting emissions into a development opportunity. A carbon price mechanism alone would negatively impact domestic income with a drop in GDP by 2050 of around −13%, meaning that only a de-growth process with a substantial reduction in production and consumption patterns might ensure the fulfilment of carbon mitigation targets [13]. In contrast, the full adoption of a sustainable transition strategy including the development and the diffusion of clean energy technologies and related infrastructures, along the whole production value chain and the consumption pattern, would ensure the system a positive GDP growth rate combined with a substantial emissions reduction.

Building on these projections, there are several key issues to be further investigated.

First, there is still uncertainty on the quantification of financial resources needed to be collected and activated by the public budget to sustain such a transition, compatible with the constraints on public finance related to the EU fiscal sustainability rules [14]. In this respect, there are some theoretical proposals mainly based on the adoption of a revenue recycling mechanism applied to carbon (and energy) taxation and practical advice suggesting the use of complementary resources directed toward technological development under investment plans to recover from the COVID-19 crisis [15,16].

The number of public resources to be invested by EU countries in this direction is substantially different if Europe's transition to full decarbonization should take place by developing green technologies in-house (the *make* option) instead of acquiring them abroad (the *buy* option). However, the choice between the two strategies is by no means neutral. For instance, undertaking the *buy* option might result in a sustainable transition process where the EU would significantly reduce the energy dependence from external sources thanks to the production of renewable energies at the expense of increasing the technological dependence from abroad if the innovative process takes place outside the EU borders. Such a possibility cannot be ruled out if one considers the marginal role historically played by technology-push instruments in climate-related policy packages planned by the EU and, more in general, the inability by past R&D policies (especially the Framework Programs and the European Fund for Strategic Investments) in fostering the innovation process at the point of bridging the scientific and technological gap of the EU vis-à-vis the United States, Japan, and, in the immediate future, China [17,18].

This last argument inevitably recalls the heated discussion on "technological sovereignty" that arose in Europe during the COVID-19 pandemic. The current crisis showed the deep technological capacity gap of Europe with both the U.S. and China in a variety of key domains, from health apparatus and vaccines to digital sectors [19,20].

According to its early notion, having "technological sovereignty" in a given area means that a country (or a group of countries) holds and preserves the ability to autonomously master knowledge in that field. Considering the globalized and interconnected nature of economies, the current meaning of "technological sovereignty" has transcended national boundaries and rather represents the appropriate level of technological capability to be held in order to avoid structural dependence with third parties [21]. This may be reached autonomously or, more likely, through mutual exchanges with other countries based on reliant alliances and partnerships [22]. In turn, technological sovereignty is a pre-condition for a plethora of other forms of sovereignty [23] such as innovation sovereignty (the ability to locally exploit technologies for the development of present and future economic activities), economic sovereignty (the ability to generate value added and prosperity through independent activities), and strategic autonomy (the ability to play an autonomous and strategic role in the geopolitical context) on issues of global importance including the green transition process.

With respect to the latter, its relevance in terms of technological sovereignty is stated by the recently published first work program for the European Innovation Council (EIC) that shapes innovation objectives on the EU's priorities for transiting to a sustainable, digital, and healthy society [24]. This calls for profound technological and innovation breakthroughs in a number of domains including green-related areas such as, as reported by the document, new pathways for green hydrogen production and engineered living materials such as advanced high-performance computing, edge computing, quantum technologies, cyber-security, artificial intelligence, block-chain, cloud infrastructure technologies and technologies for the Internet of Things, AI-driven tools for early diagnosis, point-of-care diagnostics, new approaches in cell and gene therapy, bio-processing 4.0, health intelligence services, and e-health solutions.

From a policy perspective, obtaining "technological sovereignty" in the green fields would mean exploiting the current and post-COVID-19 recovery instruments including the Next Generation EU recovery fund to devote massive resources to R&D and innovation in areas of greater scientific and technological opportunities such as green focal domains. This would require the adoption of specific green-related mission-oriented policies based on a systemic public policy toolkit that draws on frontier knowledge to attain the climate neutrality goals by contextually acquiring "technological control" over this area.

Indeed, mission-oriented innovation policies often contribute to the improvement in national competitiveness, since they are ambitious and cross-disciplinary in tackling societal and/or technological challenges [25]. Given that such policies focus on radical innovations to achieve goals of national importance (e.g., defense, infrastructure, and energy security), they create new markets and expand institutional actors of national innovation systems [26].

Seen through this lens, the EU climate strategy and technological sovereignty objectives are supposed to go hand in hand to maximize both sustainable and growth opportunities from the transformation of the economy. Therefore, the key question to be asked is whether, given the current technological state-of-the-art of Europe in the green domain, the 2050 carbon neutrality target is, as of today, compatible with the EU's existing technological capabilities (i.e., the current level of technological sovereignty in the green area) or, on the contrary, the achievement of the EU sustainable transition could led to beneficial environmental effects at the risk of triggering structural dependencies on external players, unless specific mission-oriented policies are launched.

From this point of view, assessing the current level of depth of technological sovereignty held by EU members in the green area is of paramount importance to better evaluate the potential trade-off between the make/buy options as well as to estimate the current and the desirable level of technological sovereignty to be held in environmental technologies including their costs in terms of investments, time needed, and possible efficiency losses.

To this end, our contribution aims at analyzing the technological position of the EU and other world leading countries with respect to climate change mitigation technologies (CMMTs) over the last three decades. In particular, this study is based on the analysis of patent trends, shares, and patent-related indicators in CMMTs for the period 1990–2016. Indeed, being focused on the result of the innovative process [27,28], patent data provide a wealth of information on the local presence (absence) of the scientific-technological competences needed to produce knowledge with a certain degree of autonomy, thus retrieving relevant insights with respect to the degree (lack) of technological sovereignty held by countries in a given area. Nevertheless, by reflecting the implementation of radical solutions in the field of technology, environmental-related patent data can be considered as the best proxy of a country's potential to achieve a low-emission economy through its internal knowledge resources.

The reminder of the paper is structured as follows. Section 2 describes the original database on eco-innovation built on patent data and the synthetic indexes elaborated for the analysis; Section 3 presents the main trends and evolution of CCMTs for the EU and other key players; Section 4 discusses the main relevant issues emerging from the statistical investigation that might inform the policy discourse; and Section 5 summarizes the main challenges the sustainable transition process would face in the next decades and suggests some insights for the development of an optimal policy mix design.

2. Materials and Methods

In this section, we propose a methodology based on the use of patent data to define a country's positioning in terms of technological capabilities in clean energy areas.

The dataset used for the present exercise draws information from the OECD-REGPAT database. To provide a better comparison across countries, the analysis is restricted to the most relevant patents, in other words, triadic patent families (TPFs) registered over time by the most important patent offices in the world: the United States Patent and Trademark Office (USPTO), the European Patent Office (EPO), and the Japanese Patent Office (JPO).

The tagging system for green patents follows the 4-digit Y02-Y04S classification scheme for CMMTs developed by the EPO [29]. By covering nine main categories—from energy, buildings, greenhouse gases (GHG) capture, Information & Communication Technologies (ICT), energy, transport and waste and wastewater management to smart grids—this classification provides a comprehensive overview of the technologies that, due to their technical attributes, can be referred to in the energy saving and decarbonization goals (Table 1).

The analysis was carried out on two distinct datasets that provide complementary information. The first was built by assigning the CMMT flag to any patent family (identified by a univocal id_family number) labeled with at least one of the nine CMMT subclasses. Then, in order to measure the inventive capacity of a country in decarbonization and clean energy areas, we applied the fractional counting to assign the patent application according to the inventor's place of residence. By following this criterion, if one application has more than one inventor, the application is divided equally among all of them and subsequently among their country (fractional counting), thus avoiding double counting. In the second dataset, the fractional counting was maintained, but the statistical unit was replaced by the specific CMMT subclass to which the TPF belongs. This implies that triadic patents with more than one CCMT subclass are double counted, with the final data representing the number of TPF applications filled by a country *c* in a given CMMT subclass in years. To ensure high data quality, we focused on the timeframe 1990–2016, thus dropping the last three-year period 2017–2019 as we detected a large decrease in patent applications due to time lags between the patent application and granting processes. This is mainly explained by the temporal gap in transforming the applied patent into published documents in OECD patent databases [30]. Finally, to further increase data reliability, we used a five-year moving average to compute yearly patent values.

Table 1. Climate change mitigation technology classes.

CMMTs Subclasses	Description
Y02A—Adaptation to climate change	CMMT technologies that allow adapting to the adverse effects of climate change in human, industrial (including agriculture and livestock), and economic activities.
Y02B—Buildings	CMMT technologies related to buildings, e.g., housing, house appliances or related end-user applications.
Y02C—Capture and storage of GHG	CMMT technologies for capture, storage, sequestration or disposal of greenhouse gases (GHG) included nitrous oxide (N_2O), methane (CH_4), perfluorocarbons (PFC), hydrofluorocarbons (HFC), or sulfur hexafluoride (SF6).
Y02D—ICT	CMMT technologies in information and communication technologies (ICT), i.e., information and communication technologies aiming at the reduction in their own energy use.
Y02E—Energy	CMMT technologies related to energy generation, transmission, or distribution that allow reducing greenhouse gas (GHG) emissions.
Y02P—Industry and agriculture	CMMT technologies in the production or processing of goods in any kind of industrial processing or production activity including the agri-food industry, agriculture, fishing, ranching, and the like.
Y02T—Transportation	CMMT technologies related to transportation (road transport, transportation via railways, e.g., energy recovery or reducing air resistance; aeronautics or air transport; maritime or waterways transport).
Y02W—Waste and wastewater	CMMT technologies related to solid waste management, solid waste management, enabling technologies or technologies with a potential or indirect contribution to greenhouse gas (GHG) emissions mitigation.
Y04S—Smart grids	Systems integrating technologies related to power network operation, communication, or information technologies for improving the electrical power generation, transmission, distribution, management, or usage.

To measure the actual "success" of country *c* in CCMTs, we built a set of indicators based on shares and growth rates related to patents. Therefore, we enriched the analysis with specialization and rarity indexes in order to provide complementary information with respect to the level of technological sovereignty possessed by countries in this domain. On one hand, the specialization index indicates whether a country concentrates in a specific domain with a larger share of its average innovation compared to the world average (or any other regional aggregate). On the other hand, building on the idea that technological sovereignty does not imply the need to achieve complete autonomy in a given domain but rather the ability to acquire and use technological knowledge developed elsewhere through reliable partnerships, thus avoiding unilateral dependencies [21], the rarity index provides a useful complementary measure as it is related to the number of countries specialized in it. It follows that the rarer a technology, the fewer the countries from which knowledge can be acquired, and hence the greater the risk of unilateral structural dependencies [20].

Following Hidalgo and Hausmann [31], the starting point of our analysis consists of calculating the revealed technology advantage indicator (RTA) using patent data [32]. In general, the RTA indicates whether country $c = 1, \ldots, C$ is specialized in a given technology t based on the comparison between the relative frequency of patenting in a given technology t in country c, with the relative frequency of patenting in the same technology t at the world level (Equation (1)). More specifically, the RTA was here computed on the CCMT patents filed in each country during 2000–2016. Therefore, RTA can be obtained as follows:

$$RTA_{ct} = \frac{\frac{TPF_{ct}}{\sum_{t=1}^{T} TPF_{ct}}}{\left(\frac{\sum_{t=1}^{T}(TPF_{ct})}{\sum_{t=1}^{T}\sum_{i=0}^{T} TPF_{ct}}\right)} \tag{1}$$

where TPF_{ct} is the number of triadic patent families of country c in technology t; C is the number of countries; and T is the number of technological fields. Thus, it follows that $RTA_{ct} = 1$ represents a threshold of specialization: when $RTA_{ct} > 1$, the country is said to be specialized in the generation of technology t and vice versa (Equation (2)):

$$RTA_{ct} = \begin{cases} 1\, if\ RTA_{ct} > 1 \\ 0\, if\ RTA_{ct} \leq 1 \end{cases} \tag{2}$$

Once RTA_{ct} is measured and transformed into a dichotomic specification (1, 0), it is possible to calculate the rarity index K_t, which is equal to the ratio of the number of countries with $RTA_{ct} > 1$ in CCMT domains to the total number of countries C (Equation (3)):

$$K_t = 1 - \left(\frac{\sum RTA_{c,t}}{C}\right) \tag{3}$$

As far as the RTA_{ct} is typically conceived, it could be considered as one of the potential observable outcomes stemming from a country's technological specialization pattern. Technological specialization is a process sustained, among other key enabling factors, by specific mission-oriented policy programs explicitly aimed at mastering knowledge in a given strategic field through the convergence of resources and capabilities in that direction. To put it differently, the RTA_{ct}, and to a greater extent, its dynamics over time, suggests whether or not a country is pursuing a specialization strategy in fields deemed particularly strategic, as in the case of CCMTS technologies. However, to avoid any potential bias related to the incomparability of the index on both sides of unity (given that more weight is assigned to values above 1 compared to observations below 1 when the standard RTA_{ct} is applied to quantitative analysis application), we transformed the index into a symmetrical adjusted version. According to [33], the scale of relative specialization is normalized from -100 to +100 based on a hyperbolic tangent function (Equation (4)):

$$RTA - Index = 100{\cdot}tanhln\left[\left(P_{ct}/\sum_{t=1}^{T} P_{ct}\right)\Big/\left(\sum_{t=1}^{T} P_{ct}/\sum_{c=1}^{C}\sum_{t=1}^{T} P_{ct}\right)\right] \tag{4}$$

where P_{ct} = patent applications, c = country, t = technology field. Based on this formula, a country c is said to be specialized in technology t when $RTA_{ct} - Index > 0$.

3. Results

3.1. Patent Dynamics in the CCMT: An Overview

As a starting point of the analysis, we describe the main trends of CCMT-TPFs over the last three decades distributed by the leading top inventors on the basis of world shares (Japan, USA, EU, South Korea, UK, and China) (Figure 1a) and by CMMT subclasses (Figure 1b). Although the temporal setting refers to a period where the UK was a member of the EU, we considered only the EU with 27 members, since the analysis was developed to address the features of the EU technological trajectory in CCMTs to provide elements for the future design of the climate neutrality package that will involve the 27 EU states. In line with other patent-based studies [34–36], it showed an increasing trend until 2010, which results from stringent environmental regulations [37,38] and successful technology policies, jointly implemented both on the supply and demand-side [39–41], along with a growing environmental awareness in civil society. However, similar to other contributions on this topic [42,43], we also recognize a downward trend after 2011. Following Urbaniec et al. [36], this could be due a to a combination of factors such as the financial crisis, the oversupply of some innovative energy equipment due to mass production in China, the reduction in investing incentives (as happened in Germany for subsidies for renewables), and the growing implementation of less innovative environmental practices caused by the increasing level of environmental pressure. Hence, this first evidence suggests that the ambitious environmental targets set by the EU are not coupled by a contemporaneous growth of CCMT related innovations, that, as already stated, represents a necessary condition to reduce the economic burden of climate change related policies.

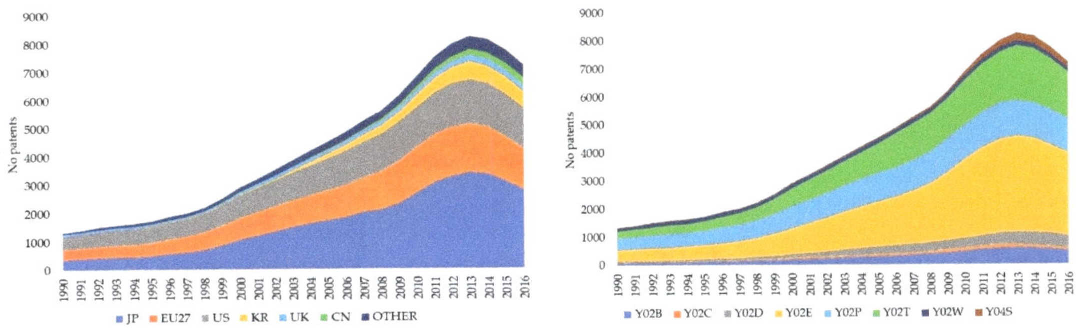

(a) (b)

Figure 1. CCMT-TPF trends in absolute values (1990–2016; 5-year MA): (**a**) by country, (**b**) by patent class.

Considering patent dynamics across countries, Japan appears by far the dominant player in the domain under scrutiny. According to Fujii and Managi [44], the Japanese supremacy in green technologies involves different causes including the new business opportunities created by the Kyoto Protocol in 2005; the efficiency improvement in patent applications due to changes in national rules governing patents (i.e., Japanese Bayh–Dole Act and the National University Corporation Act in 2003 for strengthening business–academic collaborations); and national eco-friendly policies (granted subsidies for hybrid and electric cars) as part of the emergency economic measures undertaken in response to the 2008 recession.

In the second and third positions, we observed that the EU and the U.S. had average quotas of 25% and 14%, respectively. As sustained by Waltz et al. [35], compared to the "South" of the world, both EU members and the U.S share better framework conditions for innovation as well as common ground for environmental policies and related externalities. Furthermore, Quitzow et al. [45] argue that, as environmental quality is a superior economic good with high income elasticity of demand, in high income countries such as the EU and the U.S., lead users for sustainability innovations would be typically found.

South Korea and China are in fourth and sixth places, respectively, with the UK in between. More in general, the increase in patents registered by the two Asiatic powers has been noticeable in many technological areas [46], as driven by a successful pattern of economic development built on competitive advance, export, and innovation [36]. For these reasons, both the CNIPA (Intellectual Property Administration of the People's Republic of China) and KIPO (Korean Intellectual Property Office) have been included among the largest patent offices (together with USPO, JPO, and EPO), recording an increase in applications from 17% to 43.4% during 2009–2019 [46]. According to Lema et al. [40] and Corrocher et al. [41], the process of growth in green patenting activity in the most successful latecomer countries such as China and South Korea is the result of a catch-up process enabled by mission-oriented policy actions targeted at stimulating technological opportunities (policies for the development of local technological capabilities together with the implementation of regulation and other policies) together with increasing the domestic firms' capability building, R&D investment, and new modes of knowledge transfer. Furthermore, with specific regard to the Chinese case, new green technological opportunities have been opened up by the acquisition of companies in countries that are leaders in this field [47,48].

CMMTs are not a homogeneous technological domain but a constellation of different technological sectors with differentiated technological pace over time and different degree of strategic relevance. By disentangling the CMMTs across three-digit Y02–Y04 subclasses (Figure 1b), we observed several differences among the various technology domains. Patents for decarbonization technologies applied in the energy (Y02E), transport (Y02T), and construction (Y02B) sectors are dominant during the whole period. Such evidence in rather obvious as, traditionally, the key targets of environmental policies worldwide have been represented by the most polluting industries characterized by energy and row material intensive production processes. However, new green technological patterns are emerging. For example, the systematic growth found for technologies related to environmentally-friendly ICT technologies (Y02D class) in the years 1993–2016 demonstrates the increasing pervasiveness of the digital paradigm in all areas of the technological progress. Moreover, a similar argument may explain the sharp increase in Y04S (smart grids) patent applications, whose importance has emerged since 2008 [49]. These frontier technologies make use of new digital communication, for instance, artificial intelligence (AI), to create intelligent (efficient) energy systems. Thanks to the analysis of big data trend collected for weather, energy demand, generation assets, cheapest cost, and highest efficiency of energy sources, Y04S-related applications have a great potential to directly contribute to the achievement of decarbonization and sustainable targets.

3.2. Country Specialization in CCMTs

Narrowing the analysis to the period 2000–2016 and focusing on the top-10 countries to have emerged over the last years (2010–2016), several differences appear over time and across technologies. Considering the whole Y02–Y04S area, we recognize the good positioning of the EU countries that ranked second as an aggregate, below Japan and behind the U.S. However, the EU technological performance worsens over time, moving from a share of around 25% in 2000–2009 to a share of 21.4% in 2010–2016. In the last period, EU countries have increased their presence only in the Y02W domain, a more traditional technology field related to water and waste management, while their shares have decreased in the remaining sectors.

More importantly, the greatest weakness of EU countries was in those classes at the technological frontier such as Y02D (with an average share of 9.3%) and Y04S (with an average share of 11%), representing technologies based on the use of digital devices to achieve better environmental performance. Technological supremacy over the highly strategic Y02D class was held by the U.S., which filled about 40% of patents in this domain. At some distance, the U.S. were followed by Japan (25%) and China, which showed a share of 9.6% from an initial value of 1.5% registered in the previous period. The Y04S class was dominated by Japan, where about 50% of the inventions made in this technology sector were generated. Overall, we observed a polarizing trend in favor of Japan, which reinforces its leadership in most green technologies, while the U.S. increased its relevance only in another strategic sub-domain (Y02D). Finally, South Korea and China showed the best performance in terms of dynamics, increasing their shares in all nine CMMT areas.

Similar conclusions could be made by observing, for the same countries and periods, the RTA_{ct} referring to both total CCMT patents and subdomains. As shown in Table 2, within countries ranked according to their 2010–2016 $RTA_{ct} - Index$ values, the strongest specialization was found for South Korea ($RTA_{ct} - Index = 23.3$), which outperformed Japan and France with $RTA_{ct} - Index$ values of 7.5 and 7, respectively. South Korea's superior green specialization pattern, which is particularly notable in the Y02E sub-domain, could be explained by the growing market demand for energy-saving innovations, driven by the long-term policy strategy (2009–2050 Strategy for Green Growth) together with green finance mechanisms [48]. Japan has a strong specialization in the frontier domain of smart grids (Y04S) where it recorded a $RTA_{ct} - Index$ by 35.2, while France showed the best $RTA_{ct} - Index$ in more traditional areas such as transport and water/waste management, with $RTA_{ct} - Index$ values slightly below 40. The remaining countries displayed negative specializations with some exceptions among the CCMT subclasses. On average, the $RTA_{ct} - Index$ for total CCMT patents of EU members worsened to −7.3 from a previous value of −0.9. Although the EU preserves a specialization advantage in many subdomains (Y02B; Y02C; Y02P; Y02W), its specialization appeared particularly weak in the more advanced and strategic fields such as Y04D and Y04S. In particular, with regard to the Y04D class, we recognized a strong specialization advantage for both the U.S. and China, suggesting that the technological positioning of the two economies in the environmental area is concentrated in fields on the technological frontier.

Table 2. Shares in CCMT-TPFs; total and by CPC class (2000–2009, 2010–2016; 5-year MA), (2010–2016 top-10 patenting countries plus EU27).

Country	Period	Y02–Y04S	CCMT Subclasses							
		-	Y02B	Y02C	Y02D	Y02E	Y02P	Y02T	Y02W	Y04S
JP	2000–2009	38.0	33.8	23.8	29.3	39.4	32.0	47.6	32.3	31.9
	2010–2016	40.4	39.4	32.3	25.1	40.1	33.2	53.6	25.1	49.7
EU27	2000–2009	24.5	27.5	26.2	14.9	21.4	27.5	26.3	26.0	19.1
	2010–2016	21.4	22.6	19.4	9.3	20.6	25.1	21.0	30.2	11.1
US	2000–2009	24.3	23.7	33.6	38.1	23.4	27.3	18.8	23.1	32.1
	2010–2016	19.5	18.4	29.2	39.7	17.0	22.0	14.6	24.5	23.6
DE	2000–2009	12.2	12.6	9.7	3.9	10.8	13.2	14.9	9.4	8.51
	2010–2016	9.3	7.9	7.7	2.7	9.6	11.0	9.2	9.3	3.9
KR	2000–2009	3.7	4.7	1.5	6.2	5.6	2.8	1.1	2.5	2.2
	2010–2016	7.5	5.2	1.9	6.7	11.6	6.2	4.3	2.4	4.4
FR	2000–2009	5.3	3.4	9.0	2.4	4.5	4.8	7.1	5.3	4.9
	2010–2016	5.6	3.9	4.9	1.8	5.2	5.4	7.4	7.5	2.4
UK	2000–2009	3.0	2.5	5.6	4.3	2.9	3.4	2.3	3.4	3.1
	2010–2016	2.8	2.5	3.3	3.0	2.8	2.8	2.5	3.1	2.1
CN	2000–2009	0.6	0.9	0.3	1.5	0.7	0.6	0.3	1.1	1.2
	2010–2016	2.6	3.6	1.2	9.6	2.1	2.8	0.9	3.1	2.1
NL	2000–2009	1.7	6.5	2.9	2.6	1.2	2.0	0.4	1.5	1.0
	2010–2016	1.4	6.0	1.3	0.9	1.0	1.7	0.3	1.4	1.5
SE	2000–2009	1.2	1.0	1.1	2.6	1.0	0.9	1.4	1.2	1.3
	2010–2016	1.1	0.7	1.0	2.0	0.7	1.1	1.4	1.8	0.5
CH	2000–2009	1.0	1.0	1.6	0.5	1.1	1.4	0.7	1.7	2.6
	2010–2016	1.0	0.9	1.7	0.2	1.1	1.4	0.6	1.9	0.7

The relevance of classes Y02D and Y04S from the perspective of technological sovereignty can also be inferred by observing the dynamics over time of the rarity index (K_t) for specific CCMT technologies (Table 3). In general, environmental technologies appeared less rare than the average (black dashed line), suggesting that the number of countries with a specialization advantage in these knowledge areas is higher than that recorded for other technologies.

Table 3. $RTA_{ct} - Index$ in CCMT-TPFs; total and by CPC class (2000–2009, 2010–2016; 5-year MA), (2010–2016 top-10 patenting countries plus EU27).

Country	Period	Y02–Y04S	CCMTs Subclasses							
-	-		Y02B	Y02C	Y02D	Y02E	Y02P	Y02T	Y02W	Y04S
KR	2000–2009 2010–2016	3.4 23.3	46.4 17.9	−38.0 −58.1	63.8 40.7	34.0 75.3	−9.4 34.6	−79.8 −4.5	−3.9 −51.7	20.4 1.4
JP	2000–2009 2010–2016	8.3 7.5	11.0 15.1	−21.7 −4.8	−3.6 −28.3	26.6 17.0	5.2 −1.8	40.8 43.0	6.1 −29.0	2.8 35.2
FR	2000–2009 2010–2016	0.4 7.0	−39.1 −25.4	52.8 −4.4	−51.4 −75.1	−14.9 6.8	−3.0 10.8	21.9 39.0	2.3 38.8	12.3 −61.0
DE	2000–2009 2010–2016	0.5 −0.4	6.1 −18.8	−20.2 −21.7	−76.7 −84.6	−7.0 2.0	10.7 15.2	25.0 −3.1	−22.4 0.4	−27.4 −68.3
EU27	2000–2009 2010–2016	−0.9 −7.3	4.7 8.8	20.6 16.8	−13.3 −30.3	−9.8 −8.1	23.5 22.1	−27.5 −23.3	31.6 41.8	20.4 −28.0
UK	2000–2009 2010–2016	−10.0 −6.3	−38.1 −27.3	36.3 −1.4	12.5 −7.2	−24.9 −16.4	−11.5 −13.8	−46.0 −26.0	−12.4 −5.2	−24.8 −36.9
SE	2000–2009 2010–2016	−10.0 −6.33	−38.1 −27.3	36.3 −1.4	12.5 −7.2	−24.9 −16.4	−11.5 −13.8	−46.0 −26.0	−12.4 −5.2	−24.8 −36.9
CN	2000–2009 2010–2016	−9.9 −9.0	−39.5 −37.2	−1.2 −12.6	45.8 41.5	−34.6 −58.4	−47.2 −14.8	−9.2 8.6	−19.5 3.0	38.3 −60.1
US	2000–2009 2010–2016	−3.4 −9.5	35.4 16.5	5.4 −59.8	48.3 79.3	4.1 −29.4	−14.9 −4.4	−62.9 −84.3	58.6 5.4	79.7 −37.1
NL	2000–2009 2010–2016	−6.6 −11.5	−21.6 −31.4	11.5 14.0	22.2 41.8	−23.9 −37.9	−9.1 −14.3	−42.3 −50.1	−25.8 −3.7	4.8 −6.6
CH	2000–2009 2010–2016	−17.4 −14.5	77.9 80.8	20.2 −20.1	10.3 −58.5	−51.0 −59.3	−12.9 −13.0	−94.5 −93.7	−37.7 −27.3	−21.6 −19.1

As already pointed out to explain the increasing trend observed for CCMT patent applications, this diffusion process in terms of specialization advantages can be seen as the result of international policy efforts and growing ecological awareness worldwide. However, for the majority of CCMTs, we observed a slight increase in K_t during the last years. The significant reduction in the number of countries specialized in environmental-related inventions could reasonably be attributed to the general decline in green patenting trends after 2011, caused by the aftermath of the 2008 crisis and the faster diffusion of less innovative environmentally-friendly solutions. Furthermore, the nine green technologies can be divided into groups characterized by low-rarity and high- rarity degree (Figure 2).

Unsurprisingly, the first group embraced the more traditional technologies such as those related to water/waste management (Y02W), buildings (Y02B), energy generation, transmission or distribution (Y02E), and industry (Y02P). Conversely, among the rarer environmental technologies, we found both smart grids (Y04S) and digital technologies (Y02D) together with transport related (Y02T) green patents and those applied for capturing and storing greenhouse gases (Y02C).

3.3. Climate Change Related Technologies in EU Countries

Focusing on the evolution of CCMT-TPFs within the EU countries over the last three decades (Figure 3), significant differences can be detected across countries and years both in absolute shares and $RTA_{ct} - Index$ (here calculated with the EU used as the reference territorial aggregate). With respect to the latter, a pattern of de-specialization emerged in Spain, Germany, the Scandinavian area (Denmark, Sweden, Finland) and, to a greater extent, in Eastern Europe, except for Poland, Slovakia, and Bulgaria. France was the only country to maintain a specialization advantage over the considered time span, with a positive value of $RTA_{ct} - Index$ in both 2000 and 2016. In contrast, Italy and Portugal preserved a negative $RTA_{ct} - Index$, as in the case of northern countries such as Ireland, the Netherlands, and Belgium. In terms of patent shares calculated intra-EU, Germany showed the most significant variation, decreasing its share from 53% in 2000 to 41% in 2016. Actually, after 2011, Germany experienced a significant cut in public support directed to the renewable energy sector, with a progressive reduction, especially in market-based incentives such as the feed-in-tariff scheme for solar and wind energy [50]. At the same time, for the remaining EU countries, we detected a generalized increasing trend, which is particularly relevant for France, which during 2000–2016, increased its quota by 6p.p. from an initial value of 18.9%. The third player in 2016 was Sweden with an average share of 5% (4.2% in 1990, 6.3% in 2000 and 5.8% in 2016), followed by Italy with similar quotas.

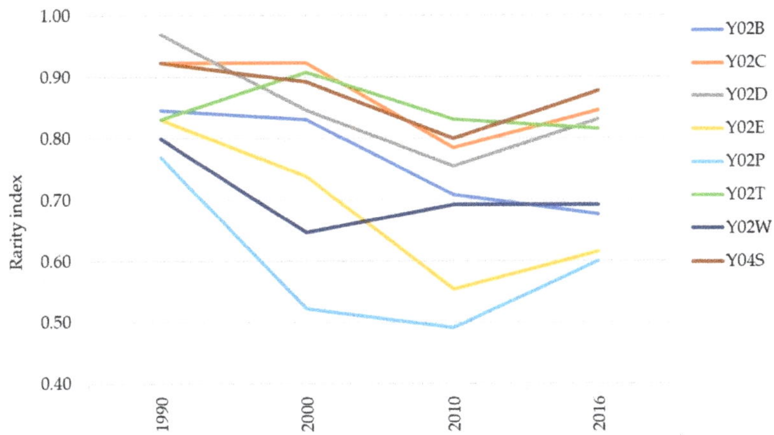

Figure 2. Rarity (K_t) in CCMT-TPFs by CPC class (2000, 2010, 2016; 5-year MA).

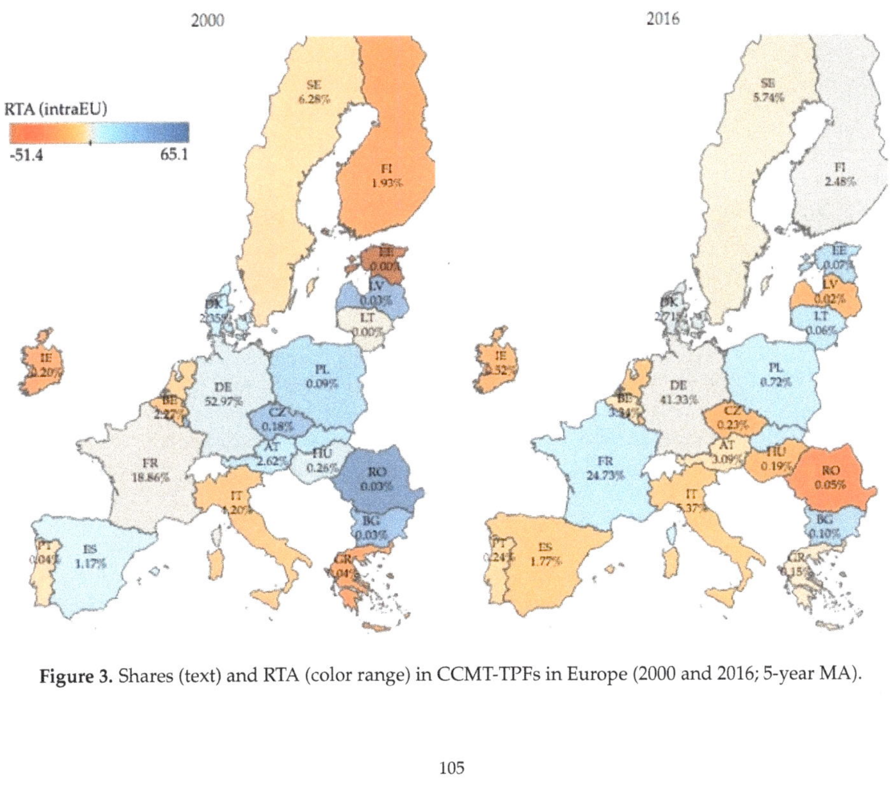

Figure 3. Shares (text) and RTA (color range) in CCMT-TPFs in Europe (2000 and 2016; 5-year MA).

Finally, to better measure the market concentration of CCMT-TPFs at the EU level, we relied on the Herfindahl–Hirschman Index (HHI), which is calculated by squaring the market share of each country and then summing up the resulting numbers. As shown in Figure 4, which reports the HHI values, the market concentration in this patenting area decreased from around 0.33 in 2000 to 0.25 in 2016. Such a reduction in the HHI could result from a cross-national policy convergence in environmental protection [51], enabled by international legal agreements that shaped national environmental policy plans, along with more diffused practices due to cross-national imitation, emulation, and learning processes [52].

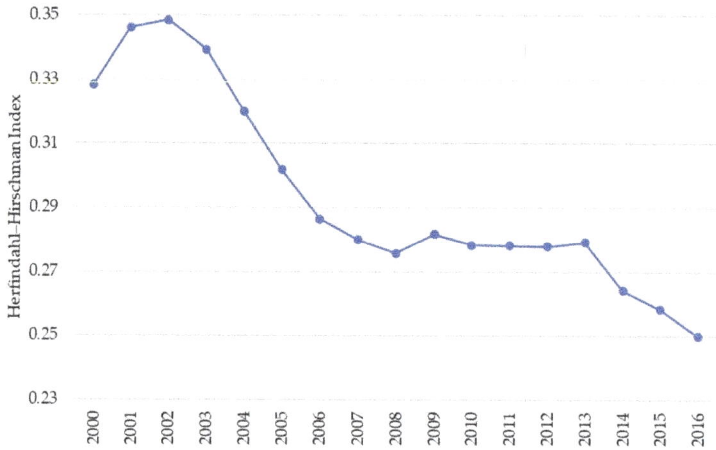

Figure 4. The HHI in CCMT-TPFs in Europe (2000–2016; 5-year moving average).

4. Discussion

In the global effort to fight climate change, the EU is at the forefront. The aim to become the world's first climate-neutral continent by 2050 is at the core of a comprehensive package of policy measures formed by the EGD and the recent Fit for 55. The climate goals represent the potential final outcome of a successful decarbonization pattern based on the development and deployment of energy-efficient, renewable, and low-carbon technologies. Accordingly, technological progress is the very engine of the European green transition. Indeed, the ability of EU countries to master technology in this direction is of paramount importance in leading the green transformation without depending on foreign partners. Furthermore, at a time of increasing geopolitical tensions and fierce global competition, having a strong technological base for developing, deploying, and using climate change mitigation technologies would represent a concrete opportunity for the EU to increase its strategical autonomy as a global economic power. In other words, ambitious environmental targets should be necessarily coupled by mission-oriented technology policies aiming at EU technological leadership in key CCMTs in order to avoid the risk of being trapped into technological structural dependences with third parties.

Based on our analysis, two main results deserve attention. First, at the global level, the pace of the generation of new green technologies as measured by patent data has slowed down in recent years. This could be explained by the existence of strict linkages between diversification patterns and existing competences. Indeed, according to Perruchas et al. [53], countries move along cumulative paths of specialization, and toward more mature green technologies. This trend, if not inverted, casts some doubts on the economic sustainability of the ambitious EU environmental targets. Second, the current EU technological positioning with respect to green domains appears to be problematic in terms of technological sovereignty, with serious risks of potential technological dependences from other countries.

In more detail, the EU shows a good ranking, being in second place for CCMT inventions filed during the last three decades. However, the dynamics over time suggests a declining trend in this area, with a significant reduction in shares. This is particularly relevant in the domains related to digital, ICT (Y02D), and smart grids (Y04S), which are considered as highly strategic in enabling speed and scale in achieving the EU's decarbonization goals, thus paving the way for the EU's "twin transition" to climate neutrality and digital leadership.

The majority of CCMT patents were applied by Japan, which in the Y04S sub-domain reached a 40% share combined with a high specialization advantage (35.2), as measured by the RTA index. The exception is the Y02D class, where the U.S. is particularly strong both in terms of quota (39.7) and specialization (79.3). In addition, we observed that the green technology arena is being reshaped by the emergence of lagging economies such as China and South Korea, which during 2010–2016 covered together about 10% of CCMT patent applications. Fueled by strong innovation-oriented policy support aimed at strengthening the whole technological environment, the increase in green patents showed by China and South Korea demonstrates that enhancing and exploiting national knowledge capabilities through policy leverage is essential to take an active role in the sustainable transition process.

Returning to the old continent, the EU is also losing ground in terms of specialization, with a share of CCMT patents on total patents that is lower than the world average, suggesting that the mission-oriented policies planned by the EU in recent years have missed the goal of concentrating innovation efforts in this area while, at the same time, past climate and energy policies have been too weak under the technology-push lever. European de-specialization was greatest in the most strategic sub-domains, namely Y02D and Y04S, which, as expected, turned out to be the rarest technologies. This has important implications in terms of technological sovereignty, as the rarity of a technology increases the risk for EU states to be structurally dependent on a few countries specializing in it. Our analysis showed that in the specific case of Y02D, with the exclusion of the U.S., the best specialization pattern was found for China (41.5): two countries on which the EU is already highly dependent for the supply of digital technologies in general [20].

Within European borders, we observed a generalized trend toward de-specialization combined with a spread of related patents, as confirmed by the dynamics of the HHI during 2000–2016. These results provide evidence of a pattern of convergence at the European level regarding the diffusion of green-related technological capabilities, a process mostly enabled by past R&D policy interventions, which appear to be successful in this respect. However, the increasing de-specialization found for CCMTs shows that Europe is still far from establishing a successful technological specialization in the green area, probably due to the lack of effective mission-oriented policy actions taken in this direction. Moreover, despite this, the massive production of green patents at the European level is mainly based in Germany and France and, to a lesser extent, in Sweden and Italy. This implies that the remaining EU members still play a rather marginal role as green innovators and further efforts in exploiting complementary skills and knowledge developed in other technological domains are required to close the gap in regional convergence [54].

The present analytical setting, despite its strength in transforming complex information into simple descriptive tools, presents a strong limitation in exploring linkages across complementary technological domains given the low degree of detail in the definition of patent families. An additional work to be developed as a future research agenda will be an in-depth analysis of cross-fertilization of inventions working with raw information on patent data.

5. Conclusions

We propose a simple analytical tool to describe the technological position of the EU and other world leading countries with respect to climate change mitigation technologies (CMMTs) over the last three decades. In particular, this study was based on the analysis of patent trends, shares, and patent-related indicators in CMMTs for the period 1990–2016. Based on our analysis, two main results appear to be relevant. First, the pace of the generation of new green technologies, as measured by patent data, has slowed down in recent years at the global level, revealing a mismatch with the urgent requirements of a rapid and radical shift in the socio-technical systems to enhance the sustainable transition. Second, the current EU technological performance in CMMTs appears to not be fully satisfactory, thus raising issues in terms of technological sovereignty, with potential technological dependences from other countries.

While the ambitious EU climate strategy represents a great opportunity for the European countries to change the development pattern, this would require additional resources to be invested for exploiting synergies and complementarities between policy actions aimed at replacing fossil fuels with cleaner alternatives. The future strategic design on the EU recovery actions should go hand in hand with the climate strategy in order to better support EU countries to compete at the technology frontier while enhancing the convergence process within European borders. In this respect, appropriate mission-oriented technology policies are required to ensure, at the same time, the meeting of ambitious environmental targets and technological independence with respect to key technologies.

Author Contributions: Conceptualization, V.C. and F.C.; Methodology, S.C., V.C. and F.C.; Software, S.C.; Validation, S.C., V.C. and F.C.; Formal analysis, S.C.; Investigation, S.C., V.C. and F.C.; Resources, V.C. and F.C.; Data curation, S.C.; Writing—original draft preparation, S.C., V.C. and F.C.; Writing—review and editing, S.C., V.C. and F.C.; Visualization, S.C., V.C. and F.C.; Funding acquisition, V.C. and F.C. All authors have read and agreed to the published version of the manuscript.

Funding: This research was funded by the Italian Ministry of University and Research (MIUR), Scientific Research Program of National Relevance PRIN2017 project "Innovation for global challenges in a connected world: the role of local resources and socioeconomic conditions". The support of Centro Economia Digitale is also gratefully acknowledged.

Institutional Review Board Statement: Not applicable.

Informed Consent Statement: Not applicable.

Data Availability Statement: Data for patents in environmental technologies are covered by copyright and can be used only by PATSTAT subscribers.

Acknowledgments: Comments, discussions, and suggestions from several persons enriched the current development of the research.

Conflicts of Interest: The authors declare no conflict of interest.

References

1. IEA (International Energy Agency). *World Energy Outlook 2020*; IEA: Paris, France, 2020.
2. Dellink, R.; Chateau, J.; Lanzi, E.; Magné, B. Long-term economic growth projections in the Shared Socioeconomic Pathways. *Glob. Environ. Chang.* **2017**, *42*, 200–214. [CrossRef]
3. EC (European Commission) Communication from the Commission. *The European Green Deal. COM (2019)640 2020*; European Commission: Bruxelles, Belgium, 2020.
4. EC (European Commission) Communication from the Commission. *Fit for 55: Delivering the EU's 2030 Climate Target on the Way to Climate Neutrality. COM (2021)550Final 2021*; European Commission: Bruxelles, Belgium, 2021.
5. Rogge, K.S.; Reichardt, K. Policy mixes for sustainability transitions: An extended concept and framework for analysis. *Res. Policy* **2016**, *45*, 1620–1635. [CrossRef]
6. Crespi, F.; Mazzanti, M.; Managi, S. Green growth, eco-innovation and sustainable transitions. *Environ. Econ. Policy Stud.* **2016**, *18*, 137–141. [CrossRef]
7. Bassi, A.M.; Costantini, V.; Paglialunga, E. Modelling the European Union Sustainability Transition: A Soft-Linking Approach. *Sustainability* **2021**, *13*, 6303. [CrossRef]
8. Keramidas, K.; Diaz Vazquez, A.; Weitzel, M.; Vandyck, T.; Tamba, M.; Tchung-Ming, S.; Soria-Ramirez, A.; Krause, J.; Van Dingenen, R.; Chai, Q.; et al. *Global Energy and Climate Outlook 2019: Electrification for the Low-Carbon Transition*; JRC119619; Institute for Energy and Transport: Seville, Spain, 2020. [CrossRef]
9. Tagliapietra, S.; Zachmann, G.; Edenhofer, O.; Glachant, J.M.; Linares, P.; Loeschel, A. The European union energy transition: Key priorities for the next five years. *Energy Policy* **2019**, *132*, 950–954. [CrossRef]

10. Costantini, V.; Crespi, F.; Palma, A. Characterizing the policy mix and its impact on eco-innovation: A patent analysis of energy-efficient technologies. *Res. Policy* **2017**, *46*, 799–819. [CrossRef]
11. Hoppmann, J.; Peters, M.; Schneider, M.; Hoffmann, V.H. The two faces of market support—How deployment policies affect technological exploration and exploitation in the solar photovoltaic industry. *Res. Policy* **2013**, *42*, 989–1003. [CrossRef]
12. Costantini, V.; Crespi, F.; Martini, C.; Pennacchio, L. Demand-pull and technology-push public support for eco-innovation: The case of the biofuels sector. *Res. Policy* **2015**, *44*, 577–595. [CrossRef]
13. Kallis, G.; Kostakis, V.; Lange, S.; Muraca, B.; Paulson, S.; Schmelzer, M. Research on degrowth. *Annu. Rev. Environ. Resour.* **2018**, *43*, 291–316. [CrossRef]
14. Speck, S.; Zoboli, R.; Paleari, S.; Marin, G.; Mazzanti, M.; Bassi, A. The Sustainability Transition in Europe in an Age of Demographic and Technological Change. Environmnetal European Agency (EEA) Report No 23/2019. Available online: https://www.eea.europa.eu/publications/sustainability-transition-in-europe (accessed on 21 June 2021).
15. Elkerbout, M.; Egenhofer, C.; Núñez Ferrer, J.; Catuti, M.; Kustova, I.; Rizos, V. The European Green Deal after Corona: Implications for EU Climate Policy. *CEPS Policy Insights* **2020**, 1–12. Available online: https://www.sipotra.it/wp-content/uploads/2020/04/The-European-Green-Deal-after-Corona-Implications-for-EU-climate-policy.pdf (accessed on 18 June 2021).
16. Sueyoshi, T.; Ryu, Y.; Yun, J.-Y. COVID-19 Response and Prospects of Clean/Sustainable Energy Transition in Industrial Nations: New Environmental Assessment. *Energies* **2021**, *14*, 1174. [CrossRef]
17. Archibugi, D.; Filippetti, A.; Frenz, M. Investment in innovation for European recovery: A public policy priority. *Sci. Public Policy* **2020**, *47*, 92–102. [CrossRef]
18. Archibugi, D.; Mariella, V. Is a European recovery possible without high-tech public corporations? *Intereconomics* **2021**, *56*, 160–166. [CrossRef] [PubMed]
19. Crespi, F.; Caravella, S.; Menghini, M.; Salvatori, C. European Technological Sovereignty: An emerging framework for policy strategy. *Intereconomics* **2021**, *6*, in press.
20. Cerra, R.; Crespi, F. *Sovranità Tecnologica: Elementi per Una Strategia Italiana ed Europea*; Centro Economia Digitale (CED): Rome, Italy, 2021. Available online: https://www.centroeconomiadigitale.com/wp-content/uploads/2021/03/CED-Sovranita-Tecnologica.pdf (accessed on 3 September 2021).
21. Edler, J.; Blind, K.; Frietsch, R.; Kimpeler, S.; Kroll, H.; Lerch, C.; Reiss, T.; Roth, F.; Schubert, T.; Schuler, J.; et al. Technology Sovereignty: From Demand to Concept. Fraunhofer Institute for Systems and Innovation Research Policy Brief No. 02/2020. Available online: https://www.econstor.eu/bitstream/10419/233462/1/policy-brief-02-2020.pdf (accessed on 3 September 2021).
22. Bauer, M.; Erixon, F. Europe's Quest for Technology Sovereignty: Opportunities and Pitfalls. ECIPE Occasional Paper, No. 02/2020. Available online: https://cdn.ecipe.org/wp-content/uploads/2020/06/2020_05_26_Event-Technology-Sovereignty-1.pdf (accessed on 3 September 2021).
23. Kelly, E.; Zubascu, F.; Naujokaityte, G.; Moran, N.; Pringle, D.; Wallace, N. *What Is 'Tech Sovereignty'?* Science Business Publishing: Brussels, Belgium, 2020. Available online: https://sciencebusiness.net/system/files/reports/2020-TECH-1.pdf (accessed on 3 September 2021).
24. EC (European Commission). *Decision on European Innovation Council (EIC) Established by the European Commission, under the Horizon Europe Programme (2021-27) C(2021)1510 2021*; European Commission: Bruxelles, Belgium, 2021.
25. Mowery, D.C.; Nelson, R.R.; Martin, B.R. Technology policy and global warming: Why new policy models are needed (or why putting new wine in old bottles won't work). *Res. Policy* **2010**, *39*, 1011–1023. [CrossRef]
26. Mazzucato, M. From market fixing to market-creating: A new framework for innovation policy. *Ind. Innov.* **2016**, *23*, 140–156. [CrossRef]
27. Albino, V.; Ardito, L.; Dangelico, R.M.; Petruzzelli, A.M. Understanding the development trends of low-carbon energy technologies: A patent analysis. *Appl. Energy* **2014**, *135*, 836–854. [CrossRef]
28. Haščič, I.; Migotto, M. *Measuring Environmental Innovation Using Patent Data*; OECD: Paris, France, 2015. [CrossRef]
29. Angelucci, S.; Hurtado-Albir, F.J.; Volpe, A. Supporting global initiatives on climate change: The EPO's "Y02-Y04S" tagging scheme. *World Pat. Inf.* **2020**, *54*, S85–S92. [CrossRef]
30. Dernis, H. *Nowcasting Patent Indicators*; Science and Technology Indicators, Working Paper 2007/3 Statistical Analysis of Science, Technology and Industry; OECD: Paris, France, 2007.
31. Hidalgo, C.A.; Hausmann, R. The building blocks of economic complexity. *Proc. Natl. Acad. Sci. USA* **2009**, *106*, 10570–10575. [CrossRef]
32. Antonelli, C.; Crespi, F.; Mongeau Ospina, C.A.; Scellato, G. Knowledge composition, Jacobs externalities and innovation performance in European regions. *Reg. Stud.* **2017**, *51*, 1708–1720. [CrossRef]
33. Grupp, H. *Foundations of the Economics of Innovation*; Edward Elgar Publishing: Cheltenham, UK; Lyme, CN, USA, 1998.
34. Karvonen, M.; Kapoor, R.; Uusitalo, A.; Ojanen, V. Technology competition in the internal combustion engine waste heat recovery: A patent landscape analysis. *J. Clean. Prod.* **2016**, *112*, 3735–3743. [CrossRef]
35. Walz, R.; Pfaff, M.; Marscheider-Weidemann, F.; Glöser-Chahoud, S. Innovations for reaching the green sustainable development goals–where will they come from? *Int. Econ. Econ. Policy* **2017**, *14*, 449–480. [CrossRef]
36. Urbaniec, M.; Tomala, J.; Martinez, S. Measurements and Trends in Technological Eco-Innovation: Evidence from Environment-Related Patents. *Resources* **2021**, *10*, 68. [CrossRef]

37. Sterlacchini, A. Trends and determinants of energy innovations: Patents, environmental policies and oil prices. *J. Econ. Policy Reform* **2020**, *23*, 49–66. [CrossRef]
38. Horbach, J.; Rammer, C.; Rennings, K. Determinants of eco-innovations by type of environmental impact—the role of regulatory push/pull, technology push and market pull. *Ecol. Econ.* **2012**, *78*, 112–122. [CrossRef]
39. Mazzucato, M. Mission-oriented innovation policies: Challenges and opportunities. *Ind. Corps. Chang.* **2018**, *27*, 803–815. [CrossRef]
40. Lema, R.; Fu, X.; Rabellotti, R. Green windows of opportunity: Latecomer development in the age of transformation toward sustainability. *Ind. Corp. Chang.* **2020**, *29*, 1193–1209. [CrossRef]
41. Corrocher, N.; Malerba, F.; Morrison, A. Technological regimes, patent growth, and catching-up in green technologies. *Ind. Corp. Chang.* **2021**, dtab025. [CrossRef]
42. Dechezleprêtre, A. How to Reverse the Dangerous Decline in Low-Carbon Innovation. Conversation. 2016. Available online: https://theconversation.com/how-to-reverse-the-dangerous-decline-in-low-carbon-innovation-66944 (accessed on 3 September 2021).
43. Cornwall, W. Clean Energy Patent Slump in US Stirs Concern. Science. 2017. Available online: https://www.sciencemag.org/news/2017/04/clean-energy-patent-slump-us-stirs-concern (accessed on 27 July 2021).
44. Fujii, H.; Managi, S. Research and development strategy for environmental technology in Japan: A comparative study of the private and public sectors. *Technol. Forecast. Soc. Chang.* **2016**, *112*, 293–302. [CrossRef]
45. Quitzow, R.; Walz, R.; Köhler, J.; Rennings, K. The concept of "lead markets" revisited: Contribution to environmental innovation theory. *Environ. Innov. Soc. Transit.* **2014**, *10*, 4–19. [CrossRef]
46. WIPO. *World Intellectual Property Indicators 2020*; World Intellectual Property Organization: Geneva, Switzerland, 2020. Available online: https://www.wipo.int/edocs/pubdocs/en/wipo_pub_941_2020.pdf (accessed on 27 July 2021).
47. Hansen, U.E.; Fold, N.; Hansen, T. Upgrading to lead firm position via international acquisition: Learning from the global biomass power plant industry. *J. Econ. Geogr.* **2016**, *16*, 131–153. [CrossRef]
48. Tolliver, C.; Fujii, H.; Keeley, A.R.; Managi, S. Green innovation and finance in Asia. *Asian Econ. Policy Rev.* **2021**, *16*, 67–87. [CrossRef]
49. Marku, E.; Zaitsava, M. Smart Grid Domain: Technology structure and innovation trends. *Int. J. Econ. Bus. Manag. Res.* **2018**, *2*, 390–440.
50. OECD. Environmental Policy: Renewable Energy Feed-in Tariffs. OECD Environment Statistics (Database). 2021. Available online: https://doi.org/10.1787/f68de84b-en (accessed on 12 October 2021).
51. Holzinger, K.; Knill, C.; Sommerer, T. Environmental policy convergence: The impact of international harmonization, transnational communication, and regulatory competition. *Int. Organ.* **2018**, *62*, 553–587. [CrossRef]
52. Busch, P.O.; Jörgens, H. The international sources of policy convergence: Explaining the spread of environmental policy innovations. *J. Eur. Public Policy* **2005**, *12*, 860–884. [CrossRef]
53. Perruchas, F.; Consoli, D.; Barbieri, N. Specialisation, diversification and the ladder of green technology development. *Res. Policy* **2020**, *49*, 103922. [CrossRef]
54. Montresor, S.; Quatraro, F. Green technologies and Smart Specialisation Strategies: A European patent-based analysis of the intertwining of technological relatedness and key enabling technologies. *Reg. Stud.* **2020**, *54*, 1354–1365. [CrossRef]

energies

Article

From Energy Audit to Energy Performance Indicators (EnPI): A Methodology to Characterize Productive Sectors. The Italian Cement Industry Case Study

Giacomo Bruni, Alessandra De Santis, Carlos Herce, Luigi Leto, Chiara Martini, Fabrizio Martini, Marcello Salvio, Federico Alberto Tocchetti and Claudia Toro *

Energy Efficiency in the Economic Sectors Laboratory (DUEE-SPS-ESE), Energy Efficiency Unit Department, Italian National Agency for New Technologies, Energy and Sustainable Economic Development (ENEA), C.R. Casaccia, Via Anguillarese 301, 00123 Rome, Italy; giacomo.bruni@enea.it (G.B.); alessandra.desantis@enea.it (A.D.S.); carlos.herce@enea.it (C.H.); luigi.leto@enea.it (L.L.); chiara.martini@enea.it (C.M.); fabrizio.martini@enea.it (F.M.); marcello.salvio@enea.it (M.S.); federicoalberto.tocchetti@enea.it (F.A.T.)

* Correspondence: claudia.toro@enea.it

Abstract: In this work, a novel methodology to assess energy performance indicators of productive and economic sectors trough the analysis of the Italian mandatory energy audits database is presented. The updating of sectoral reference energy performance indicators is fundamental for both companies and policy makers—for the formers to evaluate and compare their energy performance with competitors in order to achieve improvements and for the latter to effectively monitor the impact of energy policies. This methodology could be potentially applied to all production sectors, providing key information needed to characterize various production processes from an energy point of view. Awareness of energy efficiency and sectorial benchmarking represent the first necessary steps for companies moving towards energy transition. This paper provides details of the statistical method developed and its application to the NACE 23 division "Manufacturing of other non-metallic mineral products", with a focus on the cement industry. For this sector, results are presented in terms of specific indicators based on energy source. General results, methodological insights, and validation of the proposed case study are discussed.

Keywords: energy audits (EAs); energy performance indicators (EnPI); specific energy consumption (SEC); energy efficiency; energy management; industry; tertiary sector; cement; energy transition

Citation: Bruni, G.; De Santis, A.; Herce, C.; Leto, L.; Martini, C.; Martini, F.; Salvio, M.; Tocchetti, F.A.; Toro, C. From Energy Audit to Energy Performance Indicators (EnPI): A Methodology to Characterize Productive Sectors. The Italian Cement Industry Case Study. *Energies* **2021**, *14*, 8436. https://doi.org/10.3390/en14248436

Academic Editor: Maria Founti

Received: 3 November 2021
Accepted: 29 November 2021
Published: 14 December 2021

1. Introduction

As stated by the International Energy Agency, the cleanest and most sustainable energy is that which we do not consume. Energy efficiency was first considered hidden fuel, then first fuel, and finally as having a key role to play in a clean energy transition. Moreover, energy efficiency offers several benefits, such as improving energy security, increasing employment, and reducing CO_2 emissions [1]. However, an extended energy-efficiency gap still exists [2]; namely, profitable energy-efficiency interventions are not implemented due to several barriers pertaining to different categories [3] and involving enterprises of various sizes in different sectors [4]. A wide range of policies have been developed by different countries to address the energy-efficiency gap, as shown by Tanaka in [5]. The number of such policies steadily increased in the period from 1970 to 2011.

Energy use accounts for 75% of the EU's emissions [6]. The transformation of energy systems is central to achieving the European climate and energy goals reported in the European Green Deal [7].

The European Union stresses the need to adopt a holistic approach in which all EU actions and policies contribute to the objectives of the Green Deal itself. The Commission communication announced initiatives covering several policy areas that are all highly interlinked, including climate, environment, energy, transport, industry, agriculture, and sustainable finance.

Saving more energy and using more renewable energies is a key driver for jobs, growth, and emission reduction. In this context, energy efficiency is a milestone in the industrial and tertiary transformation process. Making production processes more efficient and rationalizing the use of energy resources are the main objectives of the European Commission's approach to the issue of energy efficiency in production processes.

The Energy Efficiency Directive 2012/27/EU (EED) [8] (and the 2018/2002 directive amendment [9]) is a key element of Europe's energy legislation. It includes a balanced set of binding measures intended to help EU Member States reach the 20% energy efficiency target by 2020. The increase in energy efficiency in the production sectors turns out to be one of the cornerstones of the new European Green Deal, introduced by the European Union to reach the challenging goal of an almost global decarbonisation of the economy by 2050. As of today, in fact, industry is still responsible for 20% of greenhouse gas emissions in the EU, and to achieve this goal, a strong paradigm shift is needed in the management of production processes (from a linear model to a circular one) and in generation, distribution, and use of energy, with particular attention paid to the efficiency.

The EED establishes a common framework of measures for the promotion of energy efficiency (EE) to ensure the achievement of European targets and to pave the way for further EE improvements beyond 2020. Article 8 of the EED introduced the obligation for large enterprises to carry out an energy audit on their production sites, starting in December 2015 and subsequently every 4 years. To this extent, the Italian definition of large enterprise is a business organization with more than 250 employees and with an annual turnover exceeding EUR 50 million and/or an annual balance-sheet total exceeding EUR 43 million. The size of the enterprise takes into consideration the core company and partner/linked enterprises within the Italian territory.

In the EED, an energy audit is defined as "a systematic procedure having the purpose of obtaining adequate knowledge of the current energy consumption profile of a building or group of buildings, of an industrial or commercial operation or installation, of a private or public service, by identifying and quantifying cost-effective energy saving opportunities, and reporting the findings" [8]. Therefore, an energy audit is the first step to characterize energetically different sites and sectors and to define a long-term strategy on energy efficiency for enterprises and policy makers.

Measuring energy-efficiency performance of equipment, processes, and factories is the first step toward effective energy management in production [10]. In order to gain a greater awareness of energy-saving opportunities, it is necessary to compare the energy performance of a site with "market references". In technical and scientific literature, it is not difficult to find references for single components (e.g., efficiency of air compressors [11] or multiple energy-efficiency measures databases, including electric motors, steam generators, cooling and refrigeration systems, heat recovery, etc., promoted by the United Nations [12], the European Commission [13], or specific countries, such as Sweden [14]). Moreover, there are multiple methods and tools available to assess the impact of energy-efficiency improvements in a single site or company. Some excellent reviews are focused on analysing energy assessment methods [15], key energy performance indicators in production [16], energy management systems in industry [17], and energy performance indicators in ISO 50001 energy management systems [18]. However, there is a lack of information on the definition of methods to evaluate the baseline of energy consumption in different economic sectors, which crucial information for the evaluation of the impact of energy-efficiency measures.

Energy performance indicators (or energy-efficiency indexes, EnPIs) can be based on economic data (i.e., value added by the production) or physical terms (i.e., tons or cubic meters of products), and at the sectoral level, they depend on the activity level of analysis, sector structure, and energy-efficiency maturity [19]. Several efforts have been made to homogenise and standardise the use of multiple energy-efficiency indicators to compare energy efficiency between countries and sectors [20]. However, currently, these methods are applied only to a limited number of energy-intensive industries (such as cement, aluminium, iron and steel, ethylene, ammonia, refining) [21] for which the variety of final product analyses is restricted and technologies are mature [22]. It is important to cite the efforts of the European IPPC Bureau to set up, review, and update BAT reference documents (BREFs), a series of sectoral analysis of more than 52,000 installations across Europe affected by the Industrial Emissions Directive (2010/75/EU) [23]. These documents are the European consumption reference for several industrial processes, providing a range of EnPIs at the European level but without specific information for each country.

The energy analysis of specific economic sectors should ideally be based on physical units of production. The information must be sufficiently disaggregated to allow for the analysis of processes and sites, while models should be the same for all economic sectors [24]. Therefore, the use of the information from energy audits is ideal to define sectoral EnPIs. Moreover, the use of linear models in energy-efficiency analysis is extensive due to their applicability in the developing benchmarks [25], determining energy savings in industries [26] (despite several production processes being not linear [27]), linking energy efficiency and productivity [28], forecasting industrial [29] and tertiary [30] energy consumption, evaluating benchmarks for plant indicators, as a basis for stochastic frontier analysis [31], modelling building consumption [32], and estimating national economic indicators of electricity consumption [33].

Another important issue is the depth level of the description of economic activities (NACE level) in order to have a compromise between availability of data and accuracy of the information. One NACE code is assigned to enterprises or production sites according to their main economic activity. The main activity is the one which contributes most to the value added of the unit. An activity, defined by a NACE code, may consist of one simple process (for example, weaving) but may also cover a whole range of sub-processes, each mentioned in different categories of the classification (for example, car manufacturing consists of specific activities, such as casting, forging, welding, assembly, painting, etc.) [34]. Therefore, for each NACE code, it is necessary to define clusters with homogeneous processes and/or products. Each NACE code is divided into four levels (section, division, group, and class), and it is recommended to carry out the definition of the sectoral indicators at a 4-digit NACE level (e.g., C23.51—manufacture of cement or D35.11—production of electricity) [35].

The analysis of energy audits to define the sectoral energy performance has been investigated in scientific literature in Germany [24,25], Sweden [26], Latvia [27], the Netherlands [28] and USA [29]. However, a high heterogeneity of the available data has been observed, as in the methodology used in the analysis and in the obtained results.

The Italian government transposed the EED in 2014 (by issuing the legislative Decree n. 102/2014, updated by legislative Decree n. 73/2020 [36]), also extending the obligation to a specific group of energy-intensive enterprises and assigning management of EED article 8 obligations to the ENEA (Italian National Agency for New Technologies, Energy, and Sustainable Economic Development).

Energy-intensive enterprises are those with large energy consumptions (more than 1 GWh of electricity) applying for tax relief on part of the purchased energy. All energy-intensive enterprises are registered in the list of "Cassa per i servizi energetici ed ambientali" (governmental agency related to electricity) [37]. Figure 1 shows schematically shows the Italian framework of mandatory energy audits. As of December 2019, the first deadline for the second compulsory cycle, 11,172 energy audits had been uploaded to the ENEA website from 6434 enterprises [38].

Over 70% of the energy audits received are complemented by a monitoring plan, in accordance with the guidelines drawn up by ENEA. The presence of monitoring systems makes the available data very useful and enables an accurate and in depth analysis of sectoral consumption (energy performance indicators, sectoral analysis, energy-consumption trend evaluation) and technologies. Monitoring also makes the energy audit itself very valuable, as it makes it possible to identify energy cost centres and therefore favours the implementation of energy-efficiency measures in order to reduce energy consumption and make production processes more efficient [39]. The Italian transposition of the NACE codes (ATECO codes) introduces two additional levels (categories and subcategories).

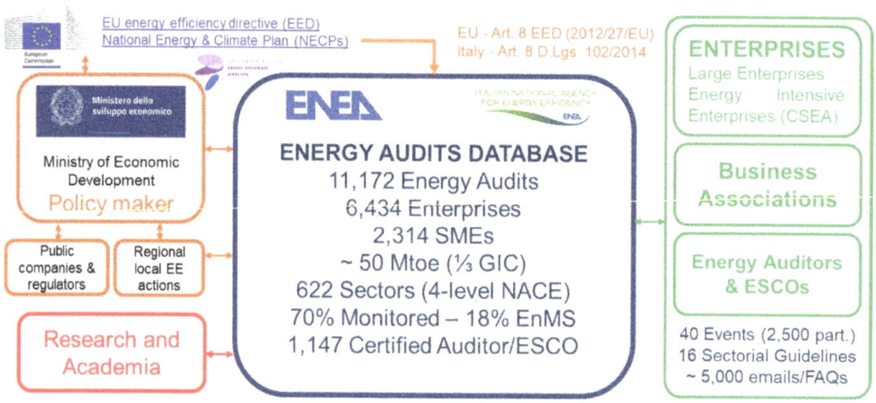

Figure 1. Italian framework of mandatory energy audits as of December 2019.

The purpose of this work is to identify a methodology that allows for the evaluation of EnPI values that is valid for the entire production process for all 4-digit NACE-level sectors (transposed in Italy as 6-digit ATECO-level sectors) and for the production process of an enterprise or production site. This methodology has been applied to more than 300 4-digit NACE-level sectors using the Italian database of mandatory energy audits from Art. 8 of EED, and it has been also applied to several subsectors at the process level.

This work has been developed within the framework of the research program National Electrical System Research (in Italian, "Piano Triennale della Ricerca del Sistema Elettrico Nazionale 2019–2021"), funded by the Italian Ministry of Economic Development, with the task "energy efficiency of industrial products and processes". The main objective of this task is the analysis of the definition of best practices and performance indicators for energy-efficiency interventions based on the information from energy audits. This task has been developed in collaboration with several universities, business associations, and sectoral experts. Detailed guidelines for six industrial sectors (glass, cement, waste valorisation, ceramic, pharma, and foundries) are available in [40].

2. Methodology

2.1. Data Collection

The Audit102 database (https://audit102.enea.it/) stores, for every audited site, production volumes representing the activity, purchase, and consumption volumes of all energy commodities. End uses are split among three main functional areas: core activity, auxiliary services, and general services. The first two levels of the tree-structure energy model are therefore available in the database. In order to limit the effort imposed on companies and conglomerates involving multiple production sites and premises, energy audits, with the analysis of energy-performance improvement actions, may be carried out on a limited number of representative sites by using a clustering strategy. Following the sampling strategy [41], focus is directed toward sites featuring a higher energy demand and those most likely to reward any energy saving measure with greater economic benefits. The database includes, for every audited site, information on energy performance improvement actions (EPIAs), both implemented and planned ones. EPIAs are described identifying the intervention category, which can be technical, (e.g., "pressure systems", "heat recovery systems and thermal plant", "inverters and other electrical machines and installations") or managerial, such as the introduction or improvement of a monitoring system or the adoption of ISO 50001 certification or training courses. Figures concerning different types of energy savings are provided, namely savings of electricity, thermal energy, transport fuels, etc.; they include achieved savings for implemented EPIAs and potential savings for planned ones. CAPEX is also covered by the database, and in the case of planned EPIAs, further economic indicators are provided: simple payback time, net present value, and actualisation rate. Based on database information, global energy savings, measured in toe, can be computed and sorted either as final or as primary energy savings, the latter referring to the technical intervention areas "cogeneration and trigeneration" and "production from renewable energy sources".

Using information on CAPEX and energy savings, it is possible to compute cost effectiveness (the cost of saving one toe of final or primary energy) for both implemented and planned EPIAs. The ATECO2007 code for the company and for every site can be used to develop sector-specific analyses, which, combined with energy consumption comparative analyses and benchmarking, could also provide useful inputs for policy making.

2.2. Definition of Energy Performance Indicators

Improving energy efficiency is considered a way to contribute to the reduction of greenhouse gas emissions. In order to achieve this goal, energy use should be reduced, and thus, energy management can be intended as a valuable means.

In international standards, energy management is based on the knowledge of energy performance indicators (EnPIs) as major key parameters to measure effects of potential energy rationalisation. EnPIs are introduced by the ISO 50001:2018 standard, where they are described as a combination of processes efficiency, energy consumption, and management of energy sources and their end use [10]. By appropriately combining the three above-mentioned terms, EnPI can measure the energy health of each manufacturing company.

ISO 50001 specifies that each organisation should identify relevant energy performance indicators and should monitor and measure its energy performance but does not define EnPIs with a precise numerical ratio. EnPIs depend on many factors, and in practice, every company should evaluate what can best meet their expectations.

ISO 50006:2014, on the other hand, provides a guideline to establish the appropriate energy performance indicator for measuring/monitoring energy efficiency and recommends the use of specific energy consumption (SEC) [25].

SEC is the numerical ratio that identifies energy consumption in a given process (it is measured as energy/unit of product), and it is frequently used in literature as EnPI [42]:

$$\text{SEC} = \frac{\text{Energy Consumption}}{\text{Production}}$$

SEC does not fully represent either conversion efficiencies or energy management (since it does not take energy-flow trim into consideration and it is not a homogeneous ratio, like efficiency). SEC represents the specific energy quantity employed by any process [19,42]. For the purpose of this work, SEC and EnPI are considered equivalent:

$$\text{EnPI} = \text{SEC}$$

Hence, SEC is certainly a part of any EnPI because it allows for the description and evaluation, as well as tracking of energy improvements or worsening, and it can therefore be used as a measure of energy health of a manufacturing company. Analysis by SECs can be considered the first step for both the adoption of an energy management strategy and the promotion of several other co-benefits that positively impact on a company's overall competitiveness. SEC can be considered a valid EnPI when it is necessary to know the overall or partial specific consumption of processes or single operations, services, or generic company activities in order to calculate the global primary energy used, electric and heat or gas consumption, and to compare, evaluate and improve energy performance. At a local level (within the same corporate structure) SEC allows for the achievement of precise information for comparison of energy performance over the years and thus improvement of energy performance. Moreover, considering numerical terms, SEC completes the essential overall footprint framework of a manufacturing company, together with pollutant emissions, water consumption, and land uses (which should be computed concurrently). At the branch level (NACE), SEC can be helpful to compare the efficiency of a single site belonging to the same production sector or to build a national reference point for the development and evaluation of energy-efficiency-related policies (e.g., adopting financial compensation mechanisms).

However, since EnPIs depend on several parameters, such as plant size, operating parameters, energy carriers used, production process, age of machines and equipment, location and environmental conditions, and business conditions, their determination at the NACE level can have some issue.

The following are some examples:

- In some companies, energy consumption may not have a clear link with production, or some companies do not have a physical production (e.g., service companies);
- The methodology must be unique for all considered sectors in order to ensure both the accuracy of the calculations and the validity of comparisons;
- The algorithm should be able to provide the actual relationship between energy consumption of a production site and its processes;
- Variables analysed with these models must not directly include economic variables (such as added value, turnover, etc.), which often have no direct relationship with the physical production processes in manufacturing [26].

The International Energy Agency, for example, recommends the development of indicators based on physical principles through models that can be applied to the levels of aggregation (from the global site level to specific technologies per department) independently to the industrial sector under study [24,43]. Hence, the production units (P.U.) depend on each analysed sector and can be defined as m^2, m^3, t, etc., in manufacturing sectors, while in service sectors, other units will be adopted (m^2 or m^3 of building, heating degree days, working hours, etc.).

In order to describe how incoming energy carriers are used by a company, ENEA created a generic representation based on energy-flow distribution in the different company areas (namely, the "Plant Energy Model" [44]). This identifies the energetic relationship between all involved processes and the final product. The algorithm provides numerical values representing the EnPIs of the site and of processes that take place within.

Two types of energy performance indicators are obtained:

- First-level index (covering all energy carriers: electricity, thermal energy, natural gas) as:
    - ratio of overall final energy consumption to the amount of service provided;
    - ratio of per-carrier energy consumption to the amount of service provided;
- Second-level index (covering the prevalent carrier(s));
    - specific consumption in the individual production department;
    - consumption by energy destination (core activities, auxiliary services, or general services)
    - consumption of characteristic technologies.

Underlined in Figure 2 is the definition of first- and secnd-level EnPIs and their representation on the site energy model. The first-Level EnPIs are computed for all manufacturing four-level NACE sectors at global plant level, including all activities and processes within the site. The second-level EnPIs are obtained only for specific energy-intensive sub-sectors (i.e., cement). These indicators are calculated for specific processes of the production cycle and for auxiliary (i.e., cogeneration or compressed-air units) and general activities (i.e., lighting, HVAC, etc.).

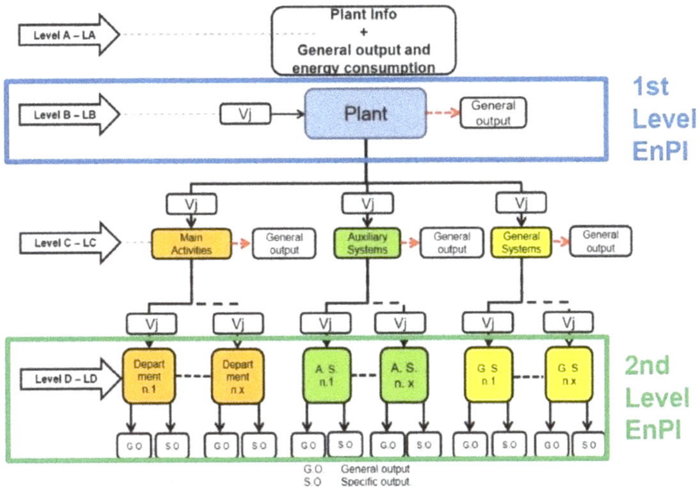

Figure 2. Site energy model and characterization of first- and second-level EnPIs [44].

In this way, the following indicators are obtained:

1. Global EnPI: includes all energy consumption;
2. Electric EnPI: includes only the electrical consumption of production processes (taken from power-grid and/or autonomous production, fossil fuels, renewable sources, and cogeneration);
3. Thermal EnPI: includes only the thermal consumption of production processes (from natural gas, heat, cold, biomass, fuel oil, coke, and other carriers);
4. Natural Gas EnPI: includes natural-gas consumption of processing activities.

Thermal EnPIs omit both automotive consumption (diesel, petrol, and LPG) and fuel consumption in cogeneration plants. Natural Gas EnPIs are part of the exploitation of previous thermal energy, omitting natural gas consumption in cogeneration or trigeneration. As mentioned above, the methodology used for calculating EnPIs allows for the measurement of energy consumption per given production.

2.3. Statistical Modelling for Energy Performance Indexes Evaluation

Starting from data stored in the Audit102 web portal, a definition of EnPI at the sectoral was carried out. The methodology used during this analysis can be outlined in the following steps:

2.3.1. Step 1—Selection of the Sample and Data Cleaning

The work in this phase consists of:

a. Qualitative and quantitative analysis of the energy consumption of every site, belonging to every four-digit NACE group stored in the ENEA database, which is the reference population.
b. Selection of a statistic sample from the reference population, with the aim of implementing the mathematical model.

With this aim, both energy-audit reports and summary spreadsheets containing the energy-consumption summaries are studied. A work of normalization was done in order to correct (whenever possible) or dismiss nonhomogeneous elements, such as:

- audits where production is measured in different units with respect to the population;
- audits lacking key information (such as the number of production units);
- audits having undergone upload issues on the Audit102 portal;

- audits referring to sites that, in terms of processes, do not actually belong to the considered NACE group;
- sites that are clearly far from the mean trend of the energy consumption vs. production ratio (outliers).

Moreover, a numeric threshold of sites was set, conventionally equal to 5, below which the modelling was not carried out because the sample would not be statistically representative. This value can be considered conservative according to results presented in Section 3, where statistically significant EnPIs are usually obtained with samples over 10 sites. This threshold increases the accuracy of the results with an increase in the required time for the analysis. During the first phase, some NACE groups are also separated, depending on the characteristic processes or products of the sites, or grouped in smaller intervals of production units, with the aim of obtaining a better description of the population. Finally, whenever a site is spotted whose assignment, in terms of NACE group, is clearly wrong, that site (and its audit report) is assigned to the right group.

This first normalization phase is a very time-consuming process since it requires a one-by-one audit analysis by highly specialized personnel.

2.3.2. Step 2—Calculation of the Real Sectoral EnPI (Mean +/− Standard Deviation)

For each NACE group, whenever possible, a single EnPI ($EnPI_r$) referring to the entire group and the whole production range was calculated, or as many EnPIs as the number of partitions the NACE group has been split into. The standard deviation for $EnPI_r$ is also calculated: this is used to determine the variation range of the EnPI. In other words, from the statistics standpoint, the dispersion of the points was assessed, with respect to a positional index, or standard deviation of the observed points referring to the sample arithmetic mean. Once the mean EnPI and its standard deviation are calculated, the coefficient of variation (CV, or relative standard deviation, RSD) of the EnPI (for a given production range) is also computed as the ratio between the standard deviation and the mean value. We will define "reliability index" of the real EnPI, as a function of the CV values, as follows:

a. «high» if it is less than 20%
b. «medium» if it is between 20% and 60%;
c. «low» if it is between 60% and 100%;
d. «invalid» if it is equal to or greater than 100%

This indicator provides quantitative information about the mean value and variability of the EnPI. These values, however, should be used with caution. The use of the mean value does not include an indication of economy of scale in the production. Hence, this indicator provides valuable information about the standardisation potential of the EnPI for specific sectors. On the one hand, a high reliability is related to stable EnPIs with production and homogenous products and processes. On the other hand, a low value of the "reliability" of real EnPIs indicates a high variability of products and processes.

2.3.3. Step 3—Analysis of Linear Correlation between Consumption and Production

Regression analysis is a statistical technique that estimates the dependence of a variable (namely, energy consumption) as a function of one or more independent variables (e.g., production quantity, degree days, etc.) while highlighting the influence of several parameters. Linear regression models are used in specific measurement and verification campaigns to estimate energy savings of energy-efficiency projects and programs [45]. In this work, we use linear regression involving two variables: energy consumption (global, electrical, thermal, or natural gas as a function of the EnPI) and production (in its specific P.U.).

Linear regression models are also preferred for analysing savings achieved with energy performance improvement actions (EPIA) in organizations with certified energy management systems (specifically between ISO 50001 and the SEP M&V protocol) [46,47]. The ISO 50006 standard "Energy management systems—Measuring energy performance using energy baselines (EnB) and energy performance indicators (EnPI)—General principles and guidance" recommends the use of linear regression for the estimation of indicators [25].

For the above reasons, linear regression models are widely used for benchmarking analysis and energy-efficiency measures [15,26,28,29]. In fact, linear models are simple to develop, can be used both at the sectoral level and at the site level, and can be derived from the data provided in energy audits without any additional hypotheses.

Linear regression is represented by the equation (see Figure 3)

$$y = a \cdot x + b$$

where y represents the final energy consumption [MJ] as the sum of a constant term, b [MJ], not dependent on the quantity of production, x, and of a variable term, a [MJ/P.U], proportional to the amount of production, x [P.U.]:

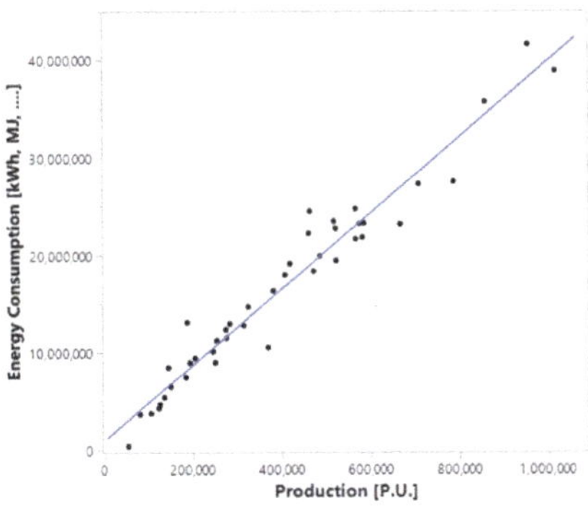

Figure 3. Linear regression between energy consumption [MJ] and production [P.U.].

In order to evaluate the statistical representativeness of the linear correlation, the following parameters were used: R2, coefficient of determination; R, coefficient of correlation (the latter to be compared with the critical Pearson correlation coefficient (Rcrit)), and finally, the *p*-value parameter. It is also necessary to define the confidence interval (CI) that may be achieved with the analysis. The statistical significance represents the range within which this value can deviate. In the context of this work, regarding the choice of the confidence intervals, it is necessary to establish the analytical acceptability of the model. The statistical significance defined for this model is confirmed when the maximum value for α is 0.05; therefore, the selected confidence interval ($CI = 1 - \alpha$) is 95%.

In literature, R^2 is used extensively in many fields as an indicator of the strength of the correlation. Linear correlation can be considered strong if $R^2 > 0.5$ and moderate if $R^2 > 0.25$ [48]. The strength of the correlation depends on the absolute value of R. The minimum value of R, which confirms the existence of a correlation between the variables, depends on the size of the population analysed, on α, and on the hypothesis concerning mono or bidirectionality. A significant correlation is considered to exist if R is greater than $R_{critic} = f(N, \alpha)$ [49].

In the end, the *p*-value, also called the probability value, is used to confirm the representativeness of the chosen sample. If the *p*-Value < α, the test is considered statistically significant (confirmation of representativeness).

Hence, the correlation is considered statistically representative if

- *p*-value < 0.05 and
- $R > Rcritic = f(N, \alpha)$ and
- $R^2 > 0.5$.

When the model did not show a significant linear relationship between energy consumption and production, the mere $EnPI_r$ calculated for every NACE group and its standard deviation were used to describe the sectors.

The analysis of these three correlation variables is only the first step. A low value certainly indicates an insignificant relationship between consumption and production (often due to the presence of economic factors). Meanwhile, a high value of the coefficient may not guarantee a significant relationship. The correlation coefficient can be very sensitive to "outliers" and may indicate association when, in fact, none exists. It is important to note that one or few anomalous items can have a large impact on R. Hence, it is necessary to inspect a plot of the data and to ensure that the data covers the range uniformly.

2.3.4. Step 4—Formulation of Specific EnPI Model (EnPIm)

As previously explained, a generic EnPI is calculated as the ratio of energy used for producing a unit of product (P.U.). Hence, following the statistical analysis of linear regression, $EnPI_m$ is calculated dividing both sides of the production function and is represented by a hyperbolic function:

$$EnPI_m\ [MJ/P.U.] = a\ [MJ/P.U.] + \frac{b\ [MJ]}{x\ [P.U.]}$$

where a and b, respectively, represent the slope and the intercept of the linear regression line.

With this approach, it is possible to represent the analytical model of EnPI (in blue in Figure 4), and the specific EnPI by site can be subsequently compared with the sectoral main value.

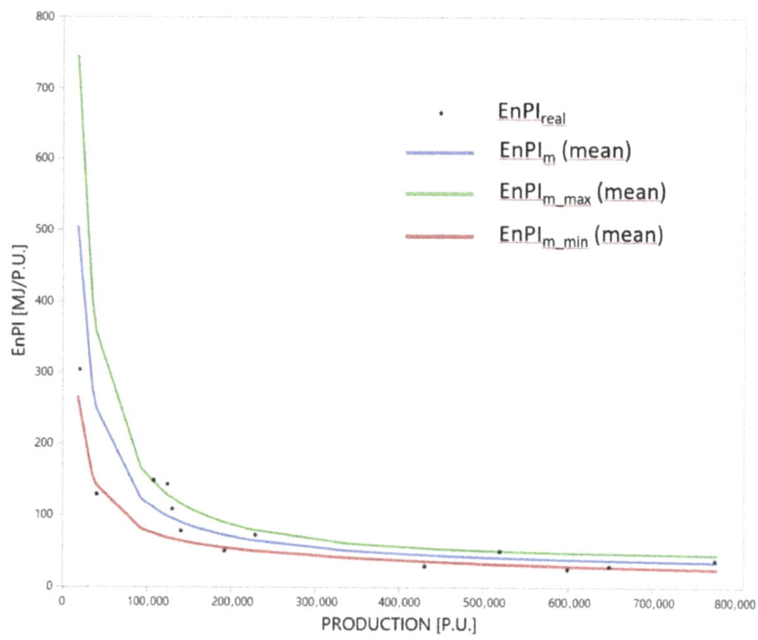

Figure 4. Mean, lower, and upper values of the EnPI model [MJ/P.U.] vs. Production [P.U.] and real EnPI for specific sites.

An analysis of uncertainty of the $EnPI_m$ has been developed to a defined significance level ($\alpha = 0.05$). The uncertainty analysis is based, in a general way, according to the central limit theorem, which, in a very simplified way, allows for the assumption that the $EnPI_m$ presents a Gaussian dispersion. Therefore, it is possible to define an upper- and lower-limit curve of statistical significance of the formula model.

$$EnPIm_{UPPER/LOWER} = EnPIm_{mean} \pm 2 \cdot \sigma_{EnPIm}$$

The uncertainty σ of the $EnPI_m$ model is calculated through the propagation of the statistical error, which is obtained on the basis of the covariance matrix, $C_{i,k}$, as follows:

$$\sigma(f) = \left(\sum_{i=1}^{n} \sum_{k=i}^{n} \left(\frac{\partial f}{\partial x_i} \frac{\partial f}{\partial x_k} (C_{i,k}) \right) \right)^{1/2}$$

where f refers to the $EnPI_m$ function and x_i e x_k are the parameters a and b estimated in the regression model. Substituting in the estimated consumption, the formula for the propagation of the statistical error becomes

$$\sigma_{EnPIm} = \left(Var(a) + \frac{Var(b)}{x^2} + \frac{2Cov(a,b)}{x} \right)^{\frac{1}{2}}$$

Figure 4 shows the mean theoretical curve (in red) of the $EnPI_m$ and its lower (red) and upper (green) limits at 2σ. Moreover, the real EnPI values of the individual sites analysed are also presented (black dots).

This graph allows for easy visualization of the effects of error propagation in the calculation of the $EnPI_m$ as a function of the production. The area contained between $EnPI_{m\_max}$ and $EnPI_{m\_min}$ represents the "variability" of the statistical case.

In practice, the confidence interval derived from the previous equations is displayed, associated with the linear regression built based on the consumption vs. production, which represents the uncertainty of the $EnPI_m$, graphically delimited by the lower and upper limit of the curve. In this way, for each diagnosis analysed, it is also possible to evaluate the value of the real EnPI with respect to the theoretical interval thus defined and to quantify the error that affected it. This model can be used as the basis for more advanced models, such as stochastic frontier analysis (SFA) [31], index decomposition analysis (IDA) [50], or data envelopment analysis (DEA) [51] for calculation of energy-efficiency impact or deviation from benchmark.

The statistical method developed and illustrated in the previous sections was applied using energy data included in the energy audits uploaded to the Audit 102 web portal at the end of December 2019.

Compared to the methodologies present in the literature [16,18,19], the methodology developed in this work has a different approach. In fact, starting from the available data set (the mandatory energy audits pursuant to art.8 directive 27/2012), we tried to find a correlation between energy consumption and production rate, assigning a level to the data set, the reliability of which is linked to the difference between the average and the standard value. The main advantage lies in its replicability in all production sectors, both in the industry and in the tertiary domains.

3. General Results of NACE 23 Division

The previously illustrated methodology was applied to more than 300 four-level NACE sectors. The quality of statistical modelling of the EnPIs is strongly dependent on each sector analysed. Specifically, it depends on the sample dimension, on the heterogeneity in terms on production range and units adopted, on the structure of energy uses, and on energy intensity of the production.

Hence, on the one hand, there are sectors with high reliability and quality of EnPI model (e.g., cement, refineries, foundries, ceramics). On the other hand, some sectors provide EnPIs with little significance (e.g., machinery manufacturing, electronics, furniture, or mining and quarrying). This section shows, as an example, the results obtained from the analysis of the NACE 23 division, "Manufacture of other non-metallic mineral products". The NACE 23 division includes 512 energy audits distributed among the different groups, as shown in Figure 5. The final global energy consumption of the NACE 23 EAs amounts to about 2.6×10^5 TJ (corresponding to about 13% of the global energy consumption of audited sites).

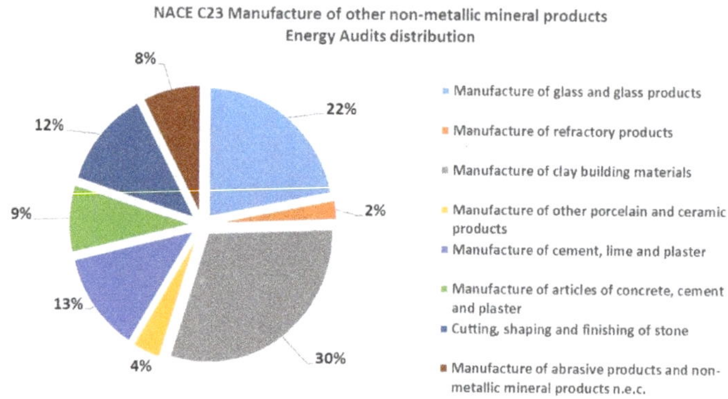

Figure 5. NACE C23 groups' energy audit percentage distribution.

The distribution of audited sites' final energy consumption among the various NACE 23 groups is reported in Figure 6, which shows that consumption is mainly attributable to glass manufacturing (47% of the total), followed by the cement industry (26%) and the manufacture of clay building materials (18%).

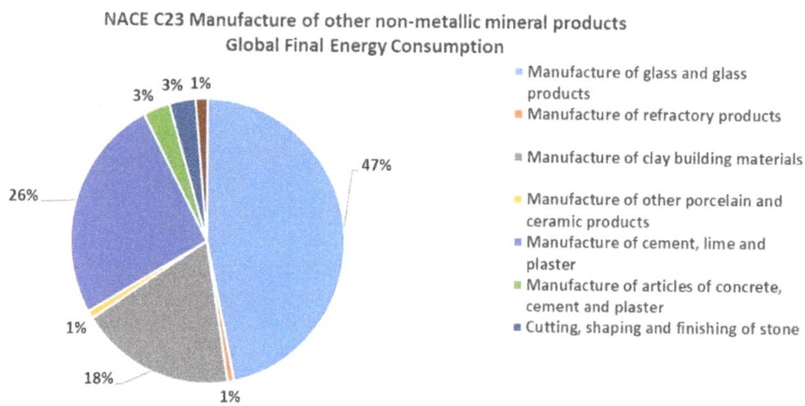

Figure 6. NACE C23 EAs groups' global final energy consumption distribution.

The NACE 23 division, in its Italian transposition (ATECO), includes 29 six-digit subcodes, and the analysis presented in the previous sections was set up for each of them. Real IPEs (see Section 2.3, Step 1) and model IPEs (see Section 2.3, Step 4) were calculated for six-digit subcodes where the number of diagnoses and the consistency of the data contained in them in terms of volumes, production processes, and units of production rates allowed it.

A total number of 41 actual EnPIs were calculated, of which 32% had high reliability, 61% had medium reliability, and the remaining 7% had poor reliability. Of these 41 EnPIs, 14 are electrical, 2 thermal, and 15 global.

Figure 7 shows, as an example, the distribution of electrical EnPIs by NACE subcode. The different colours indicate the reliability of the indexes identified. For some sub-codes, there is a higher percentage and better reliability of the identified indexes (e.g., for cement), while for others, few EnPIs have been identified and with a low reliability, mainly because of sample dimension. As shown in Figure 8, the range of the coefficient of variation for the greatest number of six-digit EnPIs (about 18%) is 10–15%. Globally, some 25% of NACE 23 four-digit EnPIs have a coefficient of variation lower than 15%, and about 50% lower than 25%.

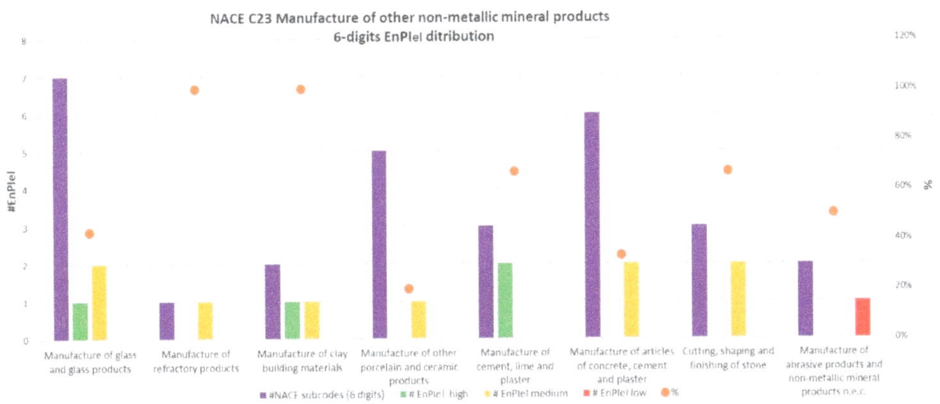

Figure 7. NACE C23 six-digit subcode EnPIel distribution.

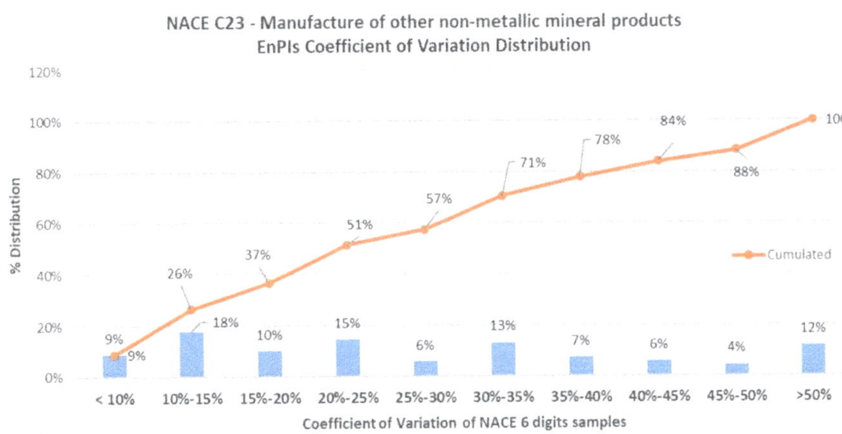

Figure 8. NACE C23 six-digit subcode ENPIs coefficient of variation distribution.

In addition to the elaboration of actual EnPIs, the linear correlation between consumption and production for NACE 23 four-digit subcodes was evaluated to develop specific ENPI models.

The main results of the correlation analysis are summarised in Figure 9. The two curves (in purple and blue) represent critical values of the Pearson correlation coefficient (R_{crit}) as a function of sample dimension for a confidence value α of 0.05 and 0.01. The indicators (triangles for electrical energy consumption, crosses for thermal energy consumption, and squares for global energy consumption) represent the correlation coefficient values (R) of the linear regressions for the different ATECO six-digit subcode sets. As explained in the dedicated paragraph, in order to regard the correlation as statistically representative, it is necessary to evaluate not only that $R > R_{critic} = f(N, \alpha)$ but also that the *p*-value is sufficiently low. In the figure, the indicators have been coloured according to the *p*-value of the linear regression. In red are the indicators for which *p* Value > 0.05; in orange, those for which 0.05 < *p* value < 0.01; and in green, the correlations with *p* value < 0.01.

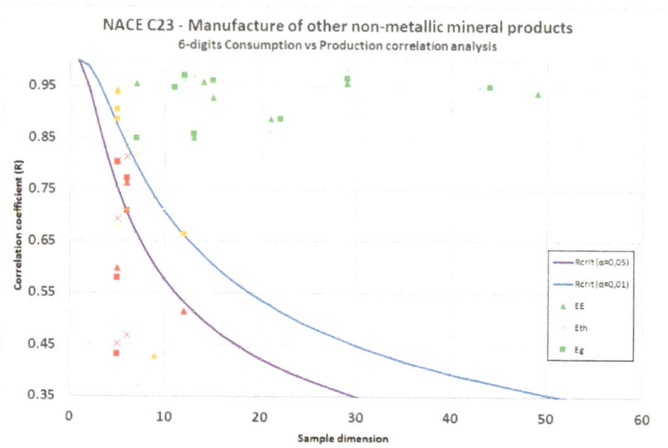

Figure 9. NACE C23 four-digit consumption vs. production correlation analysis. Purple and blue curves represent critical values of the Pearson correlation coefficient (R_{crit}) as a function of sample dimension for a confidence value α of 0.05 and 0.01. The indicators (triangles for electrical energy consumption, crosses for thermal energy consumption, and squares for global energy consumption) represent the correlation coefficient values (R) of the linear regressions for the different ATECO six-digit subcode sets. The indicators have been coloured according to the *p*-value of the linear re-gression. In red are the indicators for which *p* Value > 0.05; in orange, those for which 0.05 < *p* value < 0.01; and in green, the correlations with *p* value < 0.01.

As can be seen from Figure 9, the sample dimension has a substantial impact on the statistical significance of the correlation. The minimum sample size to be considered for the case study is approximately 8/10 EAs.

4. A Case Study: The Cement Industry

An additional focus on the cement industry (NACE 23.51) is presented in this section, as part of NACE 23 sector "Manufacture of other non-metallic mineral products". In this detailed analysis, the methodology was applied to the development of first- and second-level EnPIs. The normalization work that was pre-emptively accomplished in order to correct or dismiss nonhomogeneous elements: a partition of NACE group 23.51 depending on the characteristic processes of the sites.

In the cement sector, 47 energy audits were collected by ENEA in December 2019, in compliance with Article 8 of Italian Legislative Decree 102/2014:

- 1 energy audit related to an administrative site;
- 14 energy audits related to sites where only the grey-clinker-grinding phase is carried out;
- 30 energy audits related to sites where the complete grey-cement production cycle is carried out (in one of these sites, white cement was also produced;
- 2 energy audits related to sites where the complete white-cement production cycle is carried out.

Consumption data, gathered from 29 out of 30 energy audits related to the complete grey-cement production cycle (one site was dismissed because of non-continuous production), were subjected to the developed statistical method, with the purpose of assessing energy performance indicators of the Italian cement industry sector. The diagram in Figure 10 represents the complete grey-cement production cycle.

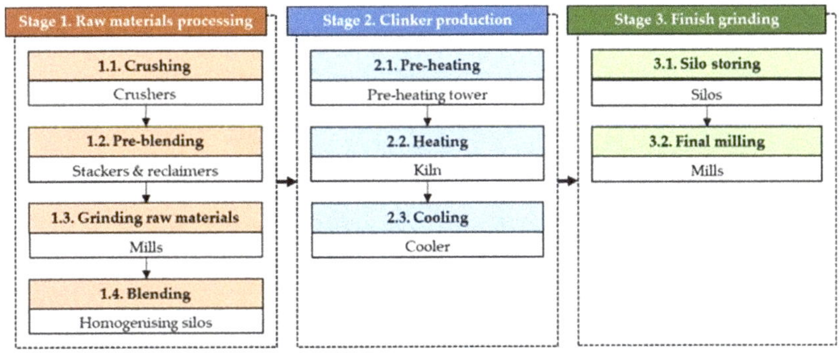

Figure 10. Complete grey-cement production cycle [3].

4.1. Energy Consumptions in Complete Grey-Cement Production Cycle

Figure 11 shows the global energy consumption distribution between electricity and heat for the grey-cement production cycle, while Figure 12 shows the global energy consumption distribution by activities (main activities, auxiliary services, and general services). Figure 13 shows the distribution of thermal consumption by fuel, while Figure 14 shows the distribution of electric consumption by activities.

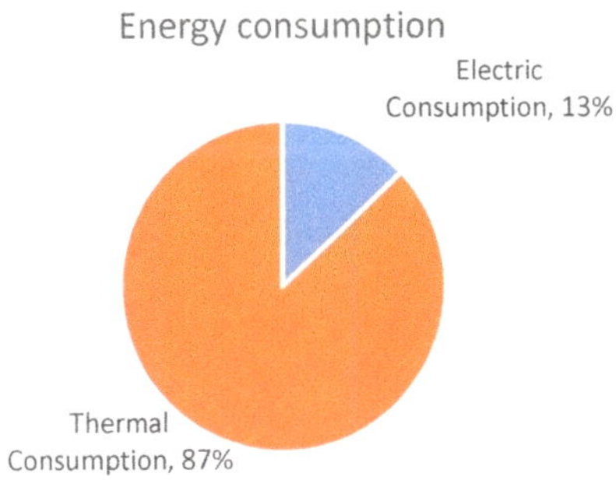

Figure 11. NACE 23.51 EA analysis: global consumption distribution between electricity and heat.

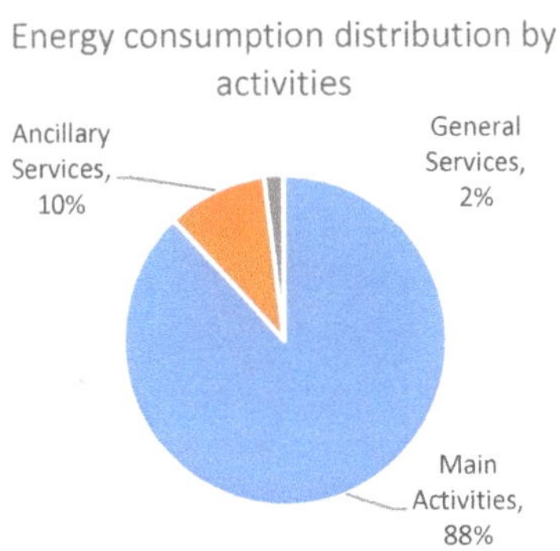

Figure 12. NACE 23.51 EA analysis: global consumption distribution by activity.

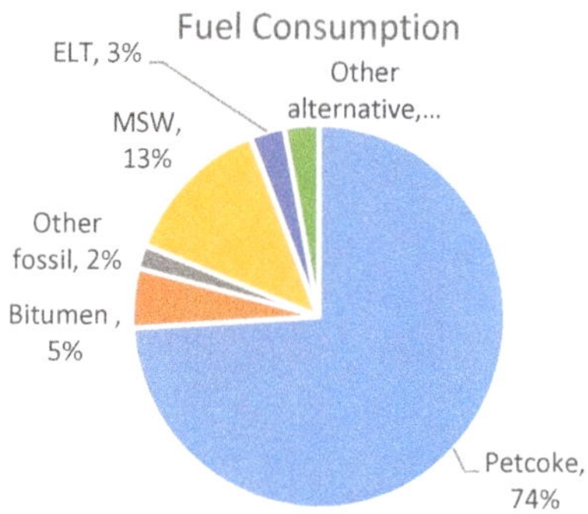

Figure 13. NACE 23.51 EA analysis: distribution of thermal consumption by fuel (ELT—end-of-life tyres, MSW—municipal solid waste).

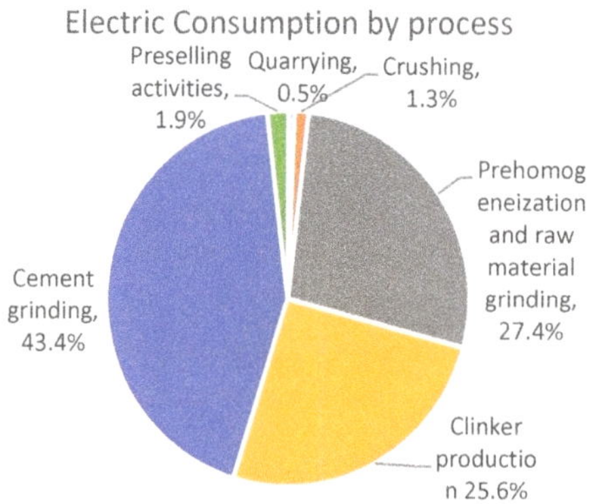

Figure 14. NACE 23.51 EA analysis: distribution of electric consumption by activity.

4.2. Electric, Thermal, and Global Energy Consumption

First, data of electric, thermal, and global consumption of each site, including those relating to auxiliary services and general services, were processed. In line with the proposed methodology, two EnPIs referring to two production ranges were calculated, and the linear regression between energy consumption (electric, thermal, and global) and production (in its specific P.U.) was analysed. Electric consumption was compared with the actual cement produced on site, expressed in tons. In order to take into account the fact that many cement production sites sell or buy clinker, take it from stocks, or put it in stock, a "virtual" electric consumption related to the complete production cycle of all cement actually produced on the site was calculated. Electric consumptions related to the quarry phase and the raw materials in the grinding phase (if carried out on site) were excluded, as these phases are not always present in the sites.

Electric consumption of clinker during shipping (where applicable) was also excluded, with the purpose of considering only electric consumption of the complete grey-cement production cycle.

Thermal consumption refers to clinker quantity produced on site (in tons) and to a "virtual" cement production calculated from the ratio between the clinker actually produced on site and the R parameter (ratio between the amount of clinker ground on site and the amount of cement actually produced). Consumption of diesel and LPG fuel, mainly used for transport (both normal vehicles and quarry vehicles, if any), was excluded from thermal consumption.

Global consumption refers to the amount of cement produced on site, expressed in tons. Additionally, for global consumption, as for electric consumption, in order to take into account the fact that many cement plants sell or buy clinker, take it from stocks, or put it in stock, an elaboration of global consumption was made in order to calculate, for each site, a "virtual" global consumption related to the complete production cycle of all cement actually produced on the site. In particular, as for electric components, the "virtual" electric consumption previously computed was considered, while for thermal components, a "virtual" thermal consumption was considered relating to the production of all the clinker necessary for cement actually produced on site. Global consumption included both electric consumption at the quarry stage, the raw materials in the grinding phase in the plant (if any), and thermal consumption of LPG and diesel fuel used for transport (both on ordinary and quarry vehicles, if any), while those of the clinker shipping phase were excluded.

Table 1 shows real electric, thermal (referring, respectively, to cement and clinker), and global EnPIs and their standard deviations. The table also shows that the reliability of EnPIs is high.

Table 1. NACE 23.51: real electric, thermal, and global EnPIs.

Energy	Production		Production Range		EnPI		Size of Population	Reliability
	Description	Units			Value	Units		
Electric	Cement	t	145,000	521,000	119 ± 10	kWh/t	15	HIGH
			521,001	1,015,000	108 ± 13		14	HIGH
Thermal	Cement	t	182,000	1,396,000	2757 ± 295	MJ/t	29	HIGH
Thermal	Clinker	t	142,000	509,000	3585 ± 264	MJ/t	16	HIGH
			509,001	1,097,000	3468 ± 230		13	HIGH
Global	Cement	t	144,803	566,000	3097 ± 220	MJ/t	18	HIGH
			566,001	1,015,000	3320 ± 384		11	HIGH

Linear regression models for electric, thermal (referring to clinker) and global consumptions are shown, respectively, in Figures 15–17, while Table 2 shows statistical regression parameters: linear regression equation, coefficient of determination (R^2), Pearson correlation coefficient (R), *p*-value, size of the population analysed (N), and Rcritic for $\alpha = 0.001$ e $\alpha = 0.005$.

The correlations are statistically representative as *p*-value < 0.0001 and R > Rcritic = f (N, α); as shown in Table 2, R2 is greater than 0.9, and the confidence value is 99% for each linear regression.

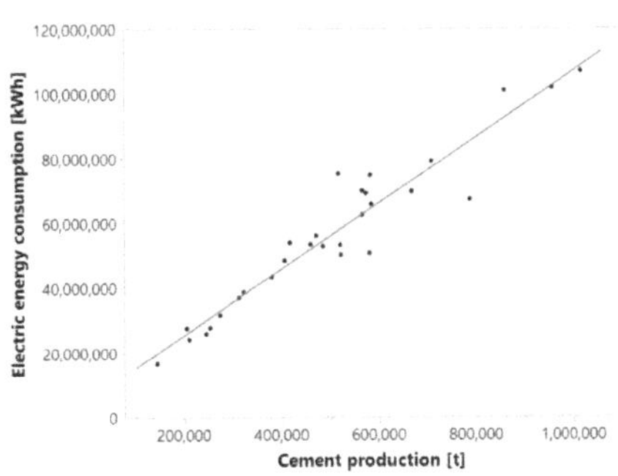

Figure 15. NACE 23.51 linear regression of site electric consumption.

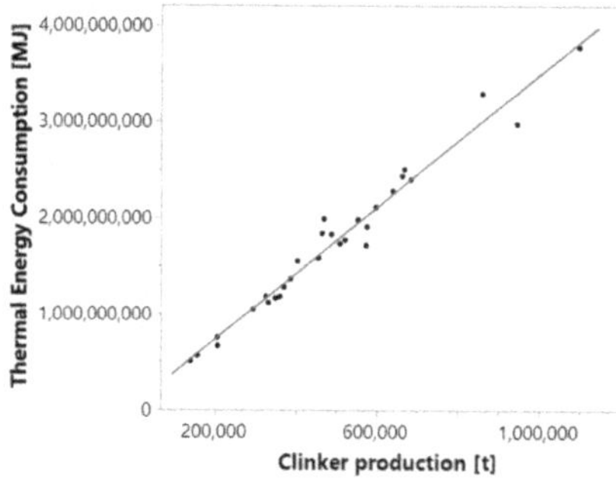

Figure 16. NACE 23.51 linear regression of site thermal consumption.

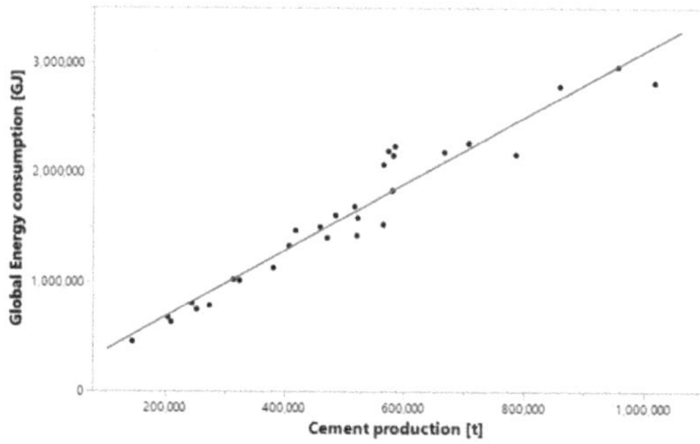

Figure 17. NACE 23.51 linear regression of site global energy consumption.

Table 2. NACE 23.51: statistical parameters of linear regression of electric, thermal, and global site consumptions.

Energy	Production		Equation	R^2	R	*p* Value	N	Confidence Value
	Description	Units						
Electric	Cement	t	EE [kWh] = $5.062 \times 10^6 + 101.9 \times t$	0.917	0.958	<0.0001	29	>99%
Thermal	Clinker	t	E_{th} [MJ] = $3.913 \times 10^7 + 3.448 \times t$	0.969	0.984	<0.0001	29	>99%
Global	Cement	t	E_g [MJ] = $7.05 \times 10^7 + 3.039 \times t$	0.933	0.966	<0.0001	29	>99%

4.3. Second-Level EnPIs—Main Phases of Grey-Cement Production Cycle

The methodology was subsequently applied to sub-process levels (second-level EnPIs, as presented in Figure 2). Hence, consumption data of main phases of the grey-cement production cycle were subjected to statistical modelling. In particular, raw materials in the grinding phase (and homogenisation of flour), the clinker grinding phase, featuring electric consumption, and the clinker cooking phase, featuring electric and thermal consumption, were considered.

- Electric consumption of raw materials in the grinding phase was compared to the amount of flour actually produced on site, expressed in tons.
- Electric consumption of clinker in the cooking phase was compared to the amount of clinker actually produced on site, expressed in tons.
- Electric consumption of clinker in the grinding phase was compared to the amount of cement actually produced on site, expressed in tons. In this case, the sites where only the clinker grinding process is carried out were also considered in the sample.
- Thermal consumption of clinker in the cooking phase was compared to the amount of clinker actually produced on site, expressed in tons.

Table 3 shows real electric and thermal EnPIs for the main phases of the grey-cement production cycle and their standard deviations. The table also shows the reliability of EnPIs.

Table 3. NACE 23.51: statistical parameters of linear regression of site consumptions for second-level EnPIs.

Energy	Phase	Production		Production Range		EnPI		Size of Population	Reliability
		Description	Units			Value	Units		
Electric	Raw materials grinding	Flour	t	219,000	740,000	23.51 ± 5.71	kWh/t	15	MEDIUM
				740,001	1,733,000	19.09 ± 4.70		11	MEDIUM
Electric	Clinker cooking	Clinker	t	141,558	597,000	31.84 ± 5.44	kWh/t	21	HIGH
				597,001	1,097,000	28.36 ± 5.99		8	MEDIUM
Electric	Clinker grinding	Cement	t	55,000	1,015,000	41.51 ± 8.74	kWh/t	42	MEDIUM
Thermal	Clinker cooking	Clinker	t	141,558	1,097,000	3505 ± 247	MJ/t	29	HIGH

Linear regression models for electric and thermal consumptions for the main phases of the grey-cement production cycle are shown, respectively, in Figures 18–21, while Table 4 shows statistical regression parameters: linear regression equation, coefficient of determination (R2), Pearson correlation coefficient (R), *p*-value, size of the population analysed (N), Rcritic for α = 0.001 e α = 0.005. Those correlations are statistically representative as *p*-value < 0.0001 and R > Rcritic = f (N, α); as shown in Table 2, R2 values are high, and the confidence value is 99% for each linear regression.

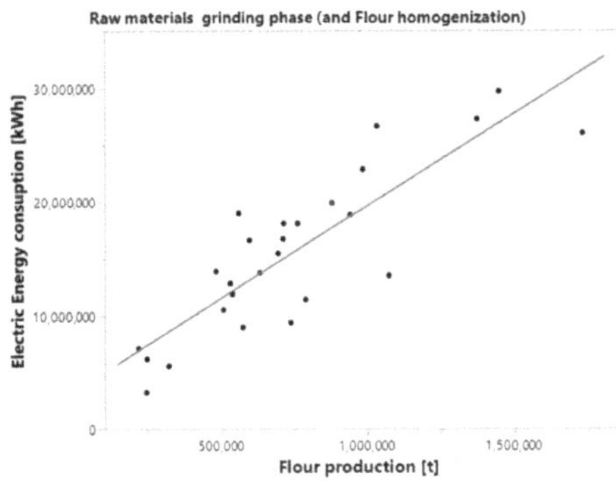

Figure 18. NACE 23.51: linear regression of electric consumption of raw material in the grinding phase (and flour homogenisation).

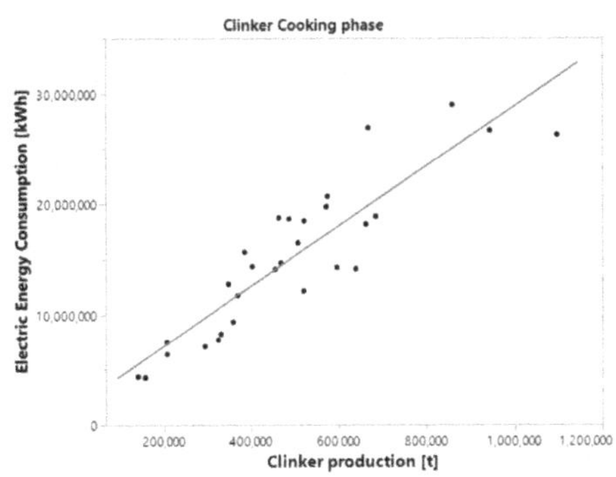

Figure 19. NACE 23.51: linear regression of electric consumption of the clinker cooking phase.

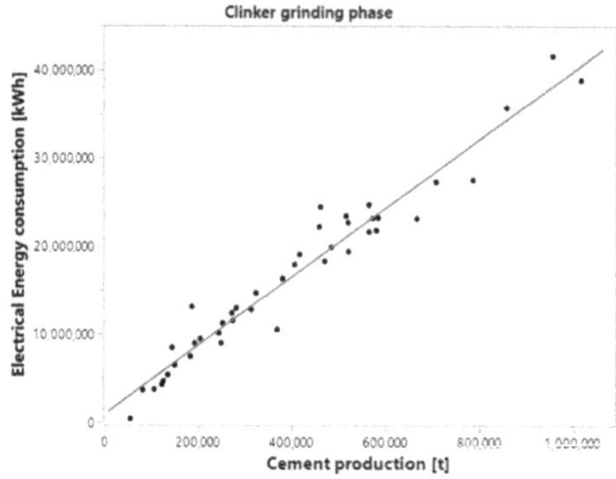

Figure 20. NACE 23.51: linear regression of electric consumption of the clinker grinding phase.

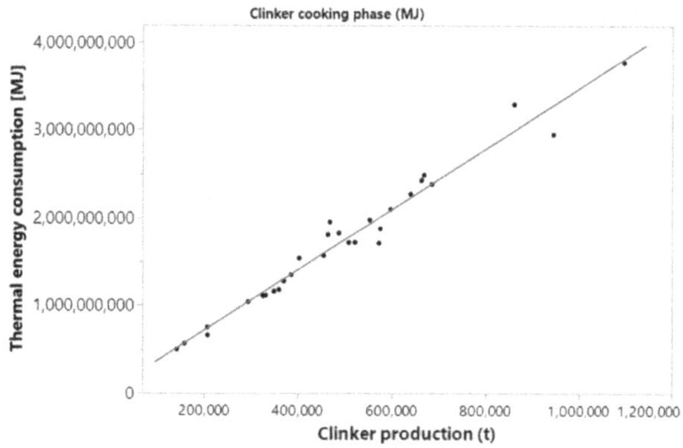

Figure 21. NACE 23.51: linear regression of thermal consumption of the clinker cooking phase.

Table 4. NACE 23.51: Statistical parameters of linear regression: phase consumptions.

Energy	Phase	Production		Equation	R^2	R	p Value	N	Confidence Value
		Description	Units						
Electric	Raw materials grinding	Flour	t	EE [kWh] = 3,395,804 + 16.24 × t	0.730	0.854	<0.0001	26	>99%
Electric	Clinker cooking	Clinker	t	EE [kWh] = 1,685,681 + 27.19 × t	0.804	0.897	<0.0001	29	>99%
Electric	Clinker grinding	Cement	t	EE [kWh] = 946,440 + 38.98 × t	0.951	0.975	<0.0001	42	>99%
Thermal	Clinker cooking	Clinker	t	E_{th} [MJ] = $2.749 \cdot 10^7$ + 3447 × t	0.969	0.984	<0.0001	29	>99%

For the clinker grinding phase, the hyperbolic function of electric EnPIs was calculated, and Figure 22 shows the theoretical (in red) curve of EnPI, its lower (blue), and upper (purple) limits at 2σ. Moreover, the real EnPI values of the individual sites analysed are also presented (blue points).

The results of the consumption vs. correlation analysis for the main production phases and auxiliary services for NACE 23.51 are summarized in Figure 23. As can be seen from the figure, although in some cases, the sample dimension is small (lower than 10 EAs), the correlations are statistically valid, given the values assumed by the correlation coefficients, R, and considering that for all of them, the p value is lower than 0.0001. The phases for which the electricity consumption vs. production correlation appears weaker are fuel transport and treatment, transport of cement to the silos, and grinding of raw material.

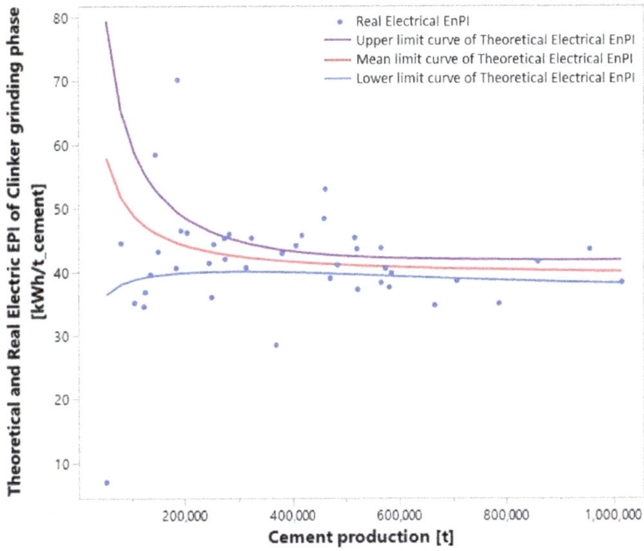

Figure 22. NACE 23.51: Theoretical and real electric EnPI of the clinker grinding phase.

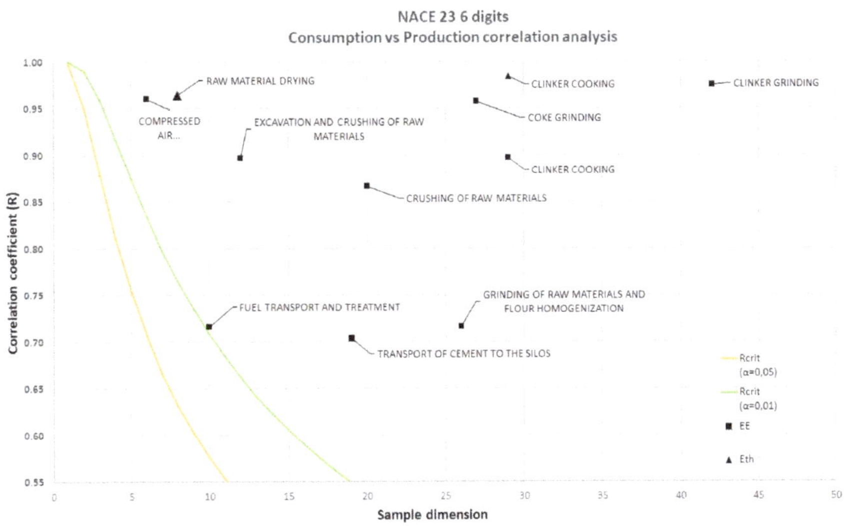

Figure 23. NACE 23.51: consumption vs. production correlation analysis for the main production phases and auxiliary services.

Numbers of Cyclones

Another application of the methodology involves energy consumption related to other variables of the plant design. Specifically, the impact of the number of cyclones on the pre-heating phase before clinker production was evaluated. In only 23 out of the 29 sites where the complete grey-cement production process is carried out, the number of cyclones of the furnace is provided.

In detail:

- in 11 of the 23 sites, the number of cyclones is equal to 4;
- in 12 of the 23 sites, the number of cyclones is equal to 5.

Table 5 shows actual thermal EnPI and its standard deviation for the entire production range in the two cases of four or five cyclones. The table also shows that the reliability of thermal EnPI for the clinker cooking phase is high in both cases.

Table 5. Real thermal EnPI of the clinker cooking phase in case of presence of four or five cyclones.

Energy	Production		Number of Cyclones	Production Range		EnPI		Size of Population	Reliability
	Description	Units				Value	Units		
Thermal	Clinker	t	4	142,000	860,000	3609 ± 291	MJ/t	11	HIGH
			5	207,000	1,097,000	3431 ± 214		12	HIGH

4.4. Validation

In Table 6, a comprehensive validation of EnPI obtained in this work with public data is provided. The data are aggregated by country or, when possible, only by dry process with multistage (three to six stages) cyclone preheaters and pre-calcining kilns.

The first level electric and thermal EnPI values is in line with the BREF sector [52] (which reports electric and thermal EnPIs values, respectively, between 90 and 150 kWh/t_cement and between 3.000 and 4.000 MJ/t_clinker) and with IEA data [53] (which report, for Italy, values, respectively, of 122 kWh/t_cement and of of 3.500 MJ/t_clinker), the most relevant benchmark for the sector in the EU. First-level EnPI values confirm that Italian cement presents a high overall energy efficiency—3.5 GJ/$t_{clinker}$—compared to the global range of 3.0–4.2 GJ/$t_{clinker}$. On the one hand, the lowest value corresponds to India, at 3.0 GJ/$t_{clinker}$ (with different product requirements and processes), and the lowest value between OECD economies correspond to Japan, at 3.4 GJ/$t_{clinker}$. On the other hand, the mean values for the EU-28 and U.S. are 3.7 GJ/$t_{clinker}$ and 3.8 GJ/$t_{clinker}$, respectively. Therefore, the Italian cement sector is very efficient in terms of energy consumption.

The information for second-level EnPIs is scarce, but the obtained electrical EnPI values for raw-material preparation, solid-fuel preparation, finish grinding, and clinker cooking are slightly higher than the best available technologies (BAT) [54] and lower than the values for the U.S. as of 1999 [55].

Ref. [56] presents values of the impact of the number of preheating cyclones on clinker thermal consumption. The values obtained in this work are in line with literature data:

- 4-cyclons preheating: 3.6 GJ/$t_{clinker}$. Overall range 3.2–3.6 GJ/$t_{clinker}$
- 5-cyclons preheating: 3.4 GJ/$t_{clinker}$. Overall range 3.1–3.5 GJ/$t_{clinker}$

Lastly, a similar analysis of thermal energy consumption as function of the clinker production is presented in [57]. Oda et al. analysed the impact of different technologies and modelled the SEC with a logarithmic function. The EnPI range is similar to that obtained in the present work; however, it cannot be directly compared due to a lack of statistical significance or correlation information.

Table 6. Validation data of calculated EnPI for the cement sector.

			This Work	EU BREF	IEA 2021	EU 2012	ASIA 95	UN 2010	EU 2019	US 99	World 2012	World 2012	EU 2012	BAT 2005	BAT UN 2010
				[52]	[53]	[58]	[59]	[22]	[60]	[55]	[61]	[57]	[58]	[54]	[22]
Clinker production	Global	GJ/t	3.5	3.0–4.0	3.5	2.8–3.7	2.1–5.4	3.3–4.2			3.0–6.5	3.0–4.5	3.0–4.2	2.85	2.9
	Electrical	kWh/t	28–32							45	60–100			22.5	
	Thermal	GJ/t	3.4–3.6							4.6					
Cement production	Global	GJ/t	3.1–3.3						2.1–5.1					2.71	
	Electrical (with griding)	kWh/t	108–111	90–150	122		93–162	109–134		150					56
	Thermal	GJ/t	2.8							4.2					
Raw-Material Preparation	Electricty	kWh/t	19–23							38	30–50			13	
Solid-Fuel Preparation	Electricty	kWh/t	13											18	
Finish Grinding	Electricty	kWh/t	25–42							52				25–31	

5. Discussion

The contribution of this work is related to the definition of a methodology to study the energy consumption of industries at a sectoral level and compute the sectoral EnPI and the related parameters that define its reliability and representativeness.

Unlike what is generally practice, sectoral EnPIs have been calculated together with their standard deviation. This helps to understand the dispersion level of site EnPIs with respect to the mean sectoral EnPI and, as a consequence, the reliability of the latter.

Moreover, this work introduces a procedure to define a model correlation between EnPI and production (or any reference parameter that has an influence on energy consumption). This same model is provided with statistical parameters that allow for the understanding not only of the correlation strength between these two variables but also of the representativeness of the sample used to produce the model.

From the results provided in this work, it was also been possible to identify the minimum sample dimension needed to identify a reliable model relating energy and production.

However, it is important to underline the practical consequences of the use of this methodology. In the first place, the definition of a reliable EnPI at the sectoral level allows different companies to compare the performances of their facilities with other homogeneous sites from the same NACE sector. This may lead to improvements in efficiency of all sites in the same NACE sector since a minimum benchmark for less efficient sites is provided. Moreover, since the methodology can be used for second-level EnPIs and, in general, for any EnPI, it makes it possible to identify benchmarks at the process level, fostering the improvement of individual processes.

However, the availability of reliable EnPIs and of EnPI models at the sectoral level is of great importance when it comes to defining new policies or enforcing obligations at the national level aimed at improving energy efficiency. A detailed characterization of the sectors, in fact, allows for the understanding of several aspects, such as:

- the different environmental impacts of certain industries with respect to others, pointing out for which sectors certain policies have to be implemented first;
- the increase in energy use related to economic growth, which allows for planning of actions to mitigate future environmental impacts;
- the need to increase efficiency in some definite sectors or the lack of the said need in other sectors where further improvements in energy efficiency are no longer possible, which helps to establish priority in actions to be taken and shows where to allocate money;
- the impact of energy-efficiency measures on future sectoral performances; such performances, in fact, can be computed from the EnPI model and from the impact of each measure, applied to each site, making it possible to predict how certain actions can lead to energy savings for the sector and the whole country.

6. Conclusions

In this work, theoretical details and applications of a novel methodology to assess global, electric, and thermal energy performance indicators (EnPIs) of productive and economic sectors trough the analysis of mandatory energy audits is presented. The methodology was applied to more than 300 four-level NACE sectors, and in this paper, the application to NACE 23 sector (Manufacture of other non-metallic mineral products) and its subsector, C23.51 (cement) is illustrated. Two types of energy performance indicators were obtained: first-level indexes covering all energy carriers consumed in the production site (electricity, thermal energy, natural gas) and second-level indexes covering prevalent carrier(s) and relative to a specific consumption in the individual production department.

It is important to underline that this methodology was applied using real data monitored by companies and collected in audits carried out by certified experts.

The available data were carefully analysed, and the application of the model led to the identification of many indexes, for which statistical representativeness was also calculated through the analysis of the coefficient of correlation and the p-value parameter.

The minimum sample size to obtain accurate EnPI models was analysed in terms of statistical significance. The study of sectoral energy consumption, therefore, allowed for the energy characterization of most of the manufacturing sectors considered. The analysis of global, electric, and thermal consumption is extremely useful in order to gather information about the general context of applicability of the achieved and potential energy and economic savings due to the energy-efficiency measures listed in the energy audits.

This methodology, applied to all production sectors, could provide key information and indications to characterize the various production processes and process phases from an energy perspective. The methodology is also capable of identifying reference sectoral consumption, and the knowledge of its current level is an enabling condition to identify the best energy-efficiency interventions at the site level, as well as to develop effective energy-efficiency incentive mechanisms at a national level. When elaborating energy-transition scenarios, information on energy performance of different sectoral production processes would allow for the careful evaluation of direct and indirect impacts of reaching long-term energy and environmental targets. For all these reasons, the methodology proposed can be considered a valuable procedure for companies, sectoral stakeholders, and policy makers to enable energy efficiency to play its key role in energy transition.

Author Contributions: Conceptualization, G.B., A.D.S., C.H., L.L., C.M., F.M., M.S., F.A.T. and C.T.; methodology, G.B., A.D.S., C.H., L.L., C.M., F.M., M.S., F.A.T. and C.T.; validation, G.B., A.D.S., C.H., L.L., C.M., F.M., M.S., F.A.T. and C.T.; formal analysis, G.B., A.D.S., C.H., L.L., C.M., F.M., M.S., F.A.T. and C.T.; data curation, G.B., A.D.S., C.H., L.L., C.M., F.M., M.S., F.A.T. and C.T.; writing—original draft preparation, G.B., A.D.S., C.H., L.L., C.M., F.M., M.S., F.A.T. and C.T.; writing—review and editing, C.H., C.M., M.S., F.A.T. and C.T.; supervision, F.M., M.S.; project administration, F.M. and M.S. All authors have read and agreed to the published version of the manuscript.

Funding: This work is funded by the National Electrical System Research Program (RdS PTR 2019-2021-WP1.6), "Piano Triennale della Ricerca del Sistema Elettrico Nazionale 2019–2021" implemented under programme agreements between the Italian Ministry for Economic Development and ENEA, CNR, and RSE S.p.A.

Institutional Review Board Statement: Not applicable.

Informed Consent Statement: Not applicable.

Data Availability Statement: Not applicable.

Conflicts of Interest: The authors declare no conflict of interest. The funders had no role in the design of the study; in the collection, analyses, or interpretation of data; in the writing of the manuscript, or in the decision to publish the results.

References

1. IEA. *Capturing the Multiple Benefits of Energy Efficiency*; IEA: Paris, France, 2015. Available online: https://www.iea.org/reports/capturing-the-multiple-benefits-of-energy-efficiency (accessed on 22 November 2021).
2. Backlund, S.; Thollander, P.; Palm, J.; Ottosson, M. Extending the energy efficiency gap. *Energy Policy* **2012**, *51*, 392–396. [CrossRef]
3. Cagno, E.; Worrell, E.; Trianni, A.; Pugliese, G. A novel approach for barriers to industrial energy efficiency. *Renew. Sustain. Energy Rev.* **2013**, *19*, 290–308. [CrossRef]
4. European Investment Bank. *Going Green. Who Is Investing in Energy Efficiency, and Why It Matters*; EIB Investment Surveys; European Investment Bank: Luxembourg, 2020. [CrossRef]
5. Tanaka, K. Review of policies and measures for energy efficiency in industry sector. *Energy Policy* **2011**, *39*, 6532–6550. [CrossRef]
6. Communication from the Commission to the European Parliament, the Council, the European Economic and Social Committee and the Committee of the Regions. Empty "Fit for 55": Delivering the EU's 2030 Climate Target on the Way to Climate Neutrality. 2021. Available online: https://eur-lex.europa.eu/legal-content/EN/TXT/HTML/?uri=CELEX:52021DC0550&from=EN (accessed on 22 November 2021).

7. Communication from the Commission to the European Parliament, the European Council, the Council, the European Economic and Social Committee and the Committee of the Regions. The European Green Deal COM/2019/640 Final. 2019. Available online: https://eur-lex.europa.eu/legal-content/EN/TXT/HTML/?uri=CELEX:52019DC0640&from=EN (accessed on 22 November 2021).
8. Directive 2012/27/EU of the European Parliament and of the Council of 25 October 2012 on Energy Efficiency. 2012. Available online: https://eur-lex.europa.eu/legal-content/EN/TXT/HTML/?uri=CELEX:32012L0027&from=EN (accessed on 22 November 2021).
9. Directive (Eu) 2018/2002 of the European Parliament and of the Council of 11 December 2018 Amending Directive 2012/27/EU on Energy Efficiency. 2018. Available online: https://eur-lex.europa.eu/legal-content/EN/TXT/HTML/?uri=CELEX:32018L2002&from=EN (accessed on 22 November 2021).
10. ISO. *ISO 50001:2018 Energy Management Systems—Requirements with Guidance for Use*; ISO: Geneva, Switzerland, 2018.
11. Benedetti, M.; Bonfa', F.; Bertini, I.; Introna, V.; Ubertini, S. Explorative study on Compressed Air Systems' energy efficiency in production and use: First steps towards the creation of a benchmarking system for large and energy-intensive industrial firms. *Appl. Energy* **2018**, *227*, 436–448. [CrossRef]
12. Fawkes, S.; Oung, K.; Thorpe, D. *Best Practices and Case Studies for Industrial Energy Efficiency Improvement—An Introduction for Policy Maker*; UNEP DTU Partnership: Copenhagen, Denmark, 2016.
13. European Commission. *Reference Document on Best Available Techniques for Energy Efficiency*; European Commission, Joint Research Centre, Institute for Prospective Technological Studies: Seville, Spain, 2009. Available online: https://eippcb.jrc.ec.europa.eu/sites/default/files/2019-11/ENE_Adopted_02-2009.pdf (accessed on 22 November 2021).
14. Blomqvist, E.; Thollander, P. An integrated dataset of energy efficiency measures published as linked open data. *Energy Effic.* **2015**, *8*, 1125–1147. [CrossRef]
15. Menghi, R.; Papetti, A.; Germani, M.; Marconi, M. Energy efficiency of manufacturing systems: A review of energy assessment methods and tools. *J. Clean. Prod.* **2019**, *240*, 118276. [CrossRef]
16. May, G.; Barletta, I.; Stahl, B.; Taisch, M. Energy management in production: A novel method to develop key performance indicators for improving energy efficiency. *Appl. Energy* **2015**, *149*, 46–61. [CrossRef]
17. Schulze, M.; Nehler, H.; Ottosson, M.; Thollander, P. Energy management in industry—A systematic review of previous findings and an integrative conceptual framework. *J. Clean. Prod.* **2016**, *112*, 3692–3708. [CrossRef]
18. Chiu, T.-Y.; Lo, S.-L.; Tsai, Y.-Y. Establishing an Integration-Energy-Practice Model for Improving Energy Performance Indicators in ISO 50001 Energy Management Systems. *Energies* **2012**, *5*, 5324–5339. [CrossRef]
19. Phylipsen, G.J.M.; Blok, K.; Worrell, E. International comparisons of energy efficiency-Methodologies for the manufacturing industry. *Energy Policy* **1997**, *25*, 715–725. [CrossRef]
20. Bosseboeuf, D.; Chateau, B.; Lapillonne, B. Cross-country comparison on energy efficiency indicators: The on-going European effort towards a common methodology. *Energy Policy* **1997**, *25*, 673–682. [CrossRef]
21. Worrell, E.; Bernstein, L.; Roy, J.; Price, L.; Harnisch, J. Industrial energy efficiency and climate change mitigation. *Energy Effic.* **2008**, *2*, 109. [CrossRef]
22. UNIDO. *Global Industrial Energy Efficiency Benchmark: An Energy Policy Tool*; Working paper; United Nations Industrial Development Organization: Vienna, Austria, 2010.
23. European IPPC Bureau. BAT Reference Documents (BREFs). Available online: https://eippcb.jrc.ec.europa.eu/reference (accessed on 22 November 2021).
24. IEA. *Energy Efficiency Indicators: Essentials for Policy Making*; International Energy Agency: Paris, France, 2014.
25. ISO. *ISO 50006:2014—Energy Management Systems—Measuring Energy Performance Using Energy Baselines (EnB) and Energy Performance Indicators (EnPI)—General Principles and Guidance*; International Organization for Standardization: Geneva, Switzerland, 2014.
26. Therkelsen, P.; Rao, P.; Sholes, D.; Meffert, B.; Green, R.; Nimbalkar, S.U.; Mckane, A. The Value of Regression Models in Determining Industrial Energy Savings. In Proceedings of the 2016 ECEEE Industrial Summer Study: Industrial Efficiency, Berlin, Germany, 12 September 2016; pp. 389–399. Available online: https://www.eceee.org/library/conference_proceedings/eceee_Industrial_Summer_Study/2016/3-energy-management-the-nuts-and-bolts/the-value-of-regression-models-in-determining-industrial-energy-savings/2016/3-079-16_Therkelsen.pdf/ (accessed on 22 November 2021).
27. Giacone, E.; Mancò, S. Energy efficiency measurement in industrial processes. *Energy* **2012**, *38*, 331–345. [CrossRef]
28. Chen, Y.; Therkelsen, P. *The Effect of Linear Regression Modeling Approaches on Determining Facility Wide Energy Savings*; Energy Technologies Area, Lawrence Berkeley National Laboratory: Berkeley, CA, USA, 2019.
29. Boyd, G.A.; Pang, J.X. Estimating the linkage between energy efficiency and productivity. *Energy Policy* **2000**, *28*, 289–296. [CrossRef]
30. Timma, L.; Skudritis, R.; Blumberga, D. Benchmarking Analysis of Energy Consumption in Supermarkets. In *Energy Procedia*; Elsevier: Amsterdam, The Nethrelands, 2016; Volume 95, pp. 435–438.
31. Boyd, G.; Dutrow, E.; Tunnessen, W. The evolution of the ENERGY STAR® energy performance indicator for benchmarking industrial plant manufacturing energy use. *J. Clean. Prod.* **2008**, *16*, 709–715. [CrossRef]
32. Vesterberg, J.; Andersson, S.; Olofsson, T. Robustness of a regression approach, aimed for calibration of whole building energy simulation tools. *Energy Build.* **2014**, *81*, 430–434. [CrossRef]

33. Bianco, V.; Manca, O.; Nardini, S. Electricity consumption forecasting in Italy using linear regression models. *Energy* **2009**, *34*, 1413–1421. [CrossRef]
34. Eurostat; European Commission. *NACE Rev. 2—Statistical Classification of Economic Activites in the European Community*; Office for Official Publications of the European Communitie: Luxembourg, 2008. Available online: https://ec.europa.eu/eurostat/web/products-manuals-and-guidelines/-/ks-ra-07-015 (accessed on 22 November 2021).
35. Eichhammer, W.; Wilhelm, M. Industrial energy efficiency: Indicators for a European cross-country comparison of energy efficiency in the manufacturing industry. *Energy Policy* **1997**, *25*, 759–772. [CrossRef]
36. Decreto Legislativo 14 Luglio 2020, n. 73 Attuazione della Direttiva (UE) 2018/2002 che Modifica la Direttiva 2012/27/UE Sull'efficienza Energetica. (20G00093) (GU Serie Generale n.175 del 14-07-2020). 2020. Available online: https://www.gazzettaufficiale.it/eli/id/2020/07/14/20G00093/sg (accessed on 22 November 2021).
37. Cantini, A.; Leoni, L.; De Carlo, F.; Salvio, M.; Martini, C.; Martini, F. Technological Energy Efficiency Improvements in Cement Industries. *Sustainability* **2021**, *13*, 3810. [CrossRef]
38. ENEA. *Energy Efficiency Annual Report 2020. Analysis and Results from Energy Efficiency Policies in Italy*; ENEA: Rome, Italy, 2020. Available online: https://www.efficienzaenergetica.enea.it/component/jdownloads/?task=download.send&id=453&catid=40%20&Itemid=101 (accessed on 22 November 2021).
39. Herce, C.; Biele, E.; Martini, C.; Salvio, M.; Toro, C. Impact of Energy Monitoring and Management Systems on the Implementation and Planning of Energy Performance Improved Actions: An Empirical Analysis Based on Energy Audits in Italy. *Energies* **2021**, *14*, 4723. [CrossRef]
40. DUEE-SPS-ESE. Publications and Proceedings—Energy Audits (in Italian). Available online: https://www.efficienzaenergetica.enea.it/servizi-per/imprese/diagnosi-energetiche/pubblicazioni-e-atti.html (accessed on 22 November 2021).
41. Thollander, P.; Karlsson, M.; Rohdin, P.; Wollin, J.; Rosenqvist, J. 3—Industrial energy systems. In *Introduction to Industrial Energy Efficiency Energy Auditing, Energy Management, and Policy Issues*; Academic Press: Cambridge, MA, USA, 2020; pp. 33–47, ISBN 978-0-12-817247-6.
42. Lawrence, A.; Thollander, P.; Andrei, M.; Karlsson, M. Specific Energy Consumption/Use (SEC) in Energy Management for Improving Energy Efficiency in Industry: Meaning, Usage and Differences. *Energies* **2019**, *12*, 247. [CrossRef]
43. IEA. *Energy Efficiency Indicators: Fundamentals on Statistics*; International Energy Agency: Paris, France, 2014.
44. Santino, D.; Biele, E.; Salvio, M. *Guidelines for Energy Audits under Article 8 of the EED: Italy's Implementation Practices and Tools*; ENEA: Rome, Italy, 2019. Available online: https://www.efficienzaenergetica.enea.it/component/jdownloads/?task=download.send&id=377&catid=40&Itemid=101 (accessed on 22 November 2021).
45. US-DOE-AMO. Energy Performance Indicator Tool—U.S. Department of Energy's Advanced Manufacturing Office. Available online: https://www.energy.gov/eere/amo/articles/energy-performance-indicator-tool (accessed on 22 November 2021).
46. IPMVP. *Concepts and Options for Determining Energy and Water Savings Volume I*; International Performance Measurement & Verification Protocol: Washington, DC, USA, 2002.
47. DoE. *Guidance for the SEP 50001TM Program Measurement & Verification Protocol: 2019*; Superior Energy Performance 50001TM (SEP 50001TM); US Department of Energy: Washington, DC, USA, 2019.
48. Asuero, A.G.; Sayago, A.; González, A.G. The Correlation Coefficient: An Overview. *Crit. Rev. Anal. Chem.* **2006**, *36*, 41–59. [CrossRef]
49. Cunningham, C.J.L.; Weathington, B.L.; Pittenger, D.J. (Eds.) Appendix B: Statistical Tables. In *Understanding and Conducting Research in the Health Sciences*; Wiley Online Books: Hoboken, NJ, USA, 2013; pp. 479–520, ISBN 9781118643624.
50. Makridou, G.; Andriosopoulos, K.; Doumpos, M.; Zopounidis, C. Measuring the efficiency of energy-intensive industries across European countries. *Energy Policy* **2016**, *88*, 573–583. [CrossRef]
51. Azadeh, A.; Amalnick, M.S.; Ghaderi, S.F.; Asadzadeh, S.M. An integrated DEA PCA numerical taxonomy approach for energy efficiency assessment and consumption optimization in energy intensive manufacturing sectors. *Energy Policy* **2007**, *35*, 3792–3806. [CrossRef]
52. JRC; European Commission; Schorcht, F.; Kourti, I.; Scalet, B.M.; Roudier, S.; Sancho, L.D. *Best Available Techniques (BAT) Reference Document for the Production of Cement, Lime and Magnesium Oxide*; European Commission, Joint Research Centre, Institute for Prospective Technological Studies: Seville, Spain, 2013. Available online: https://eippcb.jrc.ec.europa.eu/sites/default/files/2019-11/CLM_Published_def_0.pdf (accessed on 22 November 2021).
53. IEA. Driving Energy Efficiency in Heavy Industries. Available online: https://www.iea.org/articles/driving-energy-efficiency-in-heavy-industries (accessed on 22 November 2021).
54. Worrell, E.; Price, L.; Neelis, M.; Galitsky, C.; Nan, Z. *World Best Practice Energy Intensity Values for Selected Industrial Sectors*; Environmental Energy Technologies Division Lawrence Berkeley National Laboratory: Berkeley, CA, USA, 2007.
55. Worrell, E.; Galitsky, C. *Energy Efficiency Improvement and Cost Saving Opportunities for Cement Making—An ENERGY STAR® Guide for Energy and Plant Managers*; Lawrence Berkeley National Laboratory: Berkeley, CA, USA, 2008.
56. Worrell, E.; Kermeli, K.; Galitsky, C. *Energy Efficiency Improvement and Cost Saving Opportunities for Cement Making—An ENERGY STAR® Guide for Energy and Plant Managers*; U.S. EPA: Washington, DC, USA, 2013.
57. Oda, J.; Akimoto, K.; Tomoda, T.; Nagashima, M.; Wada, K.; Sano, F. International comparisons of energy efficiency in power, steel, and cement industries. *Energy Policy* **2012**, *44*, 118–129. [CrossRef]

58. Pardo, N.; Moya, J.A.; Mercier, A. Prospective on the energy efficiency and CO2 emissions in the EU cement industry. *Energy* **2011**, *36*, 3244–3254. [CrossRef]
59. APERC. *Energy Efficiency Indicators a Study of Energy Efficiency Indicators for Industry in APEC Economies*; APERC: Tokyo, Japan, 2000.
60. Odysee-Mure. Energy Efficiency—Key Indicators. Available online: https://www.indicators.odyssee-mure.eu/online-indicators.html (accessed on 22 November 2021).
61. Hasanbeigi, A.; Price, L.; Lin, E. Emerging energy-efficiency and CO_2 emission-reduction technologies for cement and concrete production: A technical review. *Renew. Sustain. Energy Rev.* **2012**, *16*, 6220–6238. [CrossRef]

energies

Article

Energy-Saving Technology Opportunities and Investments of the Italian Foundry Industry

Leonardo Leoni [1], Alessandra Cantini [1], Filippo De Carlo [1], Marcello Salvio [2], Chiara Martini [2], Claudia Toro [2,*] and Fabrizio Martini [2]

[1] Department of Industrial Engineering (DIEF), University of Florence, Viale Morgagni, 40, 50134 Florence, Italy; leonardo.leoni@unifi.it (L.L.); alessandra.cantini@unifi.it (A.C.); filippo.decarlo@unifi.it (F.D.C.)

[2] DUEE-SPS-ESE Laboratory, Italian National Agency for New Technologies, Energy and Sustainable Economic Development (ENEA), Lungotevere Thaon di Revel, 76, 00196 Rome, Italy; marcello.salvio@enea.it (M.S.); chiara.martini@enea.it (C.M.); fabrizio.martini@enea.it (F.M.)

* Correspondence: claudia.toro@enea.it

Abstract: The foundry industry is regarded as one of the most energy-intensive industrial sector due to its energy consumption up to 9 MWh/ton of produced metal. As a result, many companies are trying to increase the energy efficiency of their foundry plants. Since many energy-saving technologies are proposed by manufacturers and the literature, choosing the most appropriate one is a difficult task. Moreover, being updated with the available energy-saving solutions is complicated because of the quick technology advances. Consequently, this paper aims at investigating the recent and future opportunities and investments for reducing the energy consumptions of the technologies of Italian foundry companies. Additionally, it aims at presenting a list of available technological solutions validated by Italian experts. To this end, the Energy Audits developed by 231 plants were analyzed to extract the implemented and planned interventions. Furthermore, the economic data available within the Energy Audits were studied to determine the advantages of a given technological solutions compared to the others. It emerged that the companies are strongly investing in increasing the efficiency of the auxiliary systems such as compressors and motors. The outcomes of this study can assist both researchers and energy managers in choosing the most appropriate energy-saving solutions.

Keywords: energy-saving technologies; foundry manufacturing plant; Italian overview; energy efficiency improvements

Citation: Leoni, L.; Cantini, A.; De Carlo, F.; Salvio, M.; Martini, C.; Toro, C.; Martini, F. Energy-Saving Technology Opportunities and Investments of the Italian Foundry Industry. *Energies* **2021**, *14*, 8470. https://doi.org/10.3390/en14248470

Academic Editor: Alberto-Jesus Perea-Moreno

Received: 4 November 2021
Accepted: 6 December 2021
Published: 15 December 2021

1. Introduction

The foundry industry is among the most energy demanding industrial sectors [1] because of its energy consumption that can reach 9 MWh/ton of metal produced [2]. Indeed, the energy cost of a typical foundry plant could cover up to 7% of the total operating costs [3] and up to 15% of the added value [4]. Given the foregoing considerations, along with the introduction of new legislations [5], the importance of energy management in the foundry plants have been progressively cleared up [6]. This fundamental vision has driven the foundry companies to implement more sustainable and energy efficient solutions.

To attain a higher efficiency, it is possible to either adopt new managerial policies or replace existing old technologies with more modern and energy-saving ones [7]. Particularly, the adoption of innovative technologies could help reducing the cost of production, while coping with the rising energy cost [8]. Considering the foundry process, it is possible to act on a technological level in one of the four main phases, which are: melting, molding, casting, and finishing [9,10]. Technological improvements on auxiliary systems (e.g., compressors or motors) and heat recovery systems (e.g., installation of a Rankine turbine) are viable options as well.

Even though the foundry process could vary based on the kind of metal, which could be ferrous or nonferrous, and depending on company policies, there is a common backbone for all the processes. Indeed, the metal is melted by a furnace, which could be an electric or a fuel furnace. The melting phase is the most energy-intensive phase of the foundry process since it accounts for 70% of the total energy consumption [11]. Specifically, the energy consumption of an electric furnace is between 500 and 700 kWh/ton of metal melted, while the energy consumption related to a furnace fueled by coke is in the range of 90–120 kg/ton of metal melted [12]. With regards to the fuel furnace, the adopted fuel could be coke or methane, which is the most common for nonferrous metals. In parallel with the melting phase, the foundry process is characterized by the molding phase, which consists in the preparation of the molds. A mold is the negative of the realized pieces, and it could include cores to create cavities. Moreover, a mold could be made by sand, or it could be permanent. In case sand molding is adopted, a distinct mold must be created for each produced piece. The materials composing a sand mold are usually silica, olivine sand, and sodium silica, along with other substances such as red mud and blast furnace slag [13]. By contrary, a permanent mold could be used for multiple pieces. Indeed, the mold is made out of metal, and it can be coated with graphite or TO_2-based coatings [14]. Downstream the melting and molding phase, the casting phase occurs. The molten metal is poured into the mold, where it solidifies through the heat exchange with the colder mold. For a sand mold, the molten metal is poured through gravity, while, for a permanent mold, a die-casting or a centrifugal casting could also be adopted to let the metal spread within the mold. During this phase, the melted metal must fill all the mold cavity without creating any holes that would be present in the final product. Finally, the solidified metal is extracted from the mold, and it is sent to the finishing phase, which is tasked with the removal of all the risers, feeders, burrs, and superficial sand inclusion, along with improving the surface roughness. The finishing phase could also include thermal or chemical treatments. A schematic representation of the four main phases characterizing the foundry process is illustrated in Figure 1.

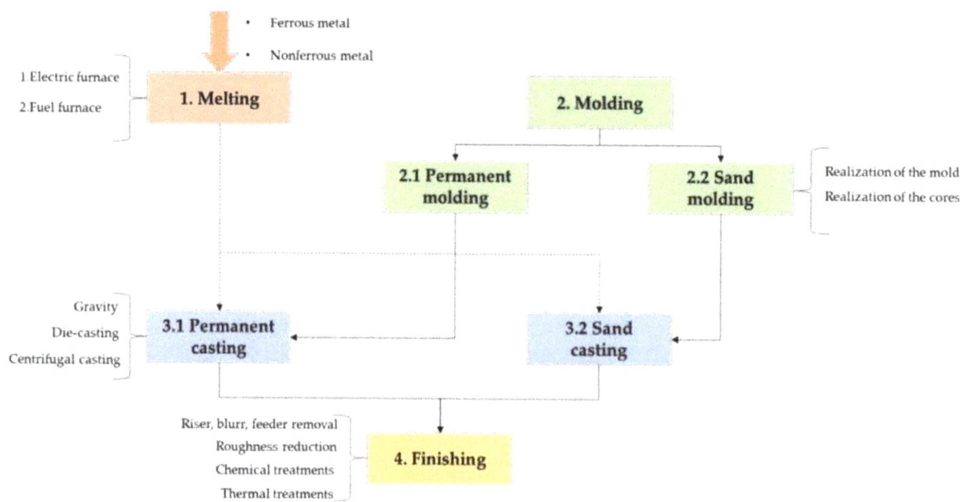

Figure 1. Main phases of the foundry production process.

Within the context of the foundry industry, the reference document to pursue a greater energy efficiency is the Best Available Techniques Reference Documents (BREFs) related to metal foundry [12]. The BREFs are documents realized by the European Union with the aim to provide a guideline for improving energy efficiency. In addition to the BREFs, several energy-efficient technological solutions are reported in the literature. For instance, technologies related to the molding and casting phase could be found in References [15–19], respectively. Specifically, the application of additive manufacturing in the foundry process is found in References [15–17].

Since reaching a higher degree of energy efficiency and environmental sustainability have become crucial requirements, there are several works on this topic for the foundry and the iron and steel industries of distinct countries. Relevant examples were Reference [20] for the foundry industry of India, while References [21–27] were related to the iron and steel industries of China, Mexico, and Taiwan, respectively. Similar studies have been conducted for the Italian foundry industry as well [1,28,29]. Both References [1,29] presented energy efficiency opportunities extracted from the analysis of Energy Audits (EAs), reporting information related to the adopted technologies, along with some relevant energy consumption data. Specifically, both of the aforementioned works underline how energy recovery could be accounted as a difficult task in a foundry plant but nevertheless, if possible, may lead to a 20% energy saving. Installing recuperative burners was also described as an effective energy-saving technology. Reference [28] was also based on an extensive analysis on the EAs of five foundry plants; however, it was more focused on auxiliary systems such as lightening, compressors, pumps, and electric motors.

Despite many researchers have been focusing their efforts on the improvements of energy consumption of energy-intensive processes, industrial technology is constantly developing [30]; thus, keeping up with the technological advances could be regarded as a tough task. Accordingly, some technologies could become obsolete and the available technological solutions for energy-saving purposes should be continuously updated. Moreover, there could be a gap between the technologies proposed by the literature and the technologies that the companies are currently investing on. In this context, this paper aims to provide an overview of the current Italian scenario and near future developments related to technological energy-saving opportunities and investments in the foundry industry, leading to an update of previous studies. Moreover, this work tries to determine the economic reasons that drive the companies towards a given solution instead of other viable options. The data required for the present study are extracted from the EAs of 231 distinct Italian foundry manufacturing sites, carried out by companies in the foundry sector to comply with Article 8 of Legislative Decree 102/2014, Italian implementation of the Energy Efficiency Directive (EED) 2012/27/EU. Indeed, consulting companies working on the field is a common practice to gather useful information and relevant feedbacks [31], along with obtaining an overview of the real situation. Finally, compared to previous similar research carried out for the Italian foundry industry [1,28,29], the present study exploits a higher number of Eas, assuring a much consistent sample. Furthermore, the same studies were mainly focused on presenting the available technologies along with some cost and energy-saving data, while an extensive analysis of all the planned and implemented interventions is neglected. This last aspect is fundamental to grasp the past and future investment trends.

It is worth mentioning that only the technological solutions are considered within this study, while the managerial solutions are disregarded. Moreover, this work does not account for the solution related to lightning, installation of sensors, and heating of the offices. By contrary, technological solutions related to heat recovery systems and auxiliary systems are considered for their pivotal role within a foundry plant. Moreover, even though the terms intervention and solution could be regarded as synonyms, the first is used for referring to something that has been implemented or planned by an Italian company, while the latter is a general word that identifies something that has been found in the literature or in the EAs.

The remainder of the present paper is organized as follows. Section 2 defines the steps of the methodology and describes the available data. Section 3 describes the obtained results related to the analysis. Subsequently, Section 4 provides a discussion on the results, and finally, in Section 5, the conclusions are presented.

2. Materials and Methods

The main objective of the present study is to investigate the actual trend of opportunities and investments in energy-saving technologies related to the Italian foundry industry. Indeed, the foundry industry is regarded as a highly energy-intensive sector; therefore, reducing energy consumption is a pivotal task to assure a sustainable and forward-looking management of the production process.

2.1. Background

In October 2012, the EED was published by the European Parliament and Council with the purpose of reaching 20% energy savings before 2020 [32]. The EED reports several legal obligations that the large companies (all companies that are not considered as small and medium enterprises) must follow to fulfill the required energy efficiency increase. Within the developed framework, Article 8 obliges the affected enterprises to produce EAs. As stated by Cantini et al. [31], an EA is a systematic document that is required to assess the current energy consumption profile and evaluate future energy-saving investments. In Italy, the EAs are collected by an agency named ENEA (Italian National Agency for New Technologies, Energy, and the Sustainable Economic Development), which is tasked with the management and control of the application of the EED's framework on Italian soil. The EAs are uploaded by the companies on the ENEA Audit 102 portal (https://audit102.enea.it/, accessed on 23 December 2019). In Italy, not only large companies but also energy-intensive enterprises are subject to the EA obligation. Energy-intensive companies are those that consume more than 1 GWh of electricity per year, that have tax relief on the electricity bill, and that are registered in the Environmental Energy Services Fund (CSEA) lists.

The EAs, which were received by ENEA in December 2019 (first expiry of the second cycle of mandatory diagnoses after 2015), contain a lot of interesting information, such as the plant location, the plant type, the type of adopted raw material, and the type of finished products manufactured by the companies. However, for the actual work, the most useful information regarded the interventions implemented by the Italian companies between 2015 and 2019, and the interventions that the companies planned to realize between 2019 and 2022. Indeed, the listed energy-saving solutions are essential to define an overview of the Italian most common opportunities and investments to limit energy consumption in the foundry sector.

To pursue the objective of the present paper, three main phases are identified as described by the following subsections.

2.2. Available Technological Solutions for Energy-Saving Purposes

At first, a literature screening on the technologies adopted by the foundry industry is conducted. Then, the obtained list is integrated with the implemented in the last five years and planned interventions found in the EAs. This activity is of prominent importance to define a comprehensive list of possible energy-saving solutions through the integration of real company information and academic studies. Subsequently, the developed list was shared with the experts of the Italian Foundry Trade Association (Assofond) to obtain valuable comments on the applicability of the listed technological solutions. To consider expert observations, the list of detected technologies was presented to the experts during a brainstorming session. The technological solutions were screened one by one and when an expert determined as necessary to add an observation, a discussion started until a common opinion by all the experts was reached. Accordingly, the developed lists of technological opportunities and investments represent a synthetic, yet useful, tool to facilitate companies in choosing appropriate energy-saving solutions.

2.3. Analysis of the Implemented and Planned Interventions in Italy

Within the context of Italian EAs, a given intervention could be proposed by more companies. Thus, to grasp a better understanding of the Italian foundry sector, the frequencies of the energy-saving technologies extracted through the EAs are estimated. Indeed, the adoption of relevant statistical parameters is pivotal to point out the most popular interventions, along with determining possible past and future trends. Denoting by SD, the total number of manufacturing sites that produced the EA and the frequencies of the implemented and planned interventions are computed through Equations (1) and (2), respectively.

$$f_{i,\,i} = \frac{n_{i,i}}{SD}, \tag{1}$$

$$f_{p,\,i} = \frac{n_{p,i}}{SD}, \tag{2}$$

where $n_{i,i}$ identifies the number of companies that implemented the ith intervention between 2015 and 2019, while $n_{p,i}$ denotes the number of companies that proposed the ith intervention as a future development. Finally, $f_{i,\,i}$ and $f_{p,\,i}$ represents the frequency of implementation and planning associated with the ith intervention.

It is worth mentioning that the foundry process could be characterized by an electric or by a fuel furnace. Moreover, the casting phase could occur through the adoption of a permanent or a sand mold. Accordingly, some of the detected interventions could have some applicability limitations. For instance, some technologies could be related to the permanent mold casting; thus, they cannot be adopted by a plant that exploits sand casting. In light of this, new frequencies are defined to get a more truthful and accurate description of the Italian foundry industry. Particularly, given an intervention extracted from the EAs, the new frequencies consider as a sample the number of plants where the aforementioned intervention is implementable (i.e., just a portion of the original sample). Consequently, the more truthful frequencies of the implemented and planned interventions are estimated through Equations (3) and (4), respectively.

$$f_{relevant\_i,\,i} = \frac{n_{i,i}}{reference\_SD}, \tag{3}$$

$$f_{relevant\_p,\,i} = \frac{n_{p,i}}{reference\_SD}, \tag{4}$$

where $n_{i,i}$ and $n_{p,i}$ still identify the number of companies that, respectively, implemented and planned the ith intervention, while $reference\_SD$ is the reduced sample size representing all the companies that could adopt the ith intervention. Finally, $f_{relevant\_i,\,i}$ and $f_{relevant\_p,\,i}$ denote the new frequencies related to the implementation and the planning.

The estimated frequencies could be useful to detect a trend and underline the most popular interventions in the Italian foundry industry, but they are not sufficient to justify whether an intervention is better than another one.

2.4. *Quantitative Economic Analysis of the Gathered Eneregy Saving Interventions*

Some of the EAs included data regarding cost, energy savings, and payback period related to the most relevant interventions. Thus, these data were collected and analyzed to determine the reasons that led the companies to adopt a specific energy-saving solution rather than others. Compared to Section 2.2, this phase allows to analyze the adopted interventions from an economic point of view, considering both the investment cost and the expected energy savings associated with each considered intervention. Specifically, a cost-effectiveness indicator is estimated as illustrated by Equation (5)

$$cost - effectiveness\ indicator = \frac{Euro\ invested\ [€]}{\text{Ton of Oil Equivalent (toe)of energy saved [ton]}}, \tag{5}$$

3. Results

The EAs were provided by 231 different manufacturing sites scattered around all Italy, with a higher density in the Lombardy Region with a total amount of 104 plants (see Figure 2). Moreover, the northern regions contain 204 sites, while 22 sites belong to the central regions, and finally, just 5 plants are located in the South of Italy. Finally, 23% of the 231 plants are considered as big enterprises, while the remaining ones are regarded as small and medium enterprises. All 231 plants are required to procure an EA.

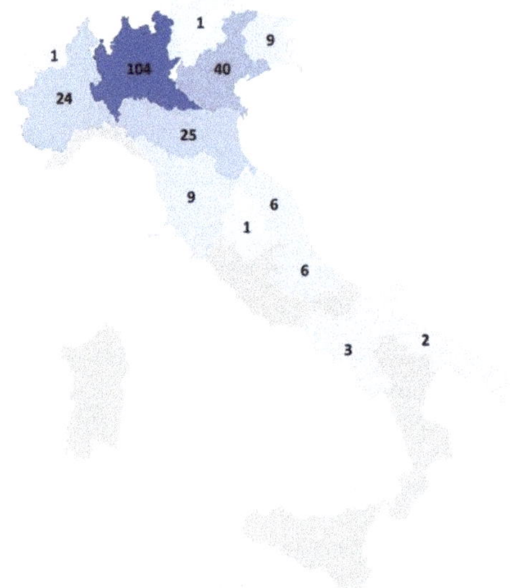

Figure 2. Geographical distribution of the 231 Italian foundry plants constituting the investigated sample.

Considering the Nomenclature of Economic Activities (NACE), the analyzed manufacturing processes comprehend the following class: (i) Casting of iron (NACE code 24.51), (ii) Casting of steel (NACE code 24.52), (iii) Casting of light material (NACE code 24.53), and (iv) Casting of non-iron ferrous material (NACE code 24.54). However, a coarser classification has been adopted for the present study by distinguish between ferrous metal casting and nonferrous metal casting, which can be identified, respectively, by the first two and the last two NACE codes.

By analyzing the 231 EAs, it emerged that 89 manufacturing sites work with ferrous metal (mainly cast iron), while 134 plants are devoted to nonferrous metal casting (mainly aluminum). Finally, seven plants realize artifacts with both ferrous and nonferrous metals. A summary of the aforementioned classification is reported by Table 1.

Table 1. Types of casted metal of the 231 manufacturing sites.

Type of Casted Metal	Manufacturing Sites	Percentage of the Sample
Ferrous	89	39%
Nonferrous	134	58%
Both	7	3%

As previously mentioned, a foundry plant could work with an electric or a fuel furnace based on management policies and choices. Accordingly, the 231 plants were classified based on the exploited type of furnace as well. It was found that 78 foundry sites operate with an electric furnace, while a fuel furnace is adopted by 132 processes. The remaining 21 sites comprehend both an electric and a gas furnace. These findings are listed by Table 2. It is worth mentioning that the analyzed steel foundries adopt only electric furnace, while 60% of the cast-iron production sites exploit an electric furnace. Half of the remaining cast-iron plants (20%) use a coke furnace, while the other half are characterized by a gas furnace. Finally, the majority of nonferrous manufacturing sites have a gas furnace.

Table 2. Types of adopted furnace by the 231 manufacturing sites.

Type of Furnace	Manufacturing Sites	Percentage of the Sample
Electric	78	33.8%
Fuel	132	57.1%
Both	21	9.1%

Finally, the last division has been identified between sites that adopt permanent casting and sites that use sand casting. Particularly, among the 231 production plants, 130 exploit permanent casting, while 91 have implemented side casting. Additionally, 10 manufacturing processes includes both permanent and side casting, as revealed by Table 3.

Table 3. Types of adopted casting by the 231 manufacturing sites.

Type of Casting	Manufacturing Sites	Percentage of the Sample
Permanent	130	56.3%
Sand	91	39.4%
Both	10	4.3%

It is worth mentioning that there is a great difference between a sand casting and a permanent casting process. Indeed, a sand mold is less expensive than a permanent mold, but it is needed to realize a new mold for each produced piece. By contrary, the investment cost related to a permanent mold is higher, but it leads to higher productivity and less finishing requirements of the casted products.

3.1. List of Energy-Saving Technologies Obtained through the Literature and EAs

A literature review was conducted to obtain a preliminary list of available technologies related to the foundry sector. Next, to provide a broader overview of the technologies that could be adopted, the implemented and planned interventions extrapolated from the EAs were integrated with the available literature. Finally, to validate the obtained output, the list of technologies was screened by two foundry experts who provided precious observations regarding the applicability of some technological solutions, along with eliminating unnecessary solutions and adding relevant ones. The two experts that took part in the process belong to the staff of the Italian foundry association and have, respectively, more than 20 and 30 years of experience.

The detected technological solutions were divided by process phases to make them more user-friendly and understandable. As an example, in Table 4, the technologies found for the melting phase are listed, while the tables related to the other phases are reported in Appendix A. In each table, the third column refers to the solution that can be implemented to improve a specific process stage, while the first and the second columns, respectively, identify the machinery and the object associated with the technological solution. Moreover, each technology found in the literature is accompanied by the related bibliographic references (fifth column), while the experts' comments are listed in the sixth column. Finally, the technological alternatives found in the Eas are reported in italics. Therefore, there could be three different types of technologies:

1. Technologies that are reported in italics and characterized by one or more bibliographic references. These technologies are found both in the literature and in the Eas.
2. Technologies that are reported in black. These technologies are only found in the literature
3. Technologies that are reported in italics and characterized by no bibliographic reference. These technologies are only found in the Eas.

To make a meaningful difference an energy hierarchy that specifies the energy approach was introduced in the fourth column. Specifically, the energy hierarchy is based on three levels similarly to what were done by Reference [33], where seven levels were used. The different energy approaches and their hierarchy are as follows:

1. Innovation: introduce a completely new technology for a part of the process or that is tasked with something that was not done before.
2. Replace: replacing a given technology with a more efficient and/or modern ones. Compared to the previous level, it introduces less changes.
3. Recover: recover thermal or electric energy.
4. Resource: change the source of energy.

Table 4. Technological energy-saving solutions for melting obtained from both the literature and Eas.

Melting					
Process Machinery	**Solution Object**	**Energy-Saving Approach**	**Energy-Saving Technological Solution**	**Reference**	**Comments from Sector Experts**
Furnace feeders	Furnace feeders	Recover	*Preheating the row material through the exhaust fumes*	[34,35]	This solution was very common in the past, but it is now less popular due to the high costs. It could be interesting in case the exhaust gas is adopted to preheat the scrap before entry the furnace.
Furnace	Burners	Innovation	*Installing recuperative burners*	[36]	
Furnace	Burners	Innovation	Installing low NO_x burners to reduce the emission	[34]	
Furnace	Burners	Replace	*Replacing the existing burners with more modern and efficient ones*	[34]	
Furnace	Burners	Innovation	*Installing regenerative burners*	[34]	
Furnace	Burners	Innovation	Installing oxy-fuel burners	[37,38]	This solution is mandatory and it is adopted in all the cast iron plants with rotating furnaces.
Furnace	Burners	Innovation	Installing a combustor for a no flame combustion	[34,38]	
Furnace	Furnace	Replace	*Replacing the existing furnace with a more modern and efficient one*	[34]	
Furnace	Furnace	Innovation	*Adopting IGBT technology for electric furnace*	[39]	
Furnace	Furnace	Innovation	Installing Ultra High Power transformer to increase the voltage of the electric arc furnace	[8]	Few companies adopt this solution, which is typical of steel foundry process characterized by smaller furnaces compared to steel production. It should be evaluated if this intervention could be convenient in a foundry plant since the furnace size and its degree of utilization are different compared to the iron and steel industry.
Furnace	Furnace	Re-source	Installing Oxy-oil technology to exploit oil as fuel and reducing the consumption of coke along with the emission	[8]	There is no similar application in Italy
Furnace	Furnace	Replace	Installing an efficient water-cooled furnace	[27]	A water-cooling system is already present.

Table 4. *Cont.*

Melting					
Process Machinery	**Solution Object**	**Energy-Saving Approach**	**Energy-Saving Technological Solution**	**Reference**	**Comments from Sector Experts**
Furnace	Furnace	Replace	*Replacing the refractory material of the furnace with a new one to reduce the heat dispersion*	[27,40]	Increasing the thickness of the refractory material leads to a reduction of the furnace production capacity, even though it assists in reducing the energy consumption. Moreover, increasing the thickness of the refractory material causes less space available for the raw material.
Furnace	Furnace	Innovation	Installing Electron Beam Furnace	[35]	
Furnace	Furnace	Innovation	Installing Solar Furnace	[35]	
Furnace	Furnace	Innovation	Installing Plasma Furnace	[35]	
Furnace	Furnace	Innovation	Installing Immersion Heaters	[35]	
Furnace	Furnace	Innovation	Installing Microwave melting technology	[35]	
Furnace	Recovery system	Innovation	Installing machines to recover metal from slag	[41]	Slag is usually selected depending on the furnace type; thus, its chemical composition is known.
Furnace	Pneumatic powder injector	Innovation	Installing a pneumatic injector to send to the furnace the dust trapped within the air filters	[42]	It leads to higher energy consumption and slag production. It assists in decreasing the environmental impact.
Furnace	Pneumatic injector lance	Innovation	*Installing a pneumatic injector lance to blow away the slag from the combustion area*	[42]	This solution inerts the slag, leading to a reduction of environmental impact. However, it could increase the energy consumption.

3.2. Frequencies of the Interventions Extracted from the Eas

After extracting the implemented and planned interventions from the Eas, the frequencies associated with each intervention was computed as explained in Section 2.3. The results of the calculation are shown in Table 5, where an italic intervention represents an intervention that is found only in the Eas.

Table 5. Number of companies that implement and plan a given intervention, along with its associated frequencies.

Process Stage	Intervention Object	Energy-Saving Approach	Intervention	$n_{i,i}$	$n_{p,i}$	$f_{i,i}$	$f_{p,i}$	$f_{relevant\_i,i}$	$f_{relevant\_p,i}$
Melting	Furnace feeders	Recover	Preheating the row material through the exhaust fumes	2	2	0.0087	0.0087	0.0087	0.0087
Melting	Burners	Innovation	Installing recuperative burners	0	7	0	0.0303	0	0.046
Melting	Burners	Replace	Replacing the existing burners with more modern and efficient ones	0	1	0	0.0043	0	0.0065
Melting	Burners	Innovation	Installing regenerative burners	0	1	0	0.0043	0	0.0065
Melting	Furnace	Replace	Replacing the existing furnace with a more modern and efficient one	6	11	0.026	0.048	0.026	0.048
Melting	Furnace	Innovation	Adopting IGBT technology for electric furnace	0	1	0	0.0043	0	0.01

Table 5. *Cont.*

Process Stage	Intervention Object	Energy-Saving Approach	Intervention	$n_{i,i}$	$n_{p,i}$	$f_{i,i}$	$f_{p,i}$	$f_{relevant\_i,i}$	$f_{relevant\_p,i}$
Melting	Furnace	Replace	Replacing the refractory material of the furnace with a new one to reduce the heat dispersion	2	1	0.0087	0.0043	0.013	0.0065
Melting	Pneumatic injector lance	Innovation	Installing a pneumatic injector lance to blow away the slag from the combustion area	0	1	0	0.0043	0	0.0043
Molding	Sand recovery system	Re-source	*Replacing the electric sand recovery system with a gas one*	0	1	0	0.0043	0	0.0099
Molding	Sand recovery system	Innovation	*Installing a sand recovery system*	1	0	0.0043	0	0.0099	0
Molding	Molding station	Replace	*Replacing the molding stations with more efficient ones*	1	0	0.0043	0	0.0099	0
Molding	Mixer	Replace	*Replacing the mixing systems with a more efficient one*	2	0	0.0087	0	0.02	0
Molding	"Hot box" molding	Innovation	Installing a pre-heating system or regenerative or recuperative burners for the furnace tasked with the production of the sand mold	1	0	0.0043	0	0.0099	0
Molding	Molding machinery	Innovation	*Installing low-pressure casting machine capable of using inorganic cores*	0	1	0	0.0043	0	0.0099
Molding	Molding machinery	Replace	*Installing an efficient filter for the sand molding process*	0	1	0	0.0043	0	0.0099
Molding	Molding machinery	Replace	Replacing the existing machinery with newer and more efficient ones	1	0	0.0043	0	0.0099	0
Casting	Casting furnace	Replace	*Replacing the casting furnace with a more efficient and newer one*	1	0	0.0043	0	0.0043	0
Casting	Cooling system	Innovation	*Installing a forced ventilation cooling system*	1	0	0.0043	0	0.0071	0
Casting	Casting machinery	Replace	Replacing the casting machineries with more efficient and newer ones	4	2	0.017	0.0087	0.029	0.014
Casting	Die-casting machinery	Replace	*Replacing the furnace where the cast waits to be poured in the mold with a more efficient and newer one*	2	3	0.0087	0.013	0.014	0.021
Casting	Die-casting machinery	Replace	*Installing a new efficient die-casting line*	0	3	0	0.0043	0	0.0099
Casting	Sand removal machinery	Replace	Replacing the machinery in charge of the sand removal process with more efficient and newer ones	1	0	0.0043	0	0.0043	0
Casting	Ladle	Innovation	Installing machines capable of scheduling an efficient preheat of the ladle	0	1	0	0.043	0	0.043
Finishing	Finishing station	Replace	*Installing a new efficient finishing line*	2	0	0.0087	0	0.0087	0
Finishing	Finishing station	Replace	*Installing high efficiency nozzles*	0	1	0	0.0043	0	0.0043
Finishing	Finishing station	Replace	Replacing old finishing machineries with more efficient and modern ones	2	1	0.0087	0.0043	0.0087	0.0043
Finishing	Heat Treatment Furnace	Replace	*Replacing the heat electric furnace with a more efficient and modern one*	0	1	0	0.0043	0	0.0043
Auxiliary Systems	Compressors	Replace	Replacing the compressor with more modern and efficient ones	21	33	0.091	0.14	0.091	0.14
Auxiliary Systems	Compressors	Innovation	Installing variable speed compressors (i.e., compressor with inverter)	0	3	0	0.013	0	0.013

Table 5. *Cont.*

Process Stage	Intervention Object	Energy-Saving Approach	Intervention	$n_{i,i}$	$n_{p,i}$	$f_{i,i}$	$f_{p,i}$	$f_{relevant\_i,i}$	$f_{relevant\_p,i}$
Auxiliary Systems	Pressure Systems	Innovation	Replacing all the equipment for pressurized air distribution with electric devices (if possible)	1	0	0.0043	0	0.0043	0
Auxiliary Systems	Suction Systems	Replace	Installing high efficiency fans	1	6	0.0043	0.026	0.0043	0.026
Auxiliary Systems	Suction Systems	Innovation	Installing variable speed fans (i.e., fans with inverter)	1	3	0.0043	0.013	0.0043	0.013
Auxiliary Systems	Suction Systems	Innovation	*Installing a forced air suction system for the furnace*	1	0	0.0043	0	0.0043	0
Auxiliary Systems	Forklifts	Replace	Replacing the forklifts with more efficient and modern ones	0	1	0	0.0043	0	0.0043
Auxiliary Systems	Forklifts' Batteries	Replace	Replacing the batteries of the forklifts	0	2	0	0.0087	0	0.0087
Auxiliary Systems	Transport Systems	Replace	*Replacing conveyor belts with more efficient and modern ones*	1	0	0.0043	0	0.0043	0
Auxiliary Systems	Transport Systems	Innovation	Installing high efficiency belt and replace V-belts with toothed belts	0	5	0	0.022	0	0.022
Auxiliary Systems	Transport Systems	Innovation	*Replacing V-belts with helicoidal belts*	0	1	0	0.0043	0	0.0043
Auxiliary Systems	Electric Transformers	Replace	*Replacing the electricity transformers with more efficient and modern ones*	1	5	0.0043	0.022	0.0043	0.022
Auxiliary Systems	Inverter	Innovation	Installing inverters on electric motors or replacing the inverters with more efficient and modern ones	11	56	0.048	0.24	0.048	0.24
Auxiliary Systems	Lift Truck	Replace	*Replacing the lift trucks with more efficient and modern ones*	0	1	0	0.0043	0	0.0043
Auxiliary Systems	Crane	Replace	*Replacing the cranes with more efficient and modern ones*	2	1	0.0087	0.0043	0.0087	0.0043
Auxiliary Systems	Passive filter	Innovation	Installing passive filters	0	10	0	0.043	0	0.043
Auxiliary Systems	Engines	Replace	Installing high efficiency electric motors (class IE2, IE3 and IE4)	3	56	0.013	0.24	0.013	0.24
Auxiliary Systems	Engines	Innovation	*Installing regenerative electric motors*	0	2	0	0.0087	0	0.0087
Auxiliary Systems	Engines	Replace	Rewinding electric motors	0	1	0	0.0043	0	0.0043
Auxiliary Systems	Engines	Innovation	Installing variable speed motors (i.e., motors with inverter)	0	2	0	0.0087	0	0.0087
Auxiliary Systems	Fluid Distribution System	Innovation	Optimize the pipelines' design leakage along with installing appropriate seal to minimize air leakage	3	66	0.013	0.29	0.013	0.29
Auxiliary Systems	Pumps	Replace	Replacing the pumps with more efficient and modern ones	2	3	0.0087	0.013	0.0087	0.013
Auxiliary Systems	Cooling Systems	Replace	*Replacing the cooling towers with more efficient and modern ones*	2	2	0.0087	0.0087	0.0087	0.0087
Heat Recovery Systems	Rankine turbine	Recover	Installing a Rankine turbine to generate electric energy through the exhaust gas	0	1	0	0.0043	0	0.0043
Heat Recovery Systems	ORC turbine	Recover	Installing a ORC turbine to generate electric energy through the exhaust gas	0	3	0	0.013	0	0.013
Heat Recovery Systems	Cogeneration	Recover	Installing cogeneration or trigeneration technologies	1	13	0.0043	0.0562	0.0043	0.056
Heat Recovery Systems	Refrigeration cycle	Recover	Installing technologies able to exploits the exhaust gas for a refrigeration cycle	0	1	0	0.0043	0	0.0043
Heat Recovery Systems	Battery	Replace	Replacing the batteries of the heat recovery systems	1	0	0.0043	0	0.0043	0

Table 5. *Cont.*

Process Stage	Intervention Object	Energy-Saving Approach	Intervention	$n_{i,i}$	$n_{p,i}$	$f_{i,i}$	$f_{p,i}$	$f_{relevant\_i,i}$	$f_{relevant\_p,i}$
Heat Recovery Systems	Evaporator	Recover	*Installing an evaporator to retrieve heat from the emulsified water*	1	0	0.0043	0	0.0043	0
Heat Recovery Systems	Sand drying system	Recover	*Installing technologies for the sand drying process*	1	0	0.0043	0	0.0043	0
Heat Recovery Systems	Exchanger	Recover	*Installing an exchanger to retrieve heat from the exhaust gas and generate hot water for the drier*	0	1	0	0.0043	0	0.0043
Heat Recovery Systems	Exchanger	Recover	*Installing an exchanger to retrieve heat from the compressors*	1	13	0.0043	0.056	0.0043	0.056
Heat Recovery Systems	Cooling system	Recover	*Installing a heat recovery system from the cooling process of the molds*	0	1	0	0.0043	0	0.0071
Heat Recovery Systems	Exchanger	Recover	*Installing an exchanger to retrieve heat and preheat the ladle*	0	4	0	0.017	0	0.017
		Total		84	336				

It is possible to state that the estimated frequencies provide a description about the adopted strategies to reduce energy consumption in the Italian foundry industry. Indeed, high values of $f_{relevant\_i,i}$ denote a common intervention exploited by the Italian companies in the last five years for energy-saving purposes. Indeed, since $f_{relevant\_i,i}$ represents the percentage of plants that implemented the ith intervention, the higher the value of $f_{relevant\_i,i}$, the more popular a given intervention has been during the past years. Moreover, $f_{relevant\_p,\ i}$ represents the portion of companies that are willing to adopt a certain technological intervention, giving a hint on possible future developments. Indeed, high values of $f_{relevant\_p,\ i}$ indicates an intervention that could be soon very popular since a high number of Italian foundry plants are planning to implement that intervention in the next future.

Figure 3 illustrates the number of interventions implemented and planned for each process phase, considering the auxiliary and the heat recovery systems as a separate phase.

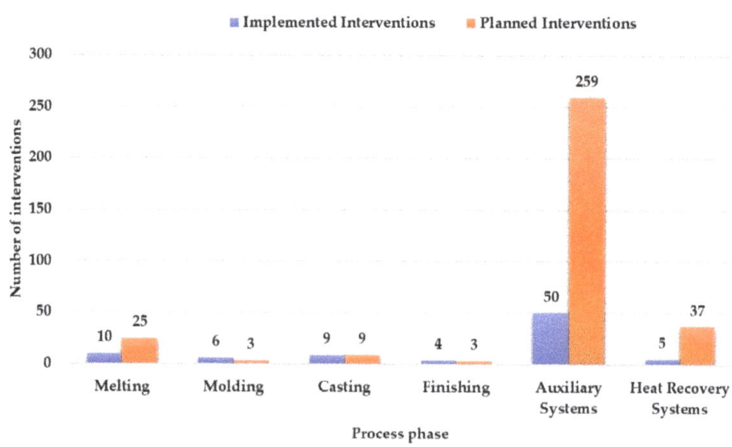

Figure 3. Number of implemented and planned interventions for each process phase.

It emerged that the companies prefer investing in auxiliary systems, which are characterized by 50 implemented interventions and 259 planned interventions. Accordingly, 59% of the implemented interventions and 77% of the planned interventions involve the auxiliary equipment. In contrast, less interest is devoted to the other process phases among which the heat recovery systems result the most considered one, with five implemented interventions and 37 planned interventions. Since the melting phase is the most energy-intensive, many efforts and investments are focused on it. Specifically, the melting phase has seen a total of 10 implemented interventions and 25 planned interventions. Finally, the finishing phase is the most neglected phase, since it is not relevant in all the manufacturing sites. For instance, the plants that adopt permanent mold casting are usually characterized by less effort on the finishing phase.

To illustrate even further the obtained results, the implemented and planned interventions for each intervention object related to the auxiliary systems, the heat recovery systems, and the melting phase are illustrated by Figures 4–6, respectively. Indeed, it is interesting to highlight the distributions of the interventions associated with the three most relevant "phases".

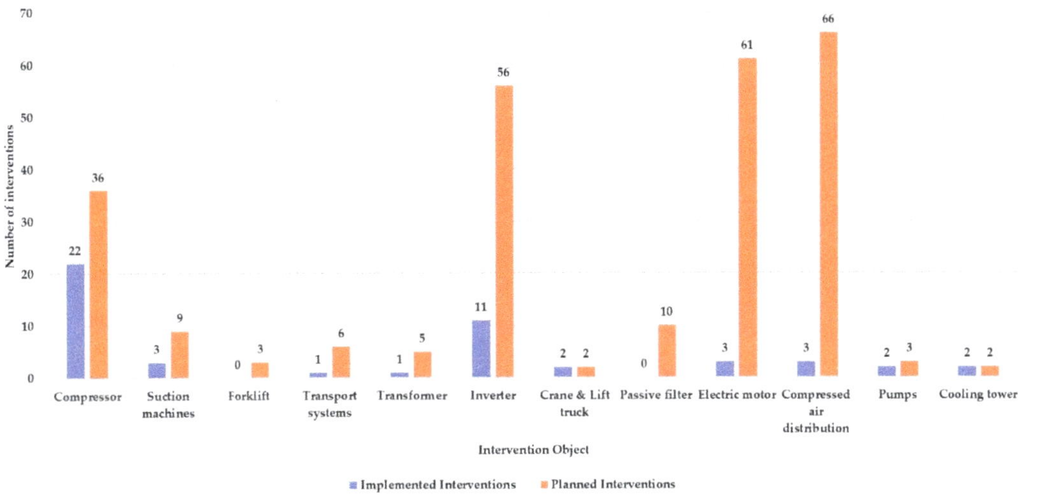

Figure 4. Number of implemented and planned interventions for the auxiliary systems.

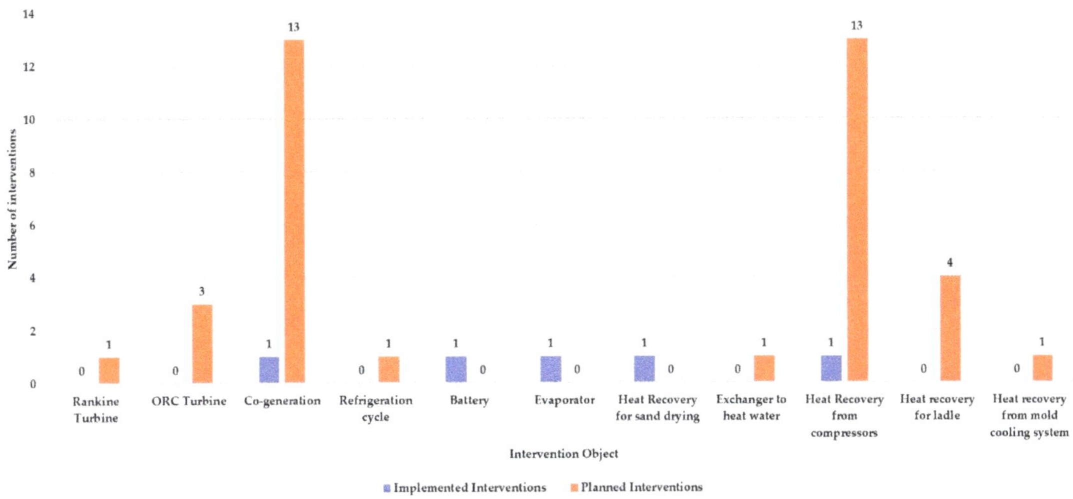

Figure 5. Number of implemented and planned interventions for the heat recovery systems.

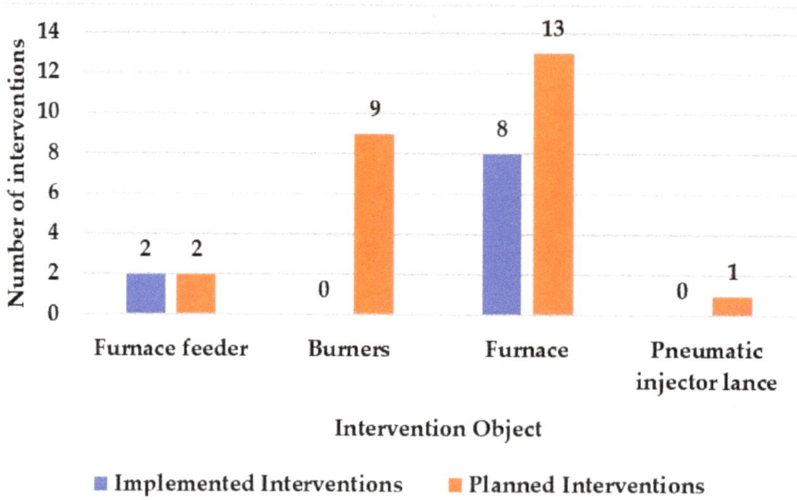

Figure 6. Number of implemented and planned interventions for the melting phase.

Considering the auxiliary systems, the most interventions were adopted for the compressors, which are characterized by 22 implemented interventions; among which, 21 interventions were regarded the replacement of old compressor with newer and more efficient ones (see Table 5). Indeed, the aforementioned intervention is associated with the highest $f_{relevant\_i,i}$, which is estimated at 0.09. This value identifies a trend of the implemented interventions, since almost 9% of the companies replaced the compressors between 2015 and 2019. Moreover, the replacement of compressors with more efficient ones has been planned by 33 enterprises, leading to a $f_{relevant\_p,i}$ equal to 0.14. Accordingly, this intervention is still regarded as one of the most beneficial. Finally, 56, 61, and 66 companies planned an intervention related to inverters, electric motors, and air distribution systems, respectively. Thus, it is expected to see an increase of the number of interventions associated with the three aforementioned intervention objects during the next years. Considering the inverters, 56 companies are willing to install new ones or replace the old ones, leading to a $f_{relevant\_p,i}$ of 0.24. However, the highest values of $f_{relevant\_p,i}$ are associated with the reduction of leaks in the air distribution systems (66 planned interventions) and the replacement of electric motors with more efficient ones (56 planned interventions). Indeed, the last two interventions yield a $f_{relevant\_p,i}$ of 0.29 and 0.24, respectively.

The heat recovery systems are mostly characterized by planned interventions; among which, the most popular one consists in installing a heat recovery system to retrieve heat from the compressor. This intervention is associated with a $f_{relevant\_p,i}$ equal to 0.056, which identifies the possibility that 6% of the companies will adopt this intervention during the next years. The installation of a cogeneration system could also become a common intervention during the next years, since it yields a $f_{relevant\_p,i}$ of 0.056 as well.

Finally, with regards to the melting phase, most of the implemented interventions is related to the furnace, which has seen eight interventions; among which, six interventions consisted of replacing the old furnace with a more efficient one, leading to a $f_{relevant\_i,i}$ of 0.026. Installing a more efficient furnace has also been planned by 11 companies, resulting in a $f_{relevant\_p,i}$ equal to 0.045.

3.3. *Cost–Benefit Quantitative Analysis*

Companies also reported quantitative data on the savings achieved by implementing technological interventions. A total of 84 implemented interventions with quantitative information were listed in the analyzed 231 EAs. Among them, the focus in this article is on technological areas of intervention related to the production process, auxiliary systems and heat recovery systems: they are summarized in Tables 6 and 7. In the tables, toe stands for ton of oil equivalent, and savings are in terms of final energy. (The column "Other savings" refers to a mix of electric and thermal savings for which the disaggregation in the two components was not available in the energy audit or to savings of other energy vectors.) Production lines determine large energy savings and the largest economic investments (both total and average). Interventions on electric or fuel furnaces are included in this area, coherently with the technological interventions shown in Table 7, and these are largely represented by replacing the existing furnace with a more modern and efficient one. Pressure systems are the second area both in terms of savings and total investment, whereas thermal power plant and heat recovery systems are the second area in terms of average investment. The average quantitative data shown in Tables 6 and 7 was computed as average of the number of production sites that reported quantitative information.

Table 6. Energy savings produced by the implemented technological interventions in the various areas of intervention. The total annual savings are calculated as the sum of thermal energy, electricity, and fuel savings.

Area of Intervention	Production Sites Reporting Quantitative Information	Electricity Savings (Toe/Year)	Thermal Energy Savings (Toe/Year)	Other Savings (Toe/Year)	Annual Savings (Toe/Year)	Annual Savings (%)	Average Annual Savings (Toe/Year)
Pressure systems	14	326	0	13	339	19%	24
Intake system	8	117	0	0	117	7%	15
Thermal power plant and heat recovery systems	3	0	59	44	103	6%	34
Engines, inverters, and other electrical installations	3	9	0	0	9	0%	3
Production lines and machines	14	348	258	610	1217	68%	87
Total	42	801	317	667	1784	100%	-

Table 7. Investments required to apply technological interventions in the various areas of intervention.

Area of Intervention	# Production Sites Reporting Quantitative Information	Total Investment (€)	Total Investment (%)	Average Investment (€)
Pressure systems	19	1,791,068	17%	94,267
Intake system	9	666,428	6%	74,487
Thermal power plant and heat recovery systems	3	615,500	6%	205,166
Engines, inverters, and other electrical installations	7	430,620	4%	71,770
Production lines and machines	20	7,347,956	68%	367,398
Total	58	10,851,572	100%	-

A cost-effectiveness indicator was calculated for each intervention, measured as Euros invested per Ton of Oil Equivalent (toe) of energy saved (see Table 8). The available information allowed to calculate it only on 11 interventions, reporting both information on energy saved and costs. The area of "Engines, inverters, and other electrical installations" shows an advantageous value of the indicator, confirming that this is a type of intervention with a large applicability, also in different industrial sectors.

Table 8. Cost-effectiveness indicator for each area of intervention.

Area of Intervention	Production Sites Reporting Quantitative Information	Cost-Effectiveness Indicator (€/toe)
Pressure systems	14	6821
Intake system	7	15,340
Thermal power plant and heat recovery systems	0	-
Engines, inverters, and other electrical installations	3	3101
Production lines and machines	11	13,900

A total of 840 planned interventions with quantitative information were identified in the EAs examined. For the purpose of this analysis, as already explained for implemented interventions, we disregard solutions related to areas of intervention not related to production process and auxiliary systems such as, for example, lighting, managerial interventions, and production from renewable sources. Tables 9 and 10 summarize the savings of final energy and investment cost indicated by those companies that proposed a feasibility study. Table 11 reports the cost-effectiveness indicators calculated for the planned interventions. Feasibility studies estimated electrical savings in all areas and thermal savings to be significant in "Thermal power plant and heat recovery systems" and "Production lines and machines" one. As in the applied interventions, also in the planned interventions the highest energy saving was associated with the production lines area, accompanied, however, by a significant investment cost (Table 10). This area shows a high cost-effectiveness indicator, and as shown in Table 4, most technological interventions are applied to furnaces; additionally, in this case, a furnace substitution represents a high share of interventions in this area. Thermal power plant and heat recovery systems have the best value of cost-effectiveness, followed by pressure systems (Table 11).

Table 9. Energy savings assessed for the planned technological interventions in the various areas of intervention. The total annual savings are calculated as the sum of thermal energy, electricity, and fuel savings.

Area of Intervention	Production Sites Reporting Quantitative Information	Annual Electricity Savings (Toe/Year)	Annual Thermal Energy Savings (Toe/Year)	Other Savings (Toe/Year)	Annual Savings (Toe/Year)	Annual Savings (%)	Average Annual Savings (Toe/Year)
Pressure systems	142	1316	47	64	1427	14%	10
Intake system	71	569	0	0	569	6%	8
Thermal power plant and heat recovery systems	56	741	917	1264	2922	29%	52
Engines, inverters, and other electrical installations	149	1072	0	128	1200	12%	600
Production lines and machines	65	2075	1380	437	3892	39%	3892
Total	483	10,009	2344	1893	14,246	100%	-

Table 10. Investments assessed for the planned interventions for distinct areas of intervention.

Area of Intervention	# Production Sites Reporting Quantitative Information	Total Investment (€)	Total Investment (%)	Average Investment (€)
Pressure systems	135	3,549,176	11%	26,290
Intake system	73	2,662,437	8%	36,472
Thermal power plant and heat recovery systems	53	7,023,606	21%	132,521
Engines, inverters, and other electrical installations	148	5,646,496	17%	48,872
Production lines and machines	68	14,568,165	44%	214,238
Total	477	30,787,442	100%	-

Table 11. Cost-effectiveness indicator for each area of intervention.

Area of Intervention	# Production Sites Reporting Quantitative Information	Cost-Effectiveness Indicator (€/toe)	Pay-Back Time (Years)
Pressure systems	133	3252	2.6
Vacuum system	71	6232	4.9
Thermal power plant and heat recovery systems	52	1935	3.4
Engines, inverters, and other electrical installations	143	6622	5.2
Production lines and machines	61	10,089	5.6

Planned technological interventions can also be analyzed distinguishing for their Payback Time class (PBT; Figure 7). In this case, 421 interventions report quantitative information: interventions with PBT between one and two years represent 8% (2.1 ktoe/year) of the total annual potential saving. Further, 26% of the potential savings is associated with interventions, having a PBT between 2 and 3 years (2.3 ktoe/year).

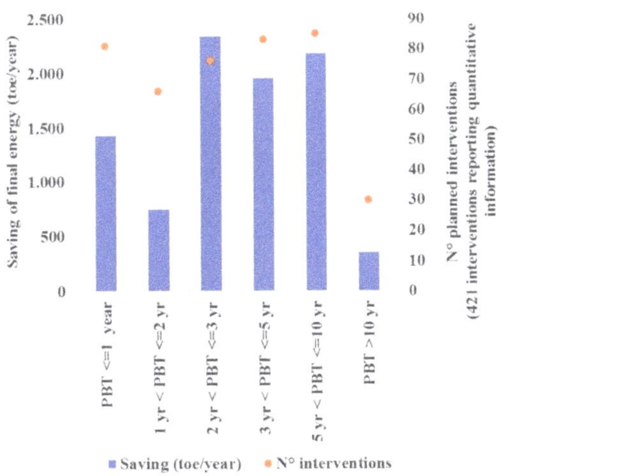

Figure 7. Annual saving and planned interventions according to the PBT classes.

4. Discussion

As depicted by Figure 3, the auxiliary systems are characterized by the highest number of implemented and planned interventions. This peculiar trend is related to the low investment cost, which results in decent values for the cost-effectiveness indicator and short payback period. Indeed, auxiliary machines are characterized by the lowest investment costs (see Table 10). Moreover, the compressors and pressure systems are associated with the lowest PBT, which is estimated at 2.6 years, and the second cost-effectiveness indicator evaluated at 3252 €/toe (see Table 11) for the planned intervention and 6821 €/toe for the implemented intervention. Furthermore, replacing old compressors with more efficient ones is regarded by the association expert as a very good strategy to reduce energy consumption. On the other side, vacuum systems and electric motors have a PBT of approximately 5 years and a cost-effectiveness indicator estimated at about 6000 €/toe for the planned intervention. Additionally, the engine sector is characterized by the best cost-effectiveness indicator among the implemented interventions (about 3000 €/toe). Another advantage of the interventions related to the auxiliary systems is the easiness of implementation, since the process remains unchanged, and the interventions are mostly characterized by the replacement of an old machine with a more efficient and modern one.

The interventions that act on the heat recovery systems are also quite popular due to the lowest cost-effectiveness indicator of 1935 €/toe (among the planned interventions). Compared to the interventions on auxiliary systems, which are always possible, the interventions related to the heat recovery systems could not always represent a viable option. Indeed, the ability to retrieve heat is limited because of the low temperature characterizing the exhaust gas, leading to some applicability restrictions. For instance, the installation of a cogeneration system is considered by the experts as difficult to implement in a foundry plant. Saying that, the installation of an Organic Rankine Cycle (ORC) turbine and the heat recovery from compressors could become popular interventions in the next future. The ORC turbine exploits lower temperature compared to the more common Rankine turbine, while the heat recovery from compressors is regarded by the experts as an emerging technology which advantages, related to energy-saving indicators, must be evaluated during the next years. Among the interventions concerning heat recovery systems, the installation of an exchanger to retrieve heat and preheat the ladle is worth mentioning. Indeed, the Italian foundry experts state that it is not a common technology, but it has great potential and margin of rationalization. It is worth mentioning that cogeneration and trigeneration interventions are not included in the heat recovery system category, but they are examined as a separate category: planned interventions reporting quantitative information are 10 and correspond to 17,586 toe/year of primary energy saving. The average cost-effectiveness indicator is 1536 €/toe of primary energy; the average PBT is 4 years, thus showing a similar value to heat recovery systems (3.4 years).

Considering the process phases, most of the interventions are implemented and planned for the melting phase, which is the most energy-intensive one. The interventions related to this area are among the most expensive ones, but they assure higher annual energy savings compared to more popular interventions. Furthermore, these kinds of interventions are usually more complex and invasive compared to the interventions related to the auxiliary systems. For instance, replacing the furnace could be very impactful on production schedule and could also lead to more strict requirements with regards to layout and spacing within the plant. However, there are some interventions that are easier to implement, such as the installation of regenerative burners, which is now mandatory and adopted by all the cast iron foundry plants, as stated by the experts.

To make a meaningful difference and resume the previous findings, Figure 8 shows that half of potential saving (4.5 ktoe/year) can be achieved by adopting interventions with PBT lower than 3 years and by mobilizing 20% of total investment associated with suggested interventions (around 5.9 million Euro). This highlights that relatively less expensive interventions are associated with a high saving potential, and such a trend appears even more significant when considering that the existing incentive mechanisms are not included within the PBT calculations.

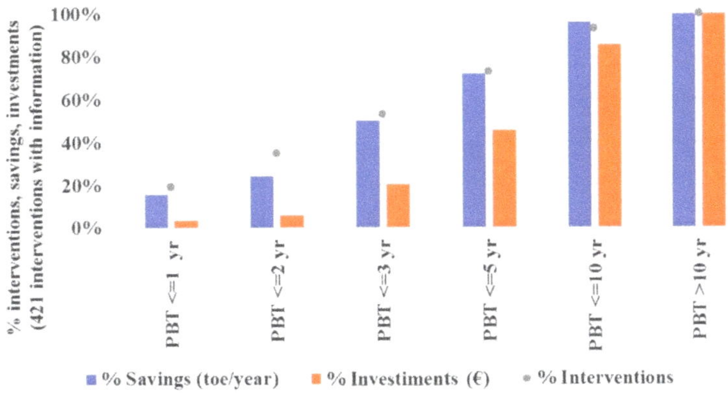

Figure 8. Cumulative saving and investment according to PBT classes.

Finally, it is worth mentioning that some technological solutions are found in the literature, but they are not currently implemented or planned by Italian companies. There could be different reasons that leads to this behavior. For instance, some technologies are emerging (e.g., 3D printing for core and mold making), which leads to uncertainty related to their application. Indeed, the costs and benefits of the emerging technology are not always clear, or the companies could be reluctant to implement a technology that is not well-known. Another reason for a low application level could be that a given technological solution is obsolete; thus, companies are not considering it anymore. Another possible reason for a scarce implementation could be that certain technologies could be disruptive for the process and could lead to major changing. As a result, there could be a quite strong resistance to change. Finally, the existing energy efficiency incentive schemes are very likely to have a role in influencing what technological solutions are more often adopted or planned: the policy coverage would evolve over time, including more promising technologies.

5. Conclusions

Given the uncertainties and difficulties that arise when planning energy-saving investments in a foundry plant, a comprehensive analysis on the interventions implemented and planned by 231 Italian manufacturing sites was conducted in this study. The in-depth study involved the EAs provided by the Italian foundry companies, along with foundry expert judgements, leading to obtain an overview of the current Italian developments on energy-saving investments. Indeed, the frequencies related to the implemented interventions provide the past trends, while the frequencies associated with the planned interventions give a hint on future trends. Moreover, as a further step of analysis, the economic data reported by the EAs were examined to determine whether there is a relationship between the adopted technological interventions and economic indicators.

The results of this study pointed out that the companies lean towards investments on the auxiliary systems and the heat recovery systems, while the melting phase has attracted most of the efforts among the process phases. Specifically, the most adopted intervention was the replacement of compressors with more efficient ones, while the most planned intervention is reducing the leakages of the air distribution systems. Among the most popular interventions that the companies are willing to implement, it is worth mentioning the following ones: replacing electric motors and furnaces with more efficient ones, installing cogeneration systems and installing or replacing the inverters. From an economic and energy-efficiency perspectives, it is possible to state that the companies prefer investing in technologies characterized by a short PBT and a decent cost-effectiveness indicator. The cost-effectiveness indicator represents the investment cost for each toe of energy saved; thus, it is strongly related to the reduction of energy consumption. The aforementioned trend is also related to the easiness of implementation of some solutions compared to others. Indeed, considering the interventions related to the heat recovery systems, it is possible to state that they guarantee the best cost-effectiveness indicators and generally higher energy-savings compared to the interventions related to pressure systems, compressors and engines. However, they are characterized by some limitations with regards to implementation.

Another fundamental outcome of this paper is the list of technologies, which have been validated by Italian foundry experts. Indeed, this output represents an up-to-date guideline for companies who are conducting a screening analysis of the possible energy-saving solutions. In other words, the aforementioned list could be exploited as a preliminary decision support tool.

During this work, energy-efficiency strategies related to lighting, heating of offices, and installation of sensors, along with the adoption of proper managerial practices, were not considered. Thus, the presented study could not account for all the possible solutions that can be adopted in a foundry plant to reduce energy consumption.

Further developments could include the analysis of other industrial sectors subjected to the EED, along with duplicating the analysis for the foundry industry of other countries. Indeed, different types of manufacturing plant could have distinct needs, leading to different choices related to energy-saving investments. Additionally, both the prices of electric energy and fuels could vary from nation to nation, resulting in diverse economic opportunities. Finally, another interesting future development could be repeating the study when the next EAs are produced to check whether the companies adopted the planned interventions and point out whether new opportunities have been identified.

Author Contributions: Conceptualization, A.C., L.L. and F.D.C.; methodology, A.C., L.L. and F.D.C.; software, C.T., C.M., M.S., F.M., A.C. and L.L.; validation, A.C., L.L., F.D.C., C.T., F.M., M.S. and C.M.; formal analysis, A.C., L.L. and C.M.; investigation, A.C., L.L. and C.M.; resources, C.M. and M.S.; data curation, A.C., L.L. and C.M.; writing—original draft preparation, L.L.; writing—review and editing, A.C., F.D.C., C.T., C.M., F.M. and M.S.; visualization, L.L.; supervision, F.D.C.; project administration, F.D.C., F.M. and M.S.; and funding acquisition, M.S. and F.M. All authors have read and agreed to the published version of the manuscript.

Funding: This work is part of the Electrical System Research (PTR 2019–2021), implemented under Programme Agreements between the Italian Ministry for Economic Development and ENEA, CNR, and RSE S.p.A.

Institutional Review Board Statement: Not applicable.

Informed Consent Statement: Not applicable.

Data Availability Statement: Data is contained within the article.

Acknowledgments: The authors gratefully acknowledge the technical support provided by the Italian Foundry Association Assofond.

Conflicts of Interest: The authors declare no conflict of interest.

Appendix A

Table A1. Technological energy-saving solutions for molding obtained from both the literature and EAs.

Molding					
Process Machinery	**Solution Object**	**Energy-Saving Approach**	**Energy-Saving Technological Solution**	**Reference**	**Comments from Sector Experts**
Molding machinery	Sand recovery system	Innovation	Installing machine capable of retrieving slag from furnace and using it as sand mold.	[43]	
Molding machinery	Sand recovery system	Re-source	*Replacing the electric sand recovery system with a gas one*	Energy Audits	This solution is in total contrast with the decarbonization requirments.
Molding machinery	Sand recovery system	Innovation	*Installing a sand recovery system*	Energy Audits	This solution generates an increase of the energy consumption; however, it could help mitigating the environmental impact. Indeed, recovering sand leads to lower level of sand waste and disposal.
Molding machinery	Molding station	Replace	*Replacing the molding stations with more efficient ones*	Energy Audits	
Molding machinery	Mixer	Replace	*Replacing the mixing systems with a more efficient one*	Energy Audits	
Molding machinery	"Hot box" molding	Innovation	*Installing a preheating system or regenerative or recuperative burners for the furnace tasked with the production of the sand mold*	[29]	
Molding machinery	Molding machinery	Replace	*Installing an efficient filter for the sand molding process*	Energy Audits	The benefits of this solution should be evaluated considering the compressed air required to clean the filter
Molding machinery	Molding machinery	Innovation	Installing a infrared system for drying the mold	[8,17]	This solution is adopted by some Italian foundry plants. It is exploited for both the refractory material of the mold and the coating of the sand mold
Molding machinery	Molding machinery	Innovation	*Installing low-pressure casting machine capable of using inorganic cores*	Energy Audits	
Molding machinery	Molding machinery	Replace	*Replacing the existing machinery with newer and more efficient ones*	[8]	
Molding machinery	3D Printer	Innovation	Installing a 3D Printer (Jet Binding) for making molds	[15,44]	
Molding machinery	3D Printer	Innovation	Installing a 3D Binder Jetting Printer for making cores. The 3D Printer blends the sand through a binder agent	[16]	Emerging techology; thus, its energy consumption should be evaluated more in-depth. More benefits for low production volumes.
Molding machinery	3D Printer	Innovation	Installing 3D Printer for making cores	[16]	Emerging techology; thus, its energy consumption should be evaluated more in-depth. More benefits for low production volumes.

Table A2. Technological energy-saving solutions for casting obtained from both the literature and EAs.

Casting					
Process Machinery	**Solution Object**	**Eenergy-Saving Approach**	**Energy-Saving Technological Solution**	**Reference**	**Comments from Sector Experts**
Casting machinery	Casting machinery	Innovation	Installing a Vacuum Suction Casting Technology	[15]	This solution allows to obtain a better surface roughness, leading to a lower energy consumption during the finishing phase. However, it is an emerging technology and it could lead to high energy consumption due to the requirments of mantaining the vaccum conditions. Before implementing this solution it is recommended to evaluate the energy consumption increase during the casting phase and the energy consumption decrease of the finishing phase.
Casting machinery	Cooling system	Innovation	Adopting the "quench casting" technique for cooling the cast	[18]	
Casting machinery	Cooling system	Innovation	Adopting the "splash casting" technique for cooling the cast	[18]	
Casting machinery	Cooling system	Innovation	*Installing a forced ventilation cooling system*	Energy Audits	
Casting machinery	Ladle	Innovation	*Installing machines capable of scheduling an efficient preheat of the ladle*	[8,38]	It is usually performed without considering energy-saving opportunities. Large margin of potential and rationalization.
Casting machinery	Ladle	Innovation	Using lid for ladle to reduce the heat loss	[8]	It is a very useful solution with large margin of potential and rationalization.
Casting machinery	Ladle	Innovation	Adopting coating material for the ladle to reduce the heat loss	[8]	It is a very useful solution with large margin of potential and rationalization
Casting machinery	Ladle	Replace	Installing ladle with a more pointed outled to reduce the porosity of the cast. It also assures a lower thermal dispersion.	[45]	It is not a common technological solution
Casting machinery	Ladle	Innovation	Installing an automatic ladle pouring system	[35]	
Casting machinery	Casting furnace	Replace	*Replacing the casting furnace with a more efficient and newer one*	Energy Audits	
Casting machinery	Casting machinery	Replace	*Replacing the casting machineries with more efficient and newer ones*	[46]	
Casting machinery	Die-casting machinery	Replace	*Installing a new efficient die-casting line*	Energy Audits	
Casting machinery	Die-casting machinery	Replace	*Replacing the furnace where the cast waits to be poured in the mold with a more efficient and newer one*	Energy Audits	
Sand removal machinery	Sand removal machinery	Replace	*Replacing the machinery in charge of the sand removal process with more efficient and newer ones*	[8]	

Table A3. Technological energy-saving solutions for finishing obtained from both the literature and EAs.

Finishing					
Process Machinery	**Solution Object**	**Energy-Saving Approach**	**Energy-Saving Technological Solution**	**Reference**	**Comments from Sector Experts**
Finishing station	Robot	Innovation	Installing a robot for finishing	[47]	The automatic operations allows to both reduce cost and save time. Moreover, they limit human interventions leading to a safer process.
Finishing station	Finishing station	Replace	*Installing a new efficient finishing line*	Energy Audits	
Finishing station	Finishing station	Replace	*Replacing old finishing machineries with more efficient and modern ones*	[8,24]	
Finishing station	Finishing station	Replace	*Installing high efficiency nozzles*	Energy Audits	
Heat Treatment Furnace	Heat Treatment Furnace	Replace	*Replacing the heat tratment furnace with a more efficient and modern one*	Energy Audits	

Table A4. Technological energy-saving solutions for auxiliary systems obtained from both the literature and EAs.

Auxiliary Systems					
Process Machinery	**Solution Object**	**Energy-Saving Approach**	**Energy-Saving Technological Solution**	**Reference**	**Comments from Sector Experts**
Compressors	Compressors	Replace	*Replacing the compressor with more modern and efficient ones*	[28]	It is a very good technological solution to reduce energy consumption
Compressors	Exchanger	Innovation	Improving the air quality through the introduction of an exchanger located at the compressor outlets	[48]	
Compressors	Filter	Innovation	Improving the air quality through the introduction of a filter located at the compressor outlets to remove volatile substances	[48]	
Compressors	Filter	Innovation	Improving the air quality through the introduction of a filter to remove oil from compressed air	[48]	
Compressors	Injection pump	Innovation	Installing a pump for better injecting and sparing the lubrificant	[48]	
Compressors	Compressors	Innovation	Installing technologies capable of isolating some sections of the system that requires specific values for the pressure of the air	[49]	
Compressors	Compressors	Replace	Optimize the size of the compressors	[49]	
Compressors	Compressors	Innovation	Installing an intercooler for the compressors	[28,50]	
Compressors	Compressors	Innovation	*Installing variable speed compressors (i.e., compressor with inverter)*	[49]	
Compressors	Compressors	Replace	Installing efficient induction motors	[49]	

Table A4. *Cont.*

Auxiliary Systems					
Process Machinery	**Solution Object**	**Energy-Saving Approach**	**Energy-Saving Technological Solution**	**Reference**	**Comments from Sector Experts**
Compressors	Compressors	Innovation	Replacing the compressors needed for creating vacuum with vacuum pumps	[49]	
Pressure systems	Pressure systems	Innovation	*Replacing all the equipment for pressurized air distribution with electric or idraulic devices (if possible)*	[49]	It is a very good technological solution to reduce energy consumption
Pressure systems	Pressure systems	Innovation	Replacing the compressor with fans, blowers or other alternative solutions (if possible)	[51]	
Lubrification system	Lubrification system		Installing technologies for electrostatic lubrification to decrease the oil contained in the air	[34]	
Suction Systems	Suction Systems	Innovation	Placed the suction systems as close as possible to the sources	[34]	It is a very good technological solution to reduce energy consumption
Suction Systems	Suction Systems	Innovation	Installing defrosters to remove condensation drops	[34]	
Suction Systems	Suction Systems	Innovation	Installing electrostatic precipitator to remove dust	[34]	
Suction Systems	Suction Systems	Innovation	Installing appropriate systes to reduce the emissions	[36]	
Suction Systems	Suction Systems	Replace	*Installing high efficiency fans*	[8]	
Suction Systems	Suction Systems	Replace	Installing fans with proper size and power	[8]	
Suction Systems	Suction Systems	Innovation	*Installing variable speed fans (i.e., fans with inverter)*	[8]	
Suction Systems	Suction Systems	Innovation	*Installing a forced air suction system for the furnace*	Energy Audits	
Transport Systems	Forklift	Replace	*Replacing the forklifts with more efficient and modern ones*	[52]	
Transport Systems	Forklift's battery	Replace	*Replacing the batteries of the forklifts*	[52]	
Transport Systems	Transport Systems	Replace	*Replacing conveyor belts with more efficient and modern ones*	Energy Audits	
Transport Systems	Transport Systems	Innovation	Installing appropriate covers and roofings for the material transportation systems	[36]	
Transport Systems	Transport Systems	Innovation	Installing frequency converters with controlled speed for the transportation systems	[53]	
Transport Systems	Transport Systems	Innovation	Replacing pneumatic, chain and screw transport systems with belt conveying systems	[53]	
Transport Systems	Transport Systems	Replace	Replacing poly-v belts with more efficient and modern ones	[54]	
Transport Systems	Transport Systems	Innovation	*Installing high efficiency belt and replace V-belts with toothed belts*	[8]	
Transport Systems	Transport Systems	Innovation	*Replacing V-belts with helicoidal belts*	Energy Audits	

Table A4. *Cont.*

Auxiliary Systems					
Process Machinery	**Solution Object**	**Energy-Saving Approach**	**Energy-Saving Technological Solution**	**Reference**	**Comments from Sector Experts**
Transport Systems	Lift truck	Replace	*Replacing the lift trucks with more efficient and modern ones*	Energy Audits	
Transport Systems	Crane	Replace	*Replacing the cranes with more efficient and modern ones*	Energy Audits	
Electricity Transformers	Electricity Transformers	Replace	*Replacing the electricity transformers with more efficient and modern ones*	Energy Audits	
Inverters	Inverters	Innovation	*Installing inverters on eletric motors or replacing the inverters with more efficient and modern ones*	[1]	
Electric Plant	Electric Plant	Innovation	*Installing passive filters*	[34]	
Engines	Engines	Replace	*Installing high efficiency electric motors (class IE2, IE3 and IE4)*	[28,54]	It is recommended for all the motors with high working hours.
Engines	Engines	Replace	Installing electric motors correctly sized in relation to the power required by the system	[54]	
Engines	Engines	Innovation	*Installing regenerative electric motors*	Energy Audits	
Engines	Engines	Innovation	Installing motors with low starting current	[55]	It assures lower consumption and less dependence on the electricity network.
Engines	Engines	Replace	*Rewinding electric motors*	[8,55]	
Engines	Engines	Replace	Rewiring the engines	[8]	
Engines	Engines	Innovation	*Installing variable speed motors (i.e., motors with inverter)*	[55]	
Fluid distribution systems	Fluid distribution systems	Replace	Installing a closed circuit cooling system	[36]	
Fluid distribution systems	Fluid distribution systems	Innovation	*Optimize the pipelines' design leakage along with installing appropriate seal to minimize air leakage*	[49,56]	
Fluid distribution systems	Fluid distribution systems	Replace	Installing a proper insulation for the fluid distribution system to reduce heat loss	[8]	
Pumps	Pumps	Replace	*Replacing the pumps with more efficient and modern ones*	[8]	
Pumps	Pumps	Replace	Installing pumps correctly sized	[8]	
Pumps	Pumps	Replace	Trim the impellers of the pumps to reduce the energy consumption	[8]	
Pumps	Pumps	Innovation	Installing variable speed pumps (i.e., pumps with inverter)	[8]	
Cooling Systems	Cooling Systems	Replace	*Replacing the cooling towers with more efficient and modern ones*	Energy Audits	
Cooling Systems	Cooling Systems	Innovation	Installing technologies able to reuse the condensation of the cooling towers	[8]	

Table A5. Technological energy-saving solutions for heat recovery systems obtained from both the literature and EAs.

Heat Recovery Systems					
Process Machinery	**Solution Object**	**Energy-Saving Approach**	**Energy-Saving Technological Solution**	**Reference**	**Comments from Sector Experts**
Furnace	Heat Recovery Systems	Recover	Installing a heat recovery system to retrieve heat from the slag	[57]	
Furnace	Heat Recovery Systems	Recover	Installing a preheater for the combustion air	[36]	
Furnace	Heat Recovery Systems	Recover	Installing a regenerator to retrieve heat from exhaust gas	[35]	
Furnace	Heat Recovery Systems	Recover	Installing a recuperator to retrieve heat from exhaust gas	[35]	
Furnace	Heat Recovery Systems	Innovation	Installing Carbon Capture and Utilization (CCU) technology for capturing the CO_2 contained in the exhaust gas and exploit it for other purposes	[58]	This is an emerging solution; thus, it is not easily implementable for small and medium enterprises
Rankine turbine	Heat Recovery Systems	Recover	*Installing a Rankine turbine to generate electric energy through the exhaust gas*	[34,38]	
Combined cycle	Heat Recovery Systems	Recover	Installing technologies able to exploits exhaust gas to produce electric energy through a combined cycle	[59]	It is usually adopted for gas cupola furnace
ORC turbine	Heat Recovery Systems	Recover	*Installing a ORC turbine to generate electric energy through the exhaust gas*	[8,38,56]	It could be adopted for medium cupola furnaces, which are exploited for at least two shifts
Burner	Heat Recovery Systems	Innovation	Installing a burner able to capture organic volatile substances to burn them and produce heat	[5]	This is an interesting technology in case the post-combustion could be totally fueled with the volatile substances, otherwise methane is required, leading to economic unsustainability
Cogeneration	Heat Recovery Systems	Recover	*Installing cogeneration or trigeneration technologies*	[8]	This solution is not easily implementable in a foundry plant
Refrigeration cycle	Heat Recovery Systems	Recover	*Installing technologies able to exploits the exhaust gas for a refrigeration cycle*	[60]	
Battery	Heat Recovery Systems	Replace	*Replacing the batteries of the heat recovery systems*	[61]	
Evaporator	Heat Recovery Systems	Recover	*Installing an evaporator to retrieve heat from the emulsified water*	Energy Audits	It is adopted for the nonferrous metals produced through die-casting
Sand drying system	Heat Recovery Systems	Recover	*Installing technologies for the sand drying process*	Energy Audits	
Furnace	Heat Recovery Systems	Recover	*Installing an exchanger to retrieve heat from the exhaust gas and generate hot water for the drier*	Energy Audits	

Table A5. *Cont.*

Heat Recovery Systems					
Process Machinery	Solution Object	Energy-Saving Approach	Energy-Saving Technological Solution	Reference	Comments from Sector Experts
Furnace	Heat Recovery Systems	Recover	Installing a turbine to retrieve heat from the high pressure of the furnace	[24,57]	
Storage Tank	Heat Recovery Systems	Recover	Installing a heat storage tank	[56]	
Compressor	Exchanger	Recover	*Installing an exchanger to retrieve heat from the compressors*	[49,56]	
Ladle	Exchanger	Recover	*Installing an exchanger to retrieve heat and preheat the ladle*	[51]	It is not a common technology; however, it has a large margin of potential and rationalization
Casting machinery	Cooling system	Recover	*Installing a heat recovery system from the cooling process of the molds*	Energy Audits	

References

1. Noro, M.; Lazzarin, R. Energy audit experiences in foundries. *Int. J. Energy Environ. Eng.* **2016**, *7*, 409–423. [CrossRef]
2. Corelli, G. La Razionalizzazione dei Processi di Fonderia. Available online: https://www.castingitaly.com/ (accessed on 22 April 2021).
3. Kermeli, K.; Deuchler, R.; Worrell, E.; Masanet, E.R. Energy Efficiency Improvement and Cost Saving Opportunities for Metal Casting: An ENERGY STAR®Guide for Energy and Plant Managers. Environmental Protection Agency: Washington, DC, USA. 2016. Available online: https://www.energystar.gov/sites/default/files/tools/ENERGY%20STAR%20Metal%20Casting%20Energy%20Guide.pdf (accessed on 1 December 2021).
4. Thollander, P.; Backlund, S.; Trianni, A.; Cagno, E. Beyond barriers–A case study on driving forces for improved energy efficiency in the foundry industries in Finland, France, Germany, Italy, Poland, Spain, and Sweden. *Appl. Energy* **2013**, *111*, 636–643. [CrossRef]
5. Sa, A.; Paramonova, S.; Thollander, P.; Cagno, E. Classification of industrial energy management practices: A case study of a Swedish foundry. *Energy Procedia* **2015**, *75*, 2581–2588. [CrossRef]
6. Thollander, P.; Ottosson, M. Energy management practices in Swedish energy-intensive industries. *J. Clean. Prod.* **2010**, *18*, 1125–1133. [CrossRef]
7. Vinci, G.; D'Ascenzo, F.; Esposito, A.; Musarra, M.; Rapa, M.; Rocchi, A. A sustainable innovation in the Italian glass production: LCA and Eco-Care matrix evaluation. *J. Clean. Prod.* **2019**, *223*, 587–595. [CrossRef]
8. Worrell, E.; Blinde, P.; Neelis, M.; Blomen, E.; Masanet, E. *Energy Efficiency Improvement and Cost Saving Opportunities for the US Iron and Steel Industry an ENERGY STAR (R) Guide for Energy and Plant Managers*; Lawrence Berkeley National Lab. (LBNL): Berkeley, CA, USA, 2010.
9. Pastor-López, I.; Santos, I.; Santamaría-Ibirika, A.; Salazar, M.; De-la-Peña-Sordo, J.; Bringas, P.G. Machine-learning-based surface defect detection and categorisation in high-precision foundry. In Proceedings of the 2012 7th IEEE Conference on Industrial Electronics and Applications (ICIEA), Singapore, 18–20 July 2012; pp. 1359–1364.
10. Assofond. Assofond—Il Processo di Fonderia. Available online: https://www.assofond.it/il-processo-di-fonderia(accessed on 22 April 2021).
11. Arasu, M.; Jeffrey, L.R. Energy consumption studies in cast iron foundries. In Proceedings of the Transactions of 57th IFC, Kolkata, India, 13–15 February 2009; pp. 331–336.
12. Prevention, I.P. *Control Reference Document on Best Available Techniques in the Smitheries and Foundries Industry*; European Commission: Bruxelles, Belgium, 2005.
13. Sinha, N.K.; Choudhary, I.; Singh, J. Influence of Mold Material on the Mold Stability for Foundry Use. *Silicon* **2021**, 1–10. [CrossRef]
14. Hamasaiid, A.; Dargusch, M.; Davidson, C.; Tovar, S.; Loulou, T.; Rezai-Aria, F.; Dour, G. Effect of mold coating materials and thickness on heat transfer in permanent mold casting of aluminum alloys. *Metall. Mater. Trans. A* **2007**, *38*, 1303–1316. [CrossRef]
15. Rodríguez-González, P.; Robles Valero, P.; Fernández-Abia, A.; Castro-Sastre, M.; Barreiro García, J. Application of Vacuum Techniques in Shell Moulds Produced by Additive Manufacturing. *Metals* **2020**, *10*, 1090. [CrossRef]
16. Lynch, P.; Hasbrouck, C.; Wilck, J.; Kay, M.; Manogharan, G. Challenges and opportunities to integrate the oldest and newest manufacturing processes: Metal casting and additive manufacturing. *Rapid Prototyp. J.* 2020. [CrossRef]

17. Wang, J.; Sama, S.R.; Manogharan, G. Re-Thinking Design Methodology for Castings: 3D Sand-Printing and Topology Optimization. *Int. J. Met.* **2019**, *13*, 2–17. [CrossRef]
18. Acar, S.; Guler, K.A. A Preliminary Study Upon the Applicability of the Direct Water Cooling with the Lost Foam Casting Process. *Int. J. Met.* **2021**, *15*, 88–97. [CrossRef]
19. Lee, J.; Lee, Y.C.; Kim, J.T. Migration from the traditional to the smart factory in the die-casting industry: Novel process data acquisition and fault detection based on artificial neural network. *J. Mater. Process. Technol.* **2021**, *290*, 116972. [CrossRef]
20. Patange, G.; Khond, M. Some studies on energy consumptions and identification of suitable energy management techniques in Indian foundry industries. *Eur. Sci. J.* **2013**, *9*, 241–252.
21. Price, L.; Sinton, J.; Worrell, E.; Phylipsen, D.; Xiulian, H.; Ji, L. Energy use and carbon dioxide emissions from steel production in China. *Energy* **2002**, *27*, 429–446. [CrossRef]
22. Lin, B.; Wang, X. Promoting energy conservation in China's iron & steel sector. *Energy* **2014**, *73*, 465–474.
23. Wei, Y.-M.; Liao, H.; Fan, Y. An empirical analysis of energy efficiency in China's iron and steel sector. *Energy* **2007**, *32*, 2262–2270. [CrossRef]
24. Guo, Z.; Fu, Z. Current situation of energy consumption and measures taken for energy saving in the iron and steel industry in China. *Energy* **2010**, *35*, 4356–4360. [CrossRef]
25. Zhang, J.; Wang, G. Energy saving technologies and productive efficiency in the Chinese iron and steel sector. *Energy* **2008**, *33*, 525–537. [CrossRef]
26. Ozawa, L.; Sheinbaum, C.; Martin, N.; Worrell, E.; Price, L. Energy use and CO2 emissions in Mexico's iron and steel industry. *Energy* **2002**, *27*, 225–239. [CrossRef]
27. Chan, D.Y.-L.; Yang, K.-H.; Lee, J.-D.; Hong, G.-B. The case study of furnace use and energy conservation in iron and steel industry. *Energy* **2010**, *35*, 1665–1670. [CrossRef]
28. Lazzarin, R.M.; Noro, M. Energy efficiency opportunities in the service plants of cast iron foundries in Italy. *Int. J. Low-Carbon Technol.* **2017**, *12*, 96–109. [CrossRef]
29. Lazzarin, R.M.; Noro, M. Energy efficiency opportunities in the production process of cast iron foundries: An experience in Italy. *Appl. Therm. Eng.* **2015**, *90*, 509–520. [CrossRef]
30. Hasanbeigi, A.; Price, L.; Lin, E. Emerging energy-efficiency and CO2 emission-reduction technologies for cement and concrete production: A technical review. *Renew. Sustain. Energy Rev.* **2012**, *16*, 6220–6238. [CrossRef]
31. Cantini, A.; Leoni, L.; De Carlo, F.; Salvio, M.; Martini, C.; Martini, F. Technological Energy Efficiency Improvements in Cement Industries. *Sustainability* **2021**, *13*, 3810. [CrossRef]
32. European Parliament and Council, Energy Efficiency Directive 2012/27/EU (EED). 2012. Available online: https://eur-lex.europa.eu/LexUriServ/LexUriServ.do?uri=OJ:L:2012:315:0001:0056:en:PDF (accessed on 5 December 2021).
33. Enright, N. The resource EFFICIENCY Hierarchy. Available online: https://www.sustainsuccess.co.uk/the-resource-efficiency-hierarchy (accessed on 23 November 2021).
34. Roudier, S.; Sancho, L.D.; Remus, R.; Aguado-Monsonet, M. *Best Available Techniques (BAT) Reference Document for Iron and Steel Production: Industrial Emissions Directive 2010/75/EU: Integrated Pollution Prevention and Control*; Publications of the European Union Luxembourg: Luxembourg, 2013.
35. BCS Incorporated. *Advanced Melting Technologies: Energy Saving Concepts and Opportunities for the Metal Casting Industry*; BCS Incorporated: Boston, MA, USA, 2005.
36. Cusano, G.; Gonzalo, M.; Farrell, F.; Remus, R.; Roudier, S.; Sancho, L. *Best Available Techniques (BAT) Reference Document for the Non-Ferrous Metals Industries-Industrial Emissions Directive 2010/75/EU (Integrated Pollution Prevention and Control)*; Publications of the European Union: Luxembourg, 2017.
37. Pardo, N.; Moya, J.A. Prospective scenarios on energy efficiency and CO_2 emissions in the European Iron & Steel industry. *Energy* **2013**, *54*, 113–128.
38. Eronen, S.; Tapola, M.; Svensson, E.; Sommarin, P.; Helber, J.; Piatek, J.; Przybylski, J.; Caballero, P.; Kemppainen, P. *Improving the Energy Efficiency of Foundries in Europe Foundrybench—Foundry Energy Efficiency Benchmarking Project (IEE/07/585/SI2.500402)*; European Commission: Bruxelles, Belgium, 2012.
39. Grunbaum, R. Voltage source converters for maintaining of power quality and stability in power distribution. Proceedings of 2005 European Conference on Power Electronics and Applications, Dresden, Germany, 11–14 September 2005.
40. Sako, E.Y.; Orsolini, H.D.; Moreira, M.; de Meo, C.E.; Pelissari, P.I.; Salvini, V.R.; Pandolfelli, V.C. Thermal ceramic coatings as energy saving alternatives for high temperature processes. *Int. J. Appl. Ceram. Technol.* **2020**, *17*, 2492–2508. [CrossRef]
41. Sibanda, V.; Sipunga, E.; Danha, G.; Mamvura, T.A. Enhancing the flotation recovery of copper minerals in smelter slags from Namibia prior to disposal. *Heliyon* **2020**, *6*, e03135. [CrossRef]
42. Jezierski, J.; Janerka, K. Wastes utilisation in foundries and metallurgical plants with pneumatic conveying techniques–selected aspects. *Interdiscip. Environ. Rev.* **2011**, *12*, 154–165. [CrossRef]
43. Prabhushankar, N.; Balaji, N. Various Alternative Sources for Silica Sand, Binders and Additives in Sand Casting and their Properties—A Review. *IOP Conf. Ser. Mater. Sci. Eng.* **2020**, *993*, 012137. [CrossRef]
44. Sama, S.R.; Badamo, T.; Manogharan, G. Case studies on integrating 3d sand-printing technology into the production portfolio of a sand-casting foundry. *Int. J. Met.* **2020**, *14*, 12–24. [CrossRef]

45. Choi, J.; Hwang, H.; Kang, S. Effect of ladle outlet geometry on internal porosity in gravity casting automotive brackets: An experimental investigation. *China Foundry* **2020**, *17*, 56–60. [CrossRef]
46. Bonollo, F.; Gramegna, N.; Timelli, G. High-pressure die-casting: Contradictions and challenges. *JOM* **2015**, *67*, 901–908. [CrossRef]
47. Gruzman, V. Foundry production digitalization. *IOP Conf. Ser. Mater. Sci. Eng.* **2020**, *966*, 012127. [CrossRef]
48. Wrona, R.; Ziółkowski, E.; Smyksy, K.; Brzeziński, M. The quality of compressed air as the necessary condition the improving the process efficiency in foundry plants. *Arch. Foundry Eng.* **2013**, *13*, 107–111. [CrossRef]
49. Eras, J.J.C.; Gutiérrez, A.S.; Santos, V.S.; Ulloa, M.J.C. Energy management of compressed air systems. Assessing the production and use of compressed air in industry. *Energy* **2020**, *213*, 118662. [CrossRef]
50. Šarevski, M.N.; Šarevski, V.N. Thermal characteristics of high-temperature R718 heat pumps with turbo compressor thermal vapor recompression. *Appl. Therm. Eng.* **2017**, *117*, 355–365. [CrossRef]
51. Institute For Industrial Productivity. Explore Energy Efficiency Technologies across the Industrial Sectors. Available online: http://www.iipinetwork.org/wp-content/Ietd/content/institute-industrial-productivity.html (accessed on 11 January 2021).
52. Artal-Sevil, J.; Bernal-Agustín, J.; Dufo-López, R.; Domínguez-Navarro, J. Forklifts, Automated Guided Vehicles and Horizontal Order Pickers in Industrial Environments. Energy Management of an Active Hybrid Power System based on Batteries, PEM Fuel Cells and Ultracapacitors. *Renew. Energy Power Qual. J. (REPQJ)* **2017**. [CrossRef]
53. Pang, Y.; Lodewijks, G. Improving energy efficiency in material transport systems by fuzzy speed control. In Proceedings of the 3rd IEEE International Symposium on Logistics and Industrial Informatics, Budapest, Hungary, 25–27 August 2011; pp. 159–164.
54. Abdel-Hadi, A.; Salem, A.R.; Abbas, A.I.; Qandil, M.; Amano, R.S. Study of Energy Saving Analysis for Different Industries. *J. Energy Resour. Technol.* **2021**, *143*, 052101. [CrossRef]
55. Kurilin, S.; Denisov, V.; Dli, M.; Bobkov, V. Scientific and technical directions of improvement of electric motors for non-ferrous metallurgy. *Non-Ferr. Met.* **2019**, *47*, 53–58. [CrossRef]
56. Helber, J.S.M. *Good Practice Guide on Energy Saving POTENTIALS and Opportunities for foundries. Foundrybench—Foundry Energy Efficiency Benchmarking Project, (IEE/07/585/SI2.500402)*; European Commission: Bruxelles, Belgium, 2011.
57. Holappa, L. Energy efficiency and sustainability in steel production. In *Applications of Process Engineering Principles in Materials Processing, Energy and Environmental Technologies*; Springer: Cham, Switzerland, 2017; pp. 401–410.
58. Hammerschmid, M.; Müller, S.; Fuchs, J.; Hofbauer, H. Evaluation of biomass-based production of below zero emission reducing gas for the iron and steel industry. *Biomass Convers. Biorefinery* **2021**, *11*, 169–187. [CrossRef]
59. Ryzhkov, A.; Levin, E.; Filippov, P.; Abaimov, N.; Gordeev, S. Making more efficient use of blast-furnace gas at Russian metallurgical plants. *Metallurgist* **2016**, *60*, 19–30. [CrossRef]
60. Chen, L.; Yang, B.; Shen, X.; Xie, Z.; Sun, F. Thermodynamic optimization opportunities for the recovery and utilization of residual energy and heat in China's iron and steel industry: A case study. *Appl. Therm. Eng.* **2015**, *86*, 151–160. [CrossRef]
61. Zhang, Y.; Shi, Y.; Zhang, L.; Li, J.; Fu, Q.; Zhu, X.; Liao, Q. A fluidized-bed reactor for enhanced mass transfer and increased performance in thermally regenerative batteries for low-grade waste heat recovery. *J. Power Source* **2021**, *495*, 229815. [CrossRef]

energies

MDPI

Article

A Methodology for the Identification and Characterization of Low-Temperature Waste Heat Sources and Sinks in Industrial Processes: Application in the Italian Dairy Sector

Lorena Giordano * and Miriam Benedetti

Energy and Sustainable Economic Development (ENEA), Italian National Agency for New Technologies, 00123 Rome, Italy; miriam.benedetti@enea.it
* Correspondence: lorena.giordano@enea.it; Tel.: +39-063-04-834-301

Abstract: Waste heat recovery is considered as one of the most promising options to improve the efficiency and sustainability of industrial processes. Even though industrial waste heat is abundantly available and its utilization is not a new concept, the implementation rate of waste-heat recovery interventions in industrial facilities is still low, due to several real or perceived barriers. Foremost challenges are represented by technical, economic, financial and regulatory factors. An additional prominent barrier lies in the lack or incompleteness of information concerning the material and energy flows within the factories, and the types and characteristics of waste heat sources and possible sinks for their internal or external reuse. With the aim to overcome some of the information barriers and increase the willingness of companies to approach waste heat recovery and reuse, a methodology to map waste heat sources and sinks in industrial processes is proposed in this study. The approach here presented combines information from the most relevant publications on the subject and data gathered from the analysis of energy audits carried out by large and energy-intensive enterprises. In order to demonstrate its feasibility, the methodology was applied to the Italian dairy sector, because of its large energy consumption and its enormous potential for the utilization of low-temperature waste heat sources.

Keywords: energy efficiency; waste heat recovery; waste heat survey; dairy industry

Citation: Giordano, L.; Benedetti, M. A Methodology for the Identification and Characterization of Low-Temperature Waste Heat Sources and Sinks in Industrial Processes: Application in the Italian Dairy Sector. *Energies* **2022**, *15*, 155. https://doi.org/10.3390/en15010155

Academic Editors: Chiara Martini and Claudia Toro

Received: 15 November 2021
Accepted: 20 December 2021
Published: 27 December 2021

1. Introduction

Energy efficiency represents one of the key principles of the European energy policies, being considered as the most affordable and effective way to reduce greenhouse gases (GHG) emissions and energy dependency, increase the security of supply and strengthen the competitiveness and environmental sustainability of EU countries.

As part of the Clean Energy for all Europeans package [1], the revised Energy Efficiency Directive (EED) [2] set a new binding target of at least 32.5% energy efficiency by 2030, that would lead to a reduction in GHG emissions of about 45%. With the adoption of the European Green Deal in December 2019 [3], the European Commission raised the level of climate ambition, setting a new 2030 target of reducing GHG emissions by at least 55% (compared to 1990). To reach this goal, in July 2021, the European Commission has proposed a new directive on energy efficiency [4], raising the energy efficiency targets by at least 36–37% and 39–41% for primary and final energy consumption, respectively.

Despite a reduction of energy consumption driven by energy efficiency measures in the last years, industry is still one of the large energy users in EU, accounting for around 26% of the final energy consumption [5] and 21% of GHG emissions [6] in 2019. As highlighted in the Strategy for Energy System Integration [7] defined in the framework of the European Green Deal, space and process heating in industry is responsible for more than 60% of the energy demand. Even though waste heat integration and process efficiency optimization are widespread practices across the industrial sectors, a huge amount of the energy consumed by heating processes is still wasted in the form of heat. Thus, the recovery of such waste heat can have a significant impact in supporting the reduction of energy demand and the decarbonization of industrial processes.

In view of this, the Strategy for Energy System Integration [7] promoted the transition to a "more circular" energy system, where unavoidable waste streams are reused for energy purposes and synergies among enterprises are exploited.

Despite its technical and economic benefits, waste heat recovery still remains unexplored, due to the existence of several technical and non-technical barriers. Major technical barriers include low heat transfer rates, temporal, spatial and quality mismatch between sources and sinks of waste heat, process-specific constraints and inaccessibility. Non-technical barriers are represented by long payback periods due to high capital costs, the lack of economy of scale and proper subsidy policies. A further challenge is represented by informational barriers, namely the little confidence in waste heat recovery technologies and the lack or incompleteness of information regarding the material and energy flows within the industrial facilities, the origin, quality and quantity of waste heat sources and possible internal or external waste heat acceptors [8].

Challenges become even greater when considering the recovery from low (from 120 °C to 230 °C) and very-low (<120 °C) grade heat [9], which actually accounts for the greatest share of the overall theoretical (physical) waste heat production in EU industry (920 TWh) [10]. This is not only due to the low energy level, but also to the lack of maturity of certain suitable waste heat recovery technologies and the relatively limited knowledge on the subject compared to other energy efficiency measures.

Recently, a certain number of studies explored challenges and opportunities for low-temperature waste heat recovery in many respects. A research branch reviewed the most suitable technologies for capturing waste heat, depending on temperature level, type of industrial process responsible for the production of excess heat and type of waste heat utilization (direct use or heat conversion) [11–13]. In addition to the direct use of waste heat via heat-exchangers, attention was paid to heat conversion technologies for heat-upgrade (absorption, compression and chemical heat pumps) and electricity production (ORC, Kalina cycles, thermoelectric generators) [13–17]. A recent study revised the potential of using nanofluids to enhance the efficiency of heat transfer in heat-to-heat and heat-to-power technologies [18].

The technical and economic feasibility of recovering low-temperature waste heat was also investigated via experimental analyses and simulations models. Case studies mainly concerned the recovery of waste heat generated in iron and steel [19–22], ceramic [23–26], paper and pulp [27,28], textile [29–33] and food sectors [34–43].

A few studies focused on the analysis of potential industrial sources of low-grade waste heat. In this regard, waste heat sources were classified according to the temperature level and type of industrial sector in [13,44,45]. Data on temperature and mass flow rate of waste heat available per unit of mass of product were provided only for textile and paper sectors [46] and food industry, focusing on dairy, meat, canned fruit and vegetable processing facilities [47,48]. As regards to dairy sector, studies in [49,50] identified the potential sinks of waste heat, in the context of a wider investigation aimed at examining retrofit options for the energy efficiency enhancement of dairy processes.

The assessment of waste heat sources and the possible waste heat acceptors was also addressed in some recent works dealing with the development of tools for the simulation and optimization of waste heat recovery projects. In this regard, Simeone et al. [8] and Wooley et al. [51] built a decision support tool to evaluate the compatibility of waste heat sources and sinks, along with economic and environmental benefits arising from the integration of available heat exchanger technologies to recover waste heat within a manufacturing facility. Notably, the framework for the energy recovery assessment provides for a preliminary survey of waste heat sources and sinks, based on direct measurements via invasive or non-invasive devices. The European project Greenfoods [52,53] dealt with the definition of an energy audit and management tool for food and beverage industries, where waste heat sources are identified and preliminary quantified by knowing the energy inputs (natural gas, coal, electricity, etc.) and setting up mass and energy balances of manufacturing processes. The CE-HEAT project [54] implemented an on-line toolbox to perform a pre-feasibility analysis of different waste heat recovery options, assuming as input data the characteristics of waste heat sources in terms of type of emission, temperature and thermal power.

However, all these works [8,51–54] did not focus on the identification of waste heat sources released by the production processes, being regarded as input data or unknown data that need to be directly measured or indirectly evaluated by solving complex energy balances.

The literature review highlighted that there is a lack of comprehensive methodologies to systematically analyze industrial processes and identify the origin and the key characteristics of low-temperature waste heat sources and sinks. Furthermore, no methodology has attempted so far to identify waste heat sources and to relate them with potential heat sinks by using literature data for the preliminary assessment and large datasets collected through the analysis of energy audits for model refinement and validation.

This could represent a relevant leap forward an increased knowledge of waste heat recovery potentials and processes, thus facilitating the first approach of companies to this topic and reducing the amount of input data required to perform the feasibility analyses.

With the aim to fill this research gap, the current study presents a methodology for the identification and characterization of the whole set of waste heat streams within a production process and the potential waste heat acceptors, which can be effectively reproduced and applied to every industrial sector. The methodology will support stakeholders, such as academics and energy consultants, in accomplishing the assessment of waste heat sources, which represent the necessary preliminary step in evaluating the technical and the economic feasible waste heat potential within industrial processes.

Specifically, the methodology provides for the definition of a typical production process, the selection of process phases generating low-grade waste heat sources, the identification and thermodynamic characterization of waste heat sources at phase-level and, finally, the definition of possible sinks for waste heat reuse. This step-by-step approach is based on data from literature review and its validation via comparison with the information retrieved from the analysis of energy audits. Such information is periodically gathered by ENEA, the Italian National Agency for New Technologies, Energy and Sustainable Economic Development, in fulfilment of Italian Legislative Decree 102/2014 [55], which established mandatory energy audits for large and energy intensive companies every four years.

In order to assess its feasibility, the methodology was applied to some of the most common manufacturing processes of the Italian dairy industry, where low-grade waste heat is characterized by a huge potential, being generated from several process stages (pre-heating, pasteurization, sterilization, cooling, clean-in-place, etc.), in the form of steam condensate, hot water, cooling water and intermediate-product streams.

This work lies in the framework of a wider research project, funded by the Ministry of Economic Development through a three-year research plan named "Ricerca di Sistema Elettrico" ("Electric System Research"). The project aims to support industrial companies in the implementation of low-temperature waste heat recovery projects. Specifically, its main outcome will be an application software to identify industrial waste heat recovery opportunities and to select the most promising options, based on the evaluation and comparison of their energy, economic and environmental performances. Thus, the methodology proposed in this study will permit the development of a specific module of the software tool, containing data for a complete survey of waste heat sources and sinks related to three Italian industrial sectors (dairy, baked products and textile).

2. Methodology for the Identification and Characterization of Low-Temperature Waste Heat Sources within Industrial Processes

Industrial processes involve a set of unit operations to convert raw materials into finished parts or products. Each unit operation is a basic step, where a physical or a chemical transformation is performed, depending on the characteristics of inlet and outlet mass and energy flows. Waste heat energy streams are represented by outlet flows, in the form of hot air, hot exhaust gases, and hot liquids, that are released into the environment at temperatures high enough above the ambient temperature to permit the recovery of some fraction of their thermal energy for useful purposes.

As well-known, the production process of a given product can change significantly from one industry to another. This is because manufacturing plants have unique characteristics, in terms of size, structure and operating procedures, that affect the pattern of mass and energy flows.

A common approach to overcome this issue is to model a typical production process, allowing to generically analyze inlet and outlet mass and energy flows and identify waste heat sources within the process itself [56]. Depending on temperature level of waste heat sources and sinks, possible combinations and waste heat recovery technologies can be preliminary identified.

Using such approach, a step-by-step methodology was defined to accomplish a complete survey of waste heat sources and sinks within industrial processes. On the basis of data collected from technical manuals, reports and research papers, the methodology allows for the definition of a typical production process, the selection of process phases generating low-grade waste heat sources, the identification and thermodynamic characterization of waste heat sources at phase-level and, finally, the definition of possible sinks for waste heat reuse. As a result, two preliminary schemes are obtained: a process scheme and a scheme of low-temperature waste heat sources. Using data arising from the analysis of energy audits, the information collected in these schemes is validated, enriched, contextualized and, finally, merged together to form a new comprehensive scheme.

Furthermore, an additional scheme concerning waste heat sinks is generated by combining literature data with data arising from waste heat recovery projects envisaged in the framework of energy audits. The two schemes can be overlapped in order to obtain a complete description of the process, the related waste heat generated and the possible ways to reuse it.

Specifically, the methodology adopted for the definition of a "typical" production process scheme includes four main steps:

(a) Review of the technical literature related to the industrial sectors

In this step, the reference documents on the best available techniques (BREF), technical reports and scientific articles are gathered and analyzed to acquire preliminary information on the main types of production processes operated within the industrial sector of interest. Based on the type and amount of information available, production processes to be further investigated are identified.

(b) Identification of process schemes from literature

With reference to production processes defined at point (a), process schemes, indicating the set of process phases required to turn raw materials into the final product, are identified from the literature review.

(c) Comparison of process schemes from literature

Process schemes from literature concerning the same product are compared to verify the existence of any differences in terms of sequence and operating conditions of production phases.

(d) Definition of a scheme of a "typical" production process

Based on the analysis and comparison of the process schemes from literature, as indicated at points (b) and (c), a "typical" process scheme for each product is defined. The scheme indicates:

- the sequence of phases of a typical production unit, starting from the preparation of raw materials up to the manufacturing and finishing of the final product;
- the raw materials and the auxiliary services required (water, steam, compressed air, etc.);
- the phases generating low-grade waste heat.

After defining the "typical" process scheme of a certain product, a scheme of waste heat sources is obtained via a four-step procedure, as detailed below:

(e) Preliminary identification of waste heat sources

In the first step, phases within the "typical" process scheme generating low-grade waste heat (point (d)) are carefully analyzed. The aim is to carry out a preliminary survey of low-temperature waste heat sources, based on the analysis of the relative inlet and outlet material flows and energy vectors (natural gas, steam, hot water, air, etc.).

(f) Comparison and verification based on literature data

The potential waste heat sources identified at point (e) are compared with those indicated in literature studies, such as scientific articles focusing on waste heat recovery in industrial processes or technical manuals concerning the energy efficiency of industrial processes. The purpose is to validate the types of waste heat sources previously identified and to detect any additional low-grade heat loss not directly deducible from the analysis of the "typical" process scheme.

(g) Characterization of waste heat sources based on literature data

Using the information gathered from scientific articles and technical manuals concerning the efficiency of industrial processes, waste heat sources are characterized, depending on data availability, in terms of type of vector (exhaust gas, steam, condensate, etc.), temperature, pressure, mass flow rate and theoretical waste heat recoverable.

(h) Definition of a scheme of low-temperature waste heat sources

Integrating the data set characterizing the waste heat sources (temperature, pressure, flow rate, theoretical waste heat recoverable per unit of product, etc.), a preliminary scheme of low-temperature waste heat sources is defined. Depending on temperature level of waste heat sources, candidate energy sinks and waste heat recovery technologies are also preliminary identified on the basis of information gathered from the technical literature review.

Validation and Contextualization via Comparison with Data from Energy Audits and Identification of Possible Waste Heat Sinks

The energy audit is widely regarded as one of the most cost-effective instruments for analyzing energy flows and assessing energy consumption within enterprises or individual processes and exploring potential energy, cost and emission saving opportunities [57,58]. From a regulatory point of view, the energy audit is regulated by the Article 8 of EED [2], according to all large enterprises shall undertake an independent, cost-effective and high-quality energy audit at least every four years or they may implement an energy or environmental management system, which includes an energy audit. In compliance with the EU regulatory framework, Article 8 has been transposed into the Italian legislation via the Legislative Decree N. 102 of 4 July 2014 [55], recently updated by the Legislative Decree n. 73/2020 [59]. According to article 8 of such Decree, the energy audit obligation applies to large (as defined in [60]) and energy intensive industries, intended as the ones with large energy consumptions (electricity consumption more than 1 GWh/year) applying for a tax relief on part of the purchased energy. All the energy intensive enterprises are registered on the list of "Cassa per i servizi energetici ed ambientali" ("Fund for Energy and Environmental Services – CSEA"). The Italian energy audit program, including data gathering and sector analysis, is managed by ENEA, which has received on 31 December 2019, 11,172 energy audits of production sites related to 6434 companies.

In the framework of the proposed methodology, the analysis of energy audits carried out by enterprises in the industrial sectors of interest is aimed at retrieving information about product processing, types and characteristics of waste heat recovery sources at different process stages and case studies regarding the implementation of technologies or solutions for the reuse of waste heat.

The purpose of such investigation is twofold:

- validating, enriching and contextualizing at national level the schemes of product manufacturing process and waste heat sources generated as a byproduct;
- identifying possible waste heat acceptors other than those defined in technical literature.

Figure 1 summarizes the main phases of the methodology developed in this study to define the schemes of product manufacturing process and low-temperature waste heat sources.

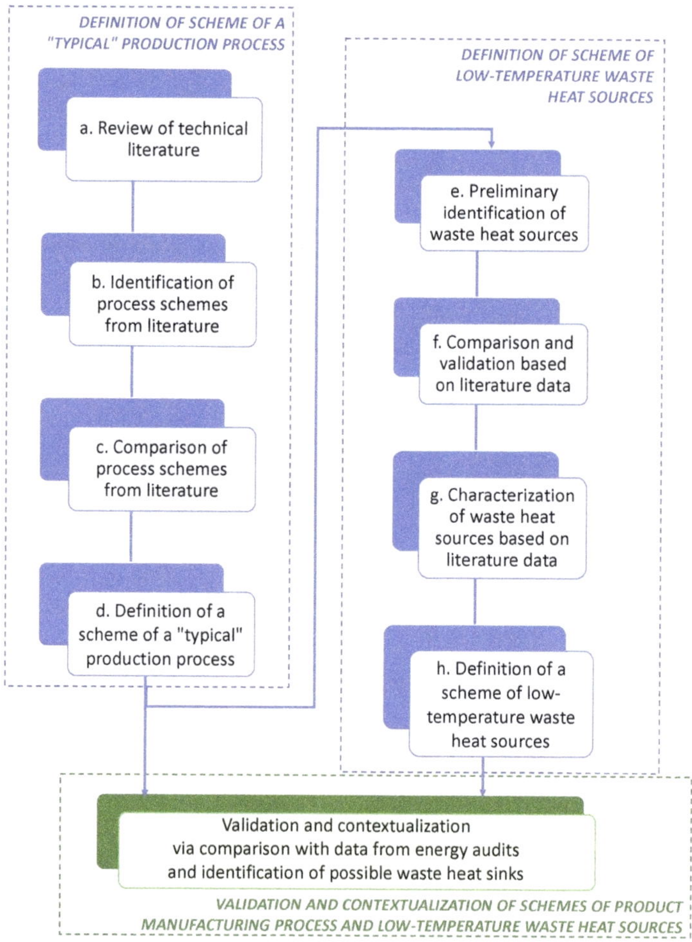

Figure 1. Methodology for the identification and characterization of the schemes of product manufacturing process (steps (**a**–**d**)) and low-temperature waste heat sources within the industrial processes (steps (**e**–**h**)).

3. Application of the Methodology to the Case of Dairy Sector

With the aim of demonstrating its feasibility, the methodology was applied to the case of dairy sector, taking as a reference the process of pasteurized milk production. The literature review allowed to identify documents containing information and process schemes regarding the pasteurized milk production. Attention was focused on the reference document on the best available techniques (BREF) for food, beverage and milk industry [61] and the study by Ramirez et al. [62], where two different process schemes were identified: the first one refers to the production process of pasteurized milk, while the second scheme outlines the process phases of the main dairy products, such as milk, powdered and condensed milk, yoghurt, butter and cheese. By analyzing and comparing these schemes, the set of stages involved in the pasteurized milk production and the temperature levels of each phase were defined. Reworking the information found in the literature, the scheme of a "typical" process for pasteurized milk production was defined. As shown in Figure 2, the new scheme:

- outlines the sequence of stages to convert raw milk into pasteurized milk;
- preliminarily identifies the main process phases responsible for low-temperature waste heat generation (indicated in black color in Figure 2), which include the phases of pasteurization and cooling of the pasteurized milk.

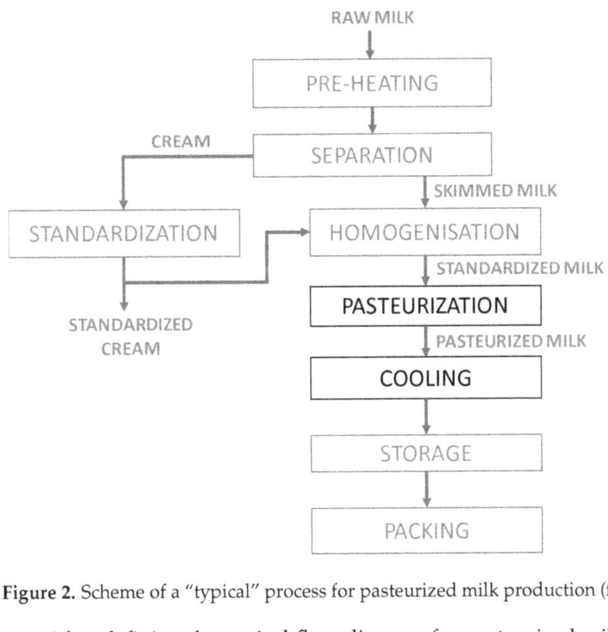

Figure 2. Scheme of a "typical" process for pasteurized milk production (from authors' elaboration).

After defining the typical flow diagram for pasteurized milk processing, a further investigation was carried out to characterize more comprehensively the process phases with low-temperature waste heat generation (pasteurization and cooling) and identify the corresponding waste heat streams.

Based on the analysis of mass and energy flows involved and the information found in [48,61–64], four low-temperature waste heat sources (S) were identified within the pasteurization phase, including:

- S1: the condensate of steam required for process water heating;
- S2: the hot process water at the pasteurizer outlet;
- S3: the pasteurizer overflow;
- S4: the water (or steam) used for the cleaning and sanitizing of the pasteurizer.

Information included in the technical literature allowed to define temperature ranges of all waste heat streams (S1–S4) [48,64]; data concerning the waste heat mass flow rate per unit of raw milk treated were found only for the pasteurizer overflow (S3) and the water (or steam) for the pasteurizer clean-up (S4) [48].

Regarding the cooling process of pasteurized milk, waste heat is released during the condensation process of the refrigerant fluid evolving within chillers for cooling water production (S5). The heat of condensation is normally removed via cooling water in a closed loop and then released into the air [65,66].

Combining the data previously collected, a preliminary scheme of low-temperature waste heat sources was outlined (Figure 3). Such diagram not only identifies and locates sources of low-temperature waste heat related to pasteurized milk processing, but it also provides a preliminary characterization of waste heat streams in terms of temperature and specific flow rates, depending on the information available in the literature.

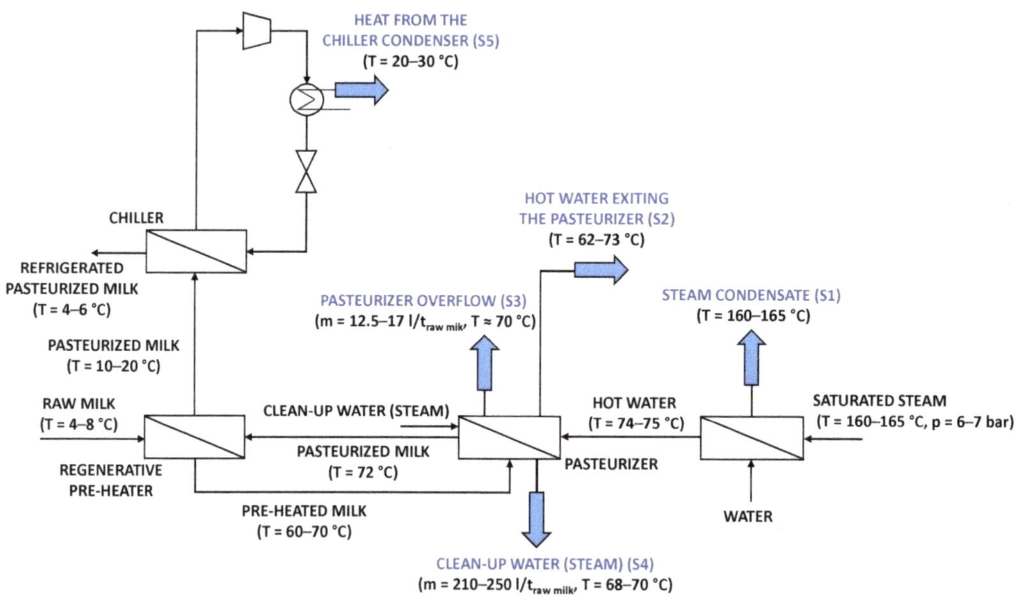

Figure 3. Scheme of low-temperature waste heat sources in pasteurized milk processing (from authors' elaboration).

3.1. Analysis of the Energy Audits of the Italian Dairy Sector

With the aim to validate, improve and contextualize the scheme of pasteurized milk production process and the corresponding scheme of low-temperature waste heat sources to the Italian dairy industry, energy audits carried out by large and energy-intensive enterprises in the year 2019, in compliance with the Legislative Decree N. 102 of 4 July 2014 [55], were examined. The paragraph outlines the main findings of this analysis, with a focus on waste heat sources identified within the manufacturing processes of dairy products and the waste heat recovery interventions envisaged by enterprises to improve their energy efficiency. It is important to emphasize that such information is rarely clearly stated in the energy audits, and the lexicon used by the enterprises is neither standardized nor always univocal. Therefore, the results presented in this paper were obtained by carefully analyzing and elaborating data retrieved from energy audits, which were then aggregated and presented in anonymous form.

3.1.1. Overview of the Italian Dairy Sector

According to the ATECO classification, the dairy sector consists of two main subsectors [67]: "10.51.10—Hygienic treatment of milk" and "10.51.20—Production of milk derivatives". The first sub-sector includes companies whose main activity is the production of fresh, pasteurized, sterilized, homogenized or ultra-high temperature (UHT) milk. The second sub-sector comprises companies involved in the production of milk-based beverages, cream made with fresh, pasteurized, sterilized, homogenized liquid milk, powdered or concentrated milk, butter, yoghurt, cheese, curd, whey, lactose and lactic ferments.

The study focused on large and energy-intensive businesses, which are obliged to undertake an energy audit in compliance with the Italian Legislative Decree 102/2014. Specifically, data provided by 79 dairy enterprises were deeply analyzed. Over 70% of companies in this sample (57) belong to the sub-sector 10.51.20, while the remainder to the sub-sector 10.51.10 (22).

The Italian dairy industry is not a uniform sector, since the production capacity can vary significantly among enterprises. In this regard, Table 1 shows the diversification of the production sites investigated in terms of annual production, varying the unit of measure adopted (tons, kilograms, litres or number of pieces). As an example, the production capacity expressed in t/year, which represents the most used unit of measure (60% of the overall sample), ranges from 450 t/year to approximately 400,000 t/year. A comparable dispersion is observed for the statistical distribution of annual productions in kg/year.

Table 1. Statistical distribution functions of the annual production of dairy enterprises investigated, varying the unit of measure.

	Unit of Measure of the Annual Production			
	t/Year	**($\times 10^3$) kg/Year**	**($\times 10^3$) L/Year**	**Pieces/Year**
Sample size	45	24	8	2
Min	450.0	42.9	4807.0	54,393.0
Lower quartile	10,905.8	3267.6	10,530.9	57,459.8
Median	33,258.7	12,531.0	18,540.8	60,526.5
Mean	60,358.2	51,622.4	20,127.4	60,526.5
Upper quartile	83,254.7	42,084.5	27,483.1	63,593.3
Max	399,319.0	400,896.2	41,943.6	66,660.0
Total	2,716,118.3	1,238,938.4	161,019.3	121,053.0

Milk processed in dairy industries is used to produce a wide variety of products: pasteurized milk, ultra-high-temperature milk, cream, butter, soft and hard cheese, yoghurt, ricotta, etc. Despite the differences in terms of production capacity and types of products, process steps involved in converting raw materials into finished products are quite similar. In this regard, Table 2 describes the sequence of unit operations required to produce some dairy products, defined on the basis of knowledge gathered from the analysis of energy audits. It is worth mentioning that the sequence of process stages for most of the products investigated has been elaborated by the authors of this paper combining all types of information available in the energy audits, including process schemes and data extrapolated by the descriptions of products manufacturing.

Table 2. Unit operations involved in manufacturing of some dairy products according to the analysis of energy audits of dairy enterprises.

Unit Operations	Butter (2 *)	Cream (3 *)	Hard Cheese (3 *)	Pasteurized Milk (2 *)	Yoghurt (3 *)
plant sterilization		X		X	
raw milk receiving			X	X	X
milk powder addition					X
filtration			X	X	
cooling and storage			X	X	
pre-heating			X	X	
cream separation/skimming			X	X	X ***
cream aging		X			
bactofugation			X		
homogeneization/ 1st homogenization				X	X ***
standardization				X	
cream receiving and storage	X				
heating and skimming	X				
regeneration	X				
cooling in storage tank					
1st pre-heating					
2nd pre-heating					
1st pasteurization/ pasteurization	X	X	X	X	X
sterilization					
thermization **					X
storage **					X
thermal treatment **					X
concentration					X
cooling/1st cooling	X	X	X	X	X
whey starter and curd addition			X		
cooking			X		
coagulation			X		
curd breaking and whey separation			X		
shaping			X		
moulds turning			X		
resting (hot chamber)			X		
transport to cold chamber			X		
salting			X		
drying and ripening			X		
washing and surface treatment			X		
2nd pasteurization	X				
steam separation (via degasser)	X				
2nd homogenization					
starter culture inoculation and aging	X				X
2nd cooling					X
churning	X				
storage in agitated tank		X			
fruits addition					X
handling and packing	X	X	X	X	X
storage	X		X		X
clean-in-place	X	X	X	X	X

* number of energy audits analyzed, ** only for processes without pasteurization; *** after pasteurization or thermal treatment.

As regard to the energy demand, natural gas is by far the most widely used fuel for the production of steam and hot water for several uses (i.e., product manufacturing, space heating, hot water production or the direct use in canteen stoves) and thus it is also the main responsible for the production of waste heat within dairy facilities. In this regard, Figure 4 shows the distribution function of the natural gas consumption of the entire sample of dairy enterprises. The yearly demand for natural gas is lower than or equal to 2100 kSm^3

for more than 80% of enterprises, while it ranges from 2100 to 4200 kSm^3 in only 10% of the sample.

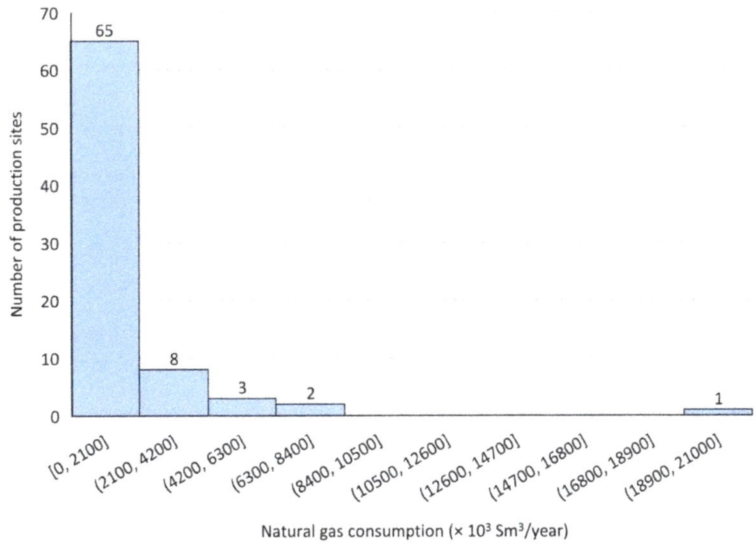

Figure 4. Distribution of yearly natural gas consumption of production sites investigated.

Based on annual data on natural gas demand and the production capacity of dairy companies investigated, the specific consumption of natural gas was also evaluated. In this regard, Figure 5 shows the trend of natural gas consumption per ton of product as a function of annual production capacity. To construct this graph, production capacities expressed in kg/year and in L/year were converted in t/year, while those in pieces/year were neglected. As shown in Figure 5, the specific consumption of natural gas decreases with the annual production according to a power law. A sharp decrease from about 300 Sm^3/t to 25 Sm^3/t is observed for annual production values up to 50,000 t/year; conversely, when that threshold is exceeded, the specific consumption undergoes a less significant decrease, stating at a minimum value of approximately 3 Sm^3/t.

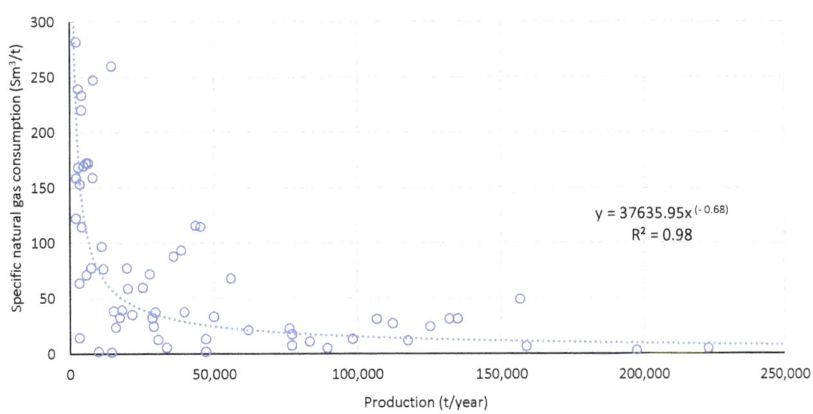

Figure 5. Trend of specific natural gas consumption as a function of the annual production capacity of dairy enterprises.

It is noted that the specific consumption of natural gas is also affected by factors other than the annual production (i.e., the product demand) which were neglected in this study, since these analyses were performed for the only purpose of sample characterization.

3.1.2. Waste Heat Recovery Interventions in the Italian Dairy Industry

As part of the energy audits, the dairy enterprises under investigation identified 365 potential energy-saving measures, which can be grouped into 15 main categories: air conditioning systems, chillers, cogeneration/trigeneration systems, compressed air systems, electric engines, electrical systems, energy management systems, heat generators, hydraulic pumps, lighting systems, process water treatment systems, production lines, renewable energy technologies, waste heat recovery, wastewater treatment plants. More than 100 retrofit interventions concerned the efficiency improvement of the lighting systems (55) and the compressed air systems (52); other energy-efficiency measures in order of importance included the replacement or refurbishment of heat generators (32), the installation of renewable energy technologies (32) and the construction or retrofitting of cogeneration and trigeneration power plants (31). Regarding the recovery and valorization of waste heat, 30 interventions were identified. As depicted in Figure 6, over two thirds of waste heat recovery interventions involved the auxiliary systems, namely the compressed air systems, the cogeneration power plants and the chiller condensing systems. The remaining interventions concerned the production lines, including the waste heat recovery from the cooling process of the whey and the milk in the aging tanks, the waste heat recovery from the sterilization and process steam condensates from the evaporation and concentration of "scotta" and from the degasser of the UHT milk.

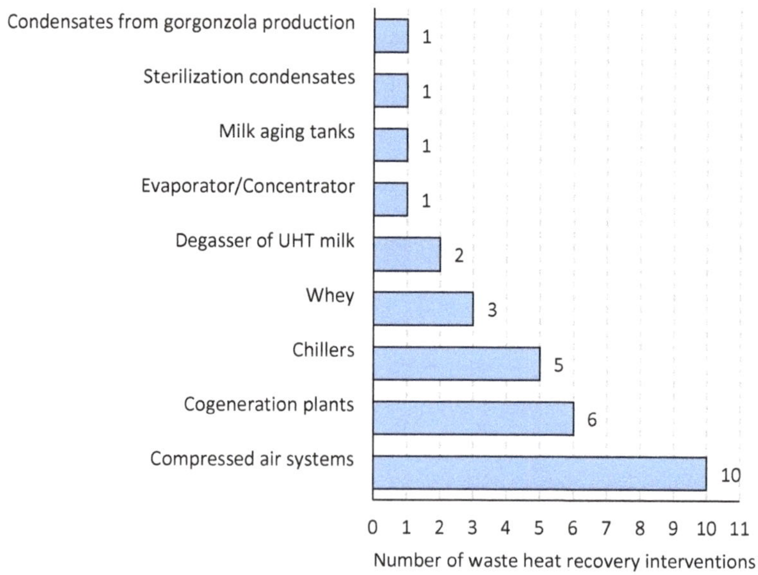

Figure 6. Number and type of waste heat recovery interventions.

Table 3 summarizes the waste heat "acceptors" according to the type of waste heat recovery intervention identified by the dairy enterprises investigated. In this regard, it is pointed out that the reuse of waste heat always occurred within the production facility itself. In the case of interventions related to the compressed air systems, the heat recovered was mainly intended for the production of domestic hot water or space heating; further uses included the heating of water of the air handling unit (AHU), the preheating of water at the boiler inlet and the preheating of cleaning water of process equipment (clean-in-place systems). The latter represented the main use of waste heat recovered from cogeneration plants, followed by the preheating of boiler make-up water, the production of chilled water and the whey heating during pasteurization. The heat recovered from the chiller condenser systems was mainly used for the preheating of water at the boiler inlet, water of clean-in-place systems and the boiler make-up water.

In the case of interventions related to the production lines, the waste heat was reused to support the product processing (i.e., for the preheating of skim milk, the reactivation of raw milk and the preheating of cleaning water) and to preheat water, air and fuel at the boiler inlet. A further type of intervention was the recovery of condensate from the sterilizers to support the production of process steam.

As outlined in Table 4, all waste heat recovery interventions envisaged by dairy enterprises were based on well-established technologies, notably heat exchangers, with or without a hot storage tank. In the case of waste heat recovery from auxiliary systems, heat exchangers were implemented in 17 cases out of 21; the remaining interventions provided for the use of canalizations or chimneys to collect the cooling air of compressors (intervention on the compressed air system) or the installation of an absorption refrigeration system (intervention on the cogeneration plant). In the case of waste recovery from the production lines, heat exchangers were virtually the only technology adopted, except in the case of steam condensate recovery from sterilizers.

3.2. Validation of the Schematic Diagram of Waste Heat Sources via Comparison with Data from Energy Audits

The analysis of energy audits allowed to validate and contextualize to the Italian dairy sector the process scheme for pasteurized milk manufacturing and the corresponding scheme of low-temperature waste heat sources.

In the case of pasteurized milk production, information was provided by the energy audits of two different production sites, including:

- the sequence of the production stages, starting from the receipt of raw materials up to the packing of the final product, including the maintenance of process equipment (sterilization and clean-in-place);
- the energy vectors used in the different production stages (steam, hot water, chilled water, etc.);
- the temperatures of raw materials, the intermediate and final products, depending on data availability.

The analysis of such information revealed that the sequence of process stages in pasteurized milk manufacturing is actually more complex than that reported in previous literature studies. Nevertheless, the limited amount of data available did not permit the further characterization of the waste heat sources in terms of mass flow rate or thermal power per unit mass of raw milk.

Combining the additional information gathered from the energy audits with the process scheme (Figure 2) and the scheme of low-temperature waste heat sources (Figure 3) obtained via the application of the proposed methodology (steps a–d and e–h, respectively), a new scheme was developed (Table 5). This provides for a more comprehensive and systematic description of the whole set of process stages involved in pasteurized milk production and the corresponding waste heat streams generated. As shown in Table 5, the scheme defines:

- the process stages involved in product manufacturing, with the identification of phases generating low-temperature waste heat;
- the energy vectors used in each stage, with the indication of temperature levels, depending on data availability;
- the low-temperature waste heat streams for each process stage, in order from the highest to the lowest temperature.

Table 3. Uses of waste heat depending on the type of waste heat recovery intervention.

Waste Heat Uses \ Waste Heat Sources	Compressed Air Systems	Cogeneration Systems	Chillers	Sterilization Condensates	Degasser of UHT Milk	Evaporator/ Concentrator of Scotta	Aging Tanks	Condensates from the Production of Gorgonzola	Whey
Milk preheating	-	-	-	-	1	-	-	-	1
Reactivation of raw milk	-	-	-	-	-	-	-	-	1
Whey heating during pasteurization	-	1	-	-	-	-	-	-	-
Condensate recovery	-	-	-	1	-	-	-	1	-
Preheating of water for clean-in-place	1	3	2	-	-	-	1	-	1
Domestic hot water production/Space heating	7	-	-	-	-	-	-	-	-
Heating of water of air handling unit (AHU)	1	-	-	-	-	-	-	-	-
Preheating of water at the boiler inlet	1	-	2	-	1	-	-	-	-
Air and fuel preheating	-	-	-	-	-	1	-	-	-
Preheating of boiler make-up water	-	1	1	-	-	-	-	-	-
Production of chilled water at 0 °C	-	1	-	-	-	-	-	-	-

Table 4. Waste heat recovery technologies depending on the type of waste heat recovery intervention.

Waste Heat Recovery Technologies \ Waste Heat Sources	Compressed Air Systems	Cogeneration Systems	Chillers	Sterilization Condensates	Degasser of UHT Milk	Evaporator/ Concentrator of Scotta	Aging Tanks	Condensates from the Production of Gorgonzola	Whey
Heat exchanger	6	4	5	-	2	1	1	1	2
Air canalizations	1	-	-	-	-	-	-	-	-
Chimney with damper	2	-	-	-	-	-	-	-	-
Heat exchanger combined with a hot water storage	1	1	-	-	-	-	-	-	1
Absorption chiller	-	1	-	-	-	-	-	-	-
Condensate recovery system	-	-	-	1	-	-	-	-	-

Table 5. Scheme of pasteurized milk manufacturing process and low-temperature waste heat sources after validation and contextualization via the analysis of energy audits.

Stage	T_p (°C)	Energy Vector	Waste Heat Sources						
plant sterilization	-	steam (T = 125 °C)	steam condensate						
raw milk receiving	4–6	-							
filtration	n/a	-							
cooling and storage	5	chilled water							chiller cooling medium
pre-heating	30	hot water or pasteurized milk	steam condensate *		hot water **				
cream separation	n/a	-							
homogeneization	n/a	-							
standardization	n/a	-							
pasteurization	75–78	hot water	steam condensate *		hot water **	pasteurized milk	pasteurizer overflow		
cooling in storage tank	3–4	chilled water							chiller cooling medium
cooling	2	chilled water							chiller cooling medium
handling and packing	n/a	compressed air		compressor cooling medium					
clean-in-place	-	hot water (T = 95 °C)	steam condensate *					clean-up water	

* only if hot water is obtained via a steam/water heat exchanger; ** without a closed loop water circuit.

Specifically, process stages involving the generation of waste heat were divided into two main categories, namely stages with "direct waste heat generation" and stages with "indirect waste heat generation". The formers (highlighted in orange color) allow for the production of low-temperature waste heat by means of an energy vector or through the intermediate or the final products obtained. By way of example, the pasteurization stage is responsible for the generation of four waste heat streams, at different pressure and temperature conditions, namely:

- the condensates from the steam used to heat the water for milk pasteurization;
- the hot water exiting the pasteurizer (only without a closed loop circuit);
- the pasteurized milk;
- the pasteurizer overflow.

Process stages with "indirect waste heat generation" (highlighted in light blue color) are responsible for the release of waste heat because of the auxiliary systems supporting the production process; in the case of the pasteurized milk processing, phases with "indirect waste heat generation" include those requiring chilled water at different temperature levels (storage, cooling after pasteurization) or compressed air (handling, packing, etc.). Indeed, the chilled water production is responsible for the release of waste heat at the chiller condenser, while the compressed air production causes the generation of waste heat that must be removed to ensure proper compressor operating conditions, as well as a compressed air temperature suitable for plant use.

Overall, seven different waste heat streams were identified, including (in order of decreasing temperature) the steam condensate, the compressor cooling medium, the hot water exiting the pasteurizer, the pasteurized milk, the pasteurizer overflow, the clean-up water and the chiller cooling medium. Finally, waste heat acceptors were identified based on the information from literature review and the analysis of waste heat recovery interventions envisaged by dairy enterprises in the context of energy audits (Section 3.1.2). In this regard, Table 6, besides providing a thermodynamic characterization of waste heat streams based on data available in literature, allows to relate waste heat streams to potential technologies for waste heat recovery and possible waste heat acceptors, identified within the process itself, the auxiliary systems and systems for other uses, including space heating, domestic hot water production, etc.

The methodology proposed in this study was also applied to dairy products other than pasteurized milk. Thus, based on data from literature and the information gathered from the analysis of energy audits of dairy enterprises, two different schemes were identified for UHT milk, pasteurized cream, butter, hard cheese and yogurt:

- a scheme of product manufacturing process and low-temperature waste heat sources;
- a scheme providing the thermodynamic characterization of waste heat streams and the identification of possible waste heat recovery technologies and waste heat acceptors.

Please refer to Appendix A of this paper for details.

Table 6. Scheme providing the characterization of waste heat sources from raw milk pasteurization and the identification of possible heat sinks.

Waste Heat Sources	Steam Condensate	Compressor Cooling Medium	Hot Water	Pasteurized Milk	Pasteurizer Overflow	Clean-Up Water	Chiller Cooling Medium
Temperature (°C)	123 [68], 100 [69], 140–150 [64]	80–95 [70]	62–73 (this study)	72 [64]	70 [48], 71.1 [71]	68–70 [64] 65 [48], 65.6 [71]	50–60 (high-grade heat), 20–30 (low-grade heat) [65]
Flow rate (L/t of raw milk)	14.7 [68], 26.5 [69], 37.3 [72], 33–42 [48], 30.3–37.8 [71]	N/A	N/A	N/A	12.5–17 [48] 11.4–15.1 [71]	163.3 [72], 210–250 [48] 189.3–227.1 [71]	N/A
Heat recovery technology	Condensate recovery system	Heat exchanger/ air canalization	Heat exchanger	Heat exchanger	Heat exchanger	Heat exchanger	Desuperheater/ heat exchanger
Heat recovery opportunities							
Boiler feedwater pre-heating	X [73]	X (this study)			X [71]	X [71]	X [65]
Boiler make-up water pre-heating		X [70]					X [65]
Clean-up water pre-heating		X (this study)	X [74]			X [75]	X [65,66]
Direct use of steam condensate for clean-in-place	X [73]						
Domestic hot water production		X (this study), [70]					X [66]
Flash steam production	X [73]						
Hot water production for air handling unit		X (this study)					
Process hot water production			X [74]				
Raw milk preheating				X (this study), [76]			
Space heating		X (this study), [70]					X [65,66]
Water source heat-pump for space heating						X [75]	X [66]

The identification and characterization of waste heat sources represents the preliminary fundamental step in the design of waste heat recovery systems. As an example, the case of an Italian dairy industry, which applied the proposed methodology to completely map the waste heat recovery sources related to the yogurt manufacturing, is here presented. Among the waste heat recovery sources identified within the process investigated, the attention was focused on two main streams: the milk coming out from the aging tank (Case 1) and the cooling medium of the compressed air system (Case 2). Thus, two waste heat recovery projects were envisaged: in Case 1, waste heat recovered via a heat exchanger was reused for heating the water intended for the clean-in-place (CIP) of process equipment; in Case 2, the waste heat recovered via a heat exchanger was collected in a storage tank and then supplied to the CIP system and the air handling unit (AHU). Characteristics of waste heat recovery sources and sinks, as well as the energy and economic performances of waste heat recovery interventions, were summarized in Table 7. All data reported in this section have been anonymized and also multiplied by a random factor due to their confidential nature.

Table 7. Energy and economic performances of two waste heat recovery projects envisaged by an Italian dairy industry that applied the methodology here proposed to map the availability of waste heat sources within its facility.

	Case 1	Case 2
Waste heat source	Milk coming out of the aging tank	Cooling medium of compressed air system
$T_{WHS,in}$ (°C)	90	90
$T_{WHS,out}$ (°C)	40	27
Waste heat recovery technology	Heat exchanger	Heat exchanger combined with a storage tank
Waste heat recovered * (%)	≈60	≈80
Waste heat sink	Cleaning water	Cleaning water/ water of air handling unit (AHU)
Natural gas saving (Sm^3/year)	60,300	23,517
Investment cost ** (€)	17,085	25,125
Pay-back time ** (years)	1.11	3.92

* assuming a heat exchanger efficiency of 95%; ** results were slightly altered using a multiplication factor to ensure data confidentiality.

4. Conclusions

The paper describes an innovative methodology for the systematic mapping of low-temperature waste heat sources and their potential acceptors in industrial manufacturing processes. The proposed approach is based on a careful analysis and elaboration of data available in technical literature. Such information is enriched and validated using data from real industrial facilities, gathered from the energy audits received by ENEA in compliance with the Legislative Decree 102/2014. The methodology is divided into four main steps: preliminary definition of a typical production process, identification of process phases with low-temperature waste-heat generation, assessment of waste heat sources and potential sinks at phase-level and, lastly, validation, enrichment and contextualization of the outcomes of the methodology via the analysis of the energy audits.

The schemes generated via the application of proposed methodology will contribute to filling the existing knowledge gap on low-temperature waste heat sources in industrial processes, thus promoting the application of waste heat recovery projects. Specifically, the schemes will facilitate to locate and characterize sources of low-temperature waste heat in industrial facilities, and to define solutions for their internal or external reuse and valorization. The schemes will also support academics and legislators in evaluating the waste heat recovery potential in specific industrial sectors, assessing the penetration rate of certain technologies and designing incentive programs for accelerating the implementation of innovative waste heat recovery solutions.

The methodology has been applied to the Italian dairy sector, because of its enormous potential for the utilization of low-temperature waste heat sources, and the resulting schemes referred to the manufacturing of the main dairy products (pasteurized milk, UHT milk, cream, butter, hard cheese and yogurt) have been produced and presented. Focusing on the case of pasteurized milk, the comparison between literature and energy audit data confirmed that main sources of waste heat are located within the pasteurization stage (steam condensate, hot water at pasteurizer exit, pasteurized milk, pasteurizer overflow). Additional sources of waste heat are related to the auxiliary systems. Namely, they are the cooling mediums of compressed air and chilled water systems, which are required to ensure the optimal operating conditions of such devices.

The analysis of energy audits, besides providing additional information on types and characteristics of low-temperature waste heat sources, highlighted that dairy manufacturers are mainly interested in implementing solutions to recover waste heat within the industrial facility itself. Specifically, waste heat is recovered within the auxiliary systems (intended for steam, compressed air and cold water production), with the aim to improve their efficiency or to support space heating or the production of hot water for equipment cleaning or domestic purposes. Heat exchangers, with or without storage systems, are by far the most common technology to capture and deliver waste heat as useful energy where it is needed. On the other side, the use of technologies to convert the waste heat into electricity still remains unexplored.

The study highlighted that the comparison with data from industrial companies is essential for improving the accuracy, the usability and the applicability of schemes of low-temperature waste heat sources and sinks.

Although those schemes were obtained on the basis of data gathered from Italian enterprises, they can be regarded as a starting point for the investigation of dairy industry in other countries. Furthermore, the methodology can be applied to any industrial sector, provided that data of sufficient quantity and/or quality from industrial companies are available.

The methodology proposed in this study is part of a wider project aimed at developing a decision support tool, which will allow industrial companies to preliminary identify low-temperature waste heat recovery opportunities and compare their performances from the energy, environmental and economic perspectives. The software tool will be supported by data regarding conversion efficiency, energy savings and costs of waste heat recovery interventions gathered from the analysis of literature and energy audits. Within this project, the schemes of low-temperature waste heat sources and sinks will be further validated via a direct interaction with industrial companies. This will provide the opportunity to collect data from smaller enterprises, thus broadening the scope of the schemes of low-temperature waste heat sources and sinks derived from the application of this methodology.

Future research directions will include:

- the modelization of waste heat sources fluctuation and intermittency in both mass or volume flow rate and temperature; such aspect has been rarely investigated in the technical literature, even though it is expected to affect to a certain extent the techno-economic performances of waste heat recovery technologies, depending on the frequency of parameters variations;
- the development of tools for the real-time optimization of waste heat recovery system performances;
- the full integration of the schemes of waste heat sources within the decision support tool under development, with aim to assist the user in the identification and characterization of available waste heat sources varying the type of industrial process investigated;
- the use of these schemes to support the knowledge dissemination and the networking among companies, with the aim to overcome technological barriers to the implementation of waste heat recovery systems.

Author Contributions: Conceptualization, L.G. and M.B.; methodology, L.G.; validation, L.G.; writing—original draft preparation, L.G.; writing—review and editing, L.G. and M.B. All authors have read and agreed to the published version of the manuscript.

Funding: This work is part of the Program "Electrical System Research 2019–2021", implemented under Program Agreements between the Italian Ministry for Economic Development and ENEA, CNR, and RSE S.p.A, Project 1.6 "Energy efficiency of industrial products and processes", WP3.

Conflicts of Interest: The authors declare no conflict of interest.

Appendix A

The appendix contains the schemes of product manufacturing process and the corresponding low-temperature waste heat sources and the schemes providing the characterization of low-temperature waste heat sources and the possible waste heat sinks related to dairy products other than pasteurized milk, such as UHT milk, cream, butter, hard cheese and yogurt production.

Table A1. Scheme of UHT milk manufacturing process and corresponding low-temperature waste heat sources.

Stage	T_p (°C)	Energy Vector	Waste Heat Sources					
raw milk receiving	10	-						
storage	10	-						
pre-heating	50	hot water or UHT milk	steam condensate *			hot water **		
separation	n/a	-						
homogenization	n/a	-						
standardization	n/a	-						
cooling in storage tank	4–5	chilled water						chiller cooling medium
1st pre-heating	27	hot water or UHT milk						
2nd pre-heating	35	steam condensate						
pasteurization	85	steam (T = 100 °C)	steam condensate					
sterilization	n/a	steam (T = 150 °C)						
steam separation (via degasser)	90	-	steam condensate	sterilized milk				
homogenization	n/a	-						
cooling	26	chilled water						chiller cooling medium
storage	26	-						
handling and packing	n/a	compressed air			compressor cooling medium			
clean-in-place	-	hot water (T = 95 °C)	steam condensate *				clean-up water	

* only if hot water is obtained via a steam/water heat exchanger; ** without a closed loop water circuit.

Table A2. Scheme providing the characterization of waste heat sources from UHT milk production and the identification of possible heat sinks.

Waste Heat Sources	Steam Condensate	Sterilized Milk	Compressor Cooling Medium	Hot Water	Clean-Up Water	Chiller Cooling Medium
Temperature (°C)	100–150 (this study)	90 (this study)	80–95 [70]	62–73 (this study)	-	50–60 (high-grade heat), 20–30 (low-grade heat) [65]
Heat recovery technology	Condensate recovery system	Heat exchanger	Heat exchanger/ air canalization	Heat exchanger	Heat exchanger	Desuperheater/ heat exchanger
Heat recovery opportunities						
Boiler feedwater pre-heating	X (this study), [73]		X (this study)		X [71]	X [65]
Boiler make-up water pre-heating			X [70]			X [65]
Clean-up water pre-heating			X (this study)	X [74]	X [75]	X [65,66]
Direct use of steam condensate for clean-in-place	X [73]					
Domestic hot water production			X (this study), [70]			X [66]
Flash steam production	X [73]					
Hot water production for air handling unit			X (this study)			
Process hot water production				X [74]		
Raw milk preheating		X (this study)				
Space heating			X (this study), [70]			X [65,66]
Water source heat-pump for space heating					X [75]	X [66]

Table A3. Scheme of pasteurized cream manufacturing process and corresponding low-temperature waste heat sources.

Stage	T_p (°C)	Energy Vector	Waste Heat Sources					
plant sterilization	-	steam (T = 125 °C) or hot water (T = 84 °C)	steam condensate *		hot water			
aging	50	-						
pasteurization	80–98	hot water	steam condensate *		hot water **	pasteurized cream		
cooling	2–5	chilled water						chiller cooling medium
storage in agitated tank	5	chilled water						chiller cooling medium
handling and packing	n/a	compressed air		compressor cooling medium				
clean-in-place	-	hot water (T = 95 °C)	steam condensate *				clean-up water	

* only if hot water is obtained via a steam/water heat exchanger; ** without a closed loop water circuit.

Table A4. Scheme providing the characterization of waste heat sources from pasteurized cream production and the identification of possible heat sinks.

Waste Heat Sources	Steam Condensate	Compressor Cooling Medium	Hot Water	Pasteurized Cream	Clean-Up Water	Chiller Cooling Medium
Temperature (°C)	123 [68], 100 [69], 140–150 [64]	80–95 [70]	62–73 (this study)	80–98 (this study)	68–70 [64], 65 [48], 65.6 [71]	50–60 (high-grade heat), 20–30 (low-grade heat) [65]
Heat recovery technology	Condensate recovery system	Heat exchanger/ air canalization	Heat exchanger	Heat exchanger	Heat exchanger	Desuperheater/ heat exchanger
Heat recovery opportunities						
Boiler feedwater pre-heating	X [73]	X (this study)			X [71]	X [65]
Boiler make-up water pre-heating		X [70]				X [65]
Clean-up water pre-heating		X (this study)	X [74]		X [75]	X [65,66]
Direct use of steam condensate for clean-in-place	X [73]					
Domestic hot water production		X (this study), [70]				X [66]
Flash steam production	X [73]					
Hot water production for air handling unit		X (this study)				
Process hot water production			X [74]			
Raw cream preheating				X (this study)		
Space heating		X (this study), [70]				X [65,66]
Water source heat-pump for space heating					X [75]	X [66]

Table A5. Scheme of butter manufacturing process and corresponding low-temperature waste heat sources.

Stage	T_p (°C)	Energy Vector	Waste Heat Sources				
cream receiving and storage	12	-					
heating and skimming	40	hot water			hot water		
storage	30	-					
regeneration	80	steam	steam condensate				
1st pasteurization	80–85	steam	steam condensate				
cooling	6–8	chilled water					chiller cooling medium
2nd pasteurization	90–100	steam					
steam separation (via degasser)	n/a	-	steam condensate				
starter culture inoculation and aging/crystallization	9	chilled water					chiller cooling medium
churning	n/a	chilled water					chiller cooling medium
handling and packing	n/a	compressed air		compressor cooling medium			
storage	4	chilled water					chiller cooling medium
clean-in-place	-	hot water	steam condensate *			clean-upwater	

* only if hot water is obtained via a steam/water heat exchanger.

Table A6. Scheme providing the characterization of waste heat sources from butter production and the identification of possible heat sinks.

Waste Heat Sources	Steam Condensate	Compressor Cooling Medium	Hot Water	Clean-Up Water	Chiller Cooling Medium
Temperature (°C)	123 [68], 100 [69], 140–150 [64]	80–95 [70]	62–73 (this study)	68–70 [64], 65 [48], 65.6 [71]	50–60 (high-grade heat), 20–30 (low-grade heat) [65]
Heat recovery technology	Condensate recovery system	Heat exchanger/ air canalization	Heat exchanger	Heat exchanger	Desuperheater/ heat exchanger
Heat recovery opportunities					
Boiler feedwater pre-heating	X [73]	X (this study)		X [71]	X [65]
Boiler make-up water pre-heating		X [70]			X [65]
Clean-up water pre-heating		X (this study)	X [74]	X [75]	X [65,66]
Direct use of steam condensate for clean-in-place	X [73]				
Domestic hot water production		X (this study), [70]			X [66]
Flash steam production	X [73]				
Hot water production for air handling unit		X (this study)			
Process hot water production			X [74]		
Space heating		X (this study), [70]			X [65,66]
Water source heat-pump for space heating				X [75]	X [66]

Table A7. Scheme of hard cheese manufacturing process and corresponding low-temperature waste heat sources.

Stage	T_p (°C)	Energy Vector	Waste Heat Sources							
raw milk receiving	4	-								
filtration	n/a	-								
cooling and storage	n/a	chilled water								chiller cooling medium
pre-heating	50–55	hot water	steam condensate *		hot water **					
skimming	n/a	-								
bactofugation	n/a	-								
thermal treatment/ pasteurization	55–72	hot water	steam condensate *		hot water **	pasteurized milk	pasteurizer overflow			
cooling and storage	10–15	chilled water								chiller cooling medium
whey starter and curd addition (in cheese kettle)	n/a	-								
cooking	20–55	steam	steam condensate							
coagulation	n/a	-								
curd breaking andwhey separation	n/a	-							whey	
shaping	n/a	-								
moulds turning	n/a	-								
resting (hot chamber)	38	steam	steam condensate							
transfer to cold chamber	n/a	-								
salting	10	chilled water								chiller cooling medium
drying and ripening	n/a	steam	steam condensate							
washing and surface treatment	n/a	-								
handling and packing	n/a	compressed air		compressor cooling medium						
storage	n/a	-								
clean-in-place	-	hot water	steam condensate *					clean-up water		

* only if hot water is obtained via a steam/water heat exchanger; ** without a closed loop water circuit.

Table A8. Scheme providing the characterization of waste heat sources from hard cheese production and the identification of possible heat sinks.

Waste Heat Sources	Steam Condensate	Compressor Cooling Medium	Hot Water	Pasteurized Milk	Pasteurizer Overflow	Clean-Up Water	Whey	Chiller Cooling Medium
Temperature (°C)	123 [68], 100 [69], 140–150 [64]	80–95 [70]	62–73 (this study)	72 [64]	70 [48], 71.1 [71]	60 [48,71]	38 [48,71]	50–60 (high-grade heat), 20–30 (low-grade heat) [65]
Flow rate (l/t of raw milk)	150 [48], 136.3 [71]	N/A	N/A	N/A	13 [48] 12.1 [71]	250–545 [48] 227.1–492.1 [71]	993 [48], 900 [71]	N/A
Heat recovery technology	Condensaterecovery system	Heat exchanger/ air canalization	Heat exchanger	Heat exchanger	Heat exchanger/ direct use	Heat exchanger	Heat exchanger	Desuperheater/heat exchanger
Heat recovery opportunities								
Boiler feedwater pre-heating	X [73]	X (this study)			X [71]	X [71]	X [71]	X [65]
Boiler make-up water pre-heating		X [70]						X [65]
Clean-up water pre-heating		X (this study)	X [74]			X [75]		X [65,66]
Direct use of steam condensate for clean-in-place	X [73]							
Domestic hot water production		X (this study), [70]						X [66]
Flash steam production	X [73]							
Hot water production for air handling unit		X (this study)						
Process hot water production			X [74]					
Raw milk preheating				X (this study), [76]				
Space heating		X (this study), [70]						X [65,66]
Water source heat-pump for space heating						X [75]		X [66]

Table A9. Scheme of yoghurt manufacturing process and corresponding low-temperature waste heat sources.

Stage	T_P (°C)	Energy Vector	Waste Heat Sources							
raw milk receiving	n/a	-								
storage	4–6	chilled water								chiller cooling medium
milk powder addition	4–6	-								
pasteurization	90	hot water	Steam condensate *		hot water **	pasteurized milk	pasteurizer overflow			
thermization ***	50	hot water	steam condensate *		hot water **					
storage ***	n/a	-								
thermal treatment ***	93	hot water	steam condensate *		hot water **	pasteurized milk	pasteurizer overflow			
skimming	n/a	-								
homogenization	n/a	-								
concentration	85	steam	steam condensate					whey concentrate		
1st cooling	30-41	chilled water								chiller cooling medium
starter culture inoculation	n/a	-								
aging	30–38	-								
2nd cooling	27–30	chilled water								chiller cooling medium
fruits addition	n/a	-								
handling and packing	n/a	compressed air		compressor cooling medium						
storage	5	chilled water								chiller cooling medium
clean-in-place	-	hot water	steam condensate *						clean-up water	

* only if hot water is obtained via a steam/water heat exchanger; ** without a closed loop water circuit; *** only for processes w/o pasteurization.

Table A10. Scheme providing the characterization of waste heat sources from yoghurt production and the identification of possible heat sinks.

Waste Heat Sources	Steam Condensate	Compressor Cooling Medium	Hot Water	Pasteurized Milk	Pasteurizer Overflow	Whey Concentrate	Clean-Up Water	Chiller Cooling Medium
Temperature (°C)	123 [68], 100 [69], 140–150 [64]	80–95 [70]	62–73 (this study)	72 [64]	70 [48], 71.1 [71]	55–68 [77]	68–70 [64], 65 [48], 65.6 [71]	50–60 (high-grade heat), 20–30 (low-grade heat) [65]
Heat recovery technology	Condensate recovery system	Heat exchanger/air canalization	Heat exchanger	Heat exchanger	Heat exchanger	Heat exchanger network	Heat exchanger	Desuperheater/heat exchanger
Heat recovery opportunities								
Boiler feedwater pre-heating	X [73]	X (this study)			X [71]		X [71]	X [65]
Boiler make-up water pre-heating		X [70]						X [65]
Clean-up water pre-heating		X (this study)	X [74]				X [75]	X [65,66]
Direct use of steam condensate for clean-in-place	X [73]							
Domestic hot water production		X (this study), [70]						X [66]
Flash steam production	X [73]							
Hot water production for air handling unit		X (this study)						
Process hot water production			X [74]					
Raw milk preheating				X (this study), [76]				
Space heating		X (this study), [70]						X [65,66]
Standardized milk pre-heating						X [77]		
Water source heat-pump for space heating							X [75]	X [66]

References

1. European Commission. Communication from the Commission to the European Parliament, the Council, the European Economic and Social Committee, the Committee of the Regions and the European Investment Bank Clean Energy for all Europeans COM/2016/0860 Final 2016. Available online: https://eur-lex.europa.eu/legal-content/en/TXT/?uri=CELEX:52016DC0860 (accessed on 5 October 2021).
2. European Commission. Directive (Eu) 2018/2002 of the European Parliament and of the Council of 11 December 2018 (Amending Directive 2012/27/EU on Energy Efficiency) 2018. Available online: https://eur-lex.europa.eu/legal-content/EN/TXT/?uri=uriserv%3AOJ.L_.2018.328.01.0210.01.ENG (accessed on 5 October 2021).
3. European Commission. Communication from the Commission to the European Parliament, the Council, the European Economic and Social Committee and the Committee of the Regions. "Fit for 55": Delivering the EU's 2030 Climate Target on the Way to Climate Neutrality 2021. Available online: https://eur-lex.europa.eu/legal-content/EN/TXT/?uri=CELEX%3A52021DC0550 (accessed on 5 October 2021).
4. European Commission. Directive of the European Parliament and of the Council on Energy Efficiency (Recast) 2021. Available online: https://eur-lex.europa.eu/legal-content/EN/TXT/?uri=CELEX%3A52021PC0558 (accessed on 5 October 2021).
5. Eurostat—Statistic Explained Energy Statistics—An Overview. Available online: https://ec.europa.eu/eurostat/statistics-explained/index.php?title=Energy_statistics_-_an_overview (accessed on 5 October 2021).
6. European Environment Agency EEA Greenhouse Gases—Data Viewer—European Environment Agency. Available online: https://www.eea.europa.eu/data-and-maps/data/data-viewers/greenhouse-gases-viewer (accessed on 5 October 2021).
7. European Commission. Communication from the Commission to the European Parliament, the Council, the European Economic and Social Committee and the Committee of the Regions. Powering a Climate-Neutral Economy: An EU Strategy for Energy System Integration 2020. Available online: https://eur-lex.europa.eu/legal-content/EN/ALL/?uri=COM:2020:299:FIN (accessed on 5 October 2021).
8. Simeone, A.; Luo, Y.; Woolley, E.; Rahimifard, S.; Boër, C. A Decision Support System for Waste Heat Recovery in Manufacturing. *CIRP Ann.* **2016**, *65*, 21–24. [CrossRef]
9. U.S. Department of Energy Waste Heat Recovery Systems. 14th Manufacturing-FocusedTechnology Assessment in Support of Chapter 6: Innovating Clean Energy Technologies in Advanced Manufacturing. Appendix to the 2015 Quadrennial Technology Review 2015. Available online: https://www.energy.gov/sites/default/files/2016/02/f30/QTR2015-6M-Waste-Heat-Recovery.pdf (accessed on 7 October 2021).
10. Bianchi, G.; Panayiotou, G.P.; Aresti, L.; Kalogirou, S.A.; Florides, G.A.; Tsamos, K.; Tassou, S.A.; Christodoulides, P. Estimating the Waste Heat Recovery in the European Union Industry. *Energ. Ecol. Environ.* **2019**, *4*, 211–221. [CrossRef]
11. Jouhara, H.; Khordehgah, N.; Almahmoud, S.; Delpech, B.; Chauhan, A.; Tassou, S.A. Waste Heat Recovery Technologies and Applications. *Therm. Sci. Eng. Prog.* **2018**, *6*, 268–289. [CrossRef]
12. Xu, Z.Y.; Wang, R.Z.; Yang, C. Perspectives for Low-Temperature Waste Heat Recovery. *Energy* **2019**, *176*, 1037–1043. [CrossRef]
13. Brückner, S.; Liu, S.; Miró, L.; Radspieler, M.; Cabeza, L.F.; Lävemann, E. Industrial Waste Heat Recovery Technologies: An Economic Analysis of Heat Transformation Technologies. *Appl. Energy* **2015**, *151*, 157–167. [CrossRef]
14. Mahmoudi, A.; Fazli, M.; Morad, M.R. A Recent Review of Waste Heat Recovery by Organic Rankine Cycle. *Appl. Therm. Eng.* **2018**, *143*, 660–675. [CrossRef]
15. Xia, L.; Liu, R.; Zeng, Y.; Zhou, P.; Liu, J.; Cao, X.; Xiang, S. A Review of Low-Temperature Heat Recovery Technologies for Industry Processes. *Chin. J. Chem. Eng.* **2019**, *27*, 2227–2237. [CrossRef]
16. Loni, R.; Najafi, G.; Bellos, E.; Rajaee, F.; Said, Z.; Mazlan, M. A Review of Industrial Waste Heat Recovery System for Power Generation with Organic Rankine Cycle: Recent Challenges and Future Outlook. *J. Clean. Prod.* **2021**, *287*, 125070. [CrossRef]
17. Champier, D. Thermoelectric Generators: A Review of Applications. *Energy Convers. Manag.* **2017**, *140*, 167–181. [CrossRef]
18. Olabi, A.G.; Elsaid, K.; Sayed, E.T.; Mahmoud, M.S.; Wilberforce, T.; Hassiba, R.J.; Abdelkareem, M.A. Application of Nanofluids for Enhanced Waste Heat Recovery: A Review. *Nano Energy* **2021**, *84*, 105871. [CrossRef]
19. Li, Y.; Xia, J.; Fang, H.; Su, Y.; Jiang, Y. Case Study on Industrial Surplus Heat of Steel Plants for District Heating in Northern China. *Energy* **2016**, *102*, 397–405. [CrossRef]
20. Fang, H.; Xia, J.; Jiang, Y. Key Issues and Solutions in a District Heating System Using Low-Grade Industrial Waste Heat. *Energy* **2015**, *86*, 589–602. [CrossRef]
21. Bonilla-Campos, I.; Nieto, N.; del Portillo-Valdes, L.; Egilegor, B.; Manzanedo, J.; Gaztañaga, H. Energy Efficiency Assessment: Process Modelling and Waste Heat Recovery Analysis. *Energy Convers. Manag.* **2019**, *196*, 1180–1192. [CrossRef]
22. Nikolaisen, M.; Andresen, T. System Impact of Heat Exchanger Pressure Loss in ORCs for Smelter Off-Gas Waste Heat Recovery. *Energy* **2021**, *215*, 118956. [CrossRef]
23. Peris, B.; Navarro-Esbrí, J.; Molés, F.; Mota-Babiloni, A. Experimental Study of an ORC (Organic Rankine Cycle) for Low Grade Waste Heat Recovery in a Ceramic Industry. *Energy* **2015**, *85*, 534–542. [CrossRef]
24. Delpech, B.; Milani, M.; Montorsi, L.; Boscardin, D.; Chauhan, A.; Almahmoud, S.; Axcell, B.; Jouhara, H. Energy Efficiency Enhancement and Waste Heat Recovery in Industrial Processes by Means of the Heat Pipe Technology: Case of the Ceramic Industry. *Energy* **2018**, *158*, 656–665. [CrossRef]

25. Jouhara, H.; Bertrand, D.; Axcell, B.; Montorsi, L.; Venturelli, M.; Almahmoud, S.; Milani, M.; Ahmad, L.; Chauhan, A. Investigation on a Full-Scale Heat Pipe Heat Exchanger in the Ceramics Industry for Waste Heat Recovery. *Energy* **2021**, *223*, 120037. [CrossRef]
26. Brough, D.; Mezquita, A.; Ferrer, S.; Segarra, C.; Chauhan, A.; Almahmoud, S.; Khordehgah, N.; Ahmad, L.; Middleton, D.; Sewell, H.I.; et al. An Experimental Study and Computational Validation of Waste Heat Recovery from a Lab Scale Ceramic Kiln Using a Vertical Multi-Pass Heat Pipe Heat Exchanger. *Energy* **2020**, *208*, 118325. [CrossRef]
27. Öhman, H. Implementation and Evaluation of a Low Temperature Waste Heat Recovery Power Cycle Using NH3 in an Organic Rankine Cycle. *Energy* **2012**, *48*, 227–232. [CrossRef]
28. Savulescu, L.; Périn-Levasseur, Z.; Benali, M. Innovative Visualization Technique for Energy Flow Analysis: Waste Heat Recovery and Energy Savings Opportunities. *Appl. Therm. Eng.* **2013**, *61*, 143–148. [CrossRef]
29. Fiaschi, D.; Manfrida, G.; Russo, L.; Talluri, L. Improvement of Waste Heat Recuperation on an Industrial Textile Dryer: Redesign of Heat Exchangers Network and Components. *Energy Convers. Manag.* **2017**, *150*, 924–940. [CrossRef]
30. Rakib, M.I.; Saidur, R.; Mohamad, E.N.; Afifi, A.M. Waste-Heat Utilization—The Sustainable Technologies to Minimize Energy Consumption in Bangladesh Textile Sector. *J. Clean. Prod.* **2017**, *142*, 1867–1876. [CrossRef]
31. Law, R.; Harvey, A.; Reay, D. A Knowledge-Based System for Low-Grade Waste Heat Recovery in the Process Industries. *Appl. Therm. Eng.* **2016**, *94*, 590–599. [CrossRef]
32. Fiaschi, D.; Talluri, L. Assessment of the Thermoelectric Conversion Potential of Low-Temperature Waste Heat from Textile Dry-Cleaning Processes. *E3S Web Conf.* **2019**, *113*, 03019. [CrossRef]
33. El Fil, B.; Garimella, S. Waste Heat Recovery in Commercial Gas-Fired Tumble Dryers. *Energy* **2021**, *218*, 119407. [CrossRef]
34. Wu, H.; Jouhara, H.; Tassou, S.A.; Karayiannis, T.G. Modelling of Energy Flows in Potato Crisp Frying Processes. *Appl. Energy* **2012**, *89*, 81–88. [CrossRef]
35. Aneke, M.; Agnew, B.; Underwood, C.; Menkiti, M. Thermodynamic Analysis of Alternative Refrigeration Cycles Driven from Waste Heat in a Food Processing Application. *Int. J. Refrig.* **2012**, *35*, 1349–1358. [CrossRef]
36. Bellocchi, S.; Guizzi, G.L.; Manno, M.; Pentimalli, M.; Salvatori, M.; Zaccagnini, A. Adsorbent Materials for Low-Grade Waste Heat Recovery: Application to Industrial Pasta Drying Processes. *Energy* **2017**, *140*, 729–745. [CrossRef]
37. Atkins, M.J.; Walmsley, M.R.W.; Neale, J.R. Integrating Heat Recovery from Milk Powder Spray Dryer Exhausts in the Dairy Industry. *Appl. Therm. Eng.* **2011**, *31*, 2101–2106. [CrossRef]
38. Aneke, M.; Agnew, B.; Underwood, C.; Wu, H.; Masheiti, S. Power Generation from Waste Heat in a Food Processing Application. *Appl. Therm. Eng.* **2012**, *36*, 171–180. [CrossRef]
39. Sapali, S.N.; Pise, S.M.; Pise, A.T.; Ghewade, D.V. Investigations of Waste Heat Recovery from Bulk Milk Cooler. *Case Stud. Therm. Eng.* **2014**, *4*, 136–143. [CrossRef]
40. Pantaleo, A.M.; Fordham, J.; Oyewunmi, O.A.; De Palma, P.; Markides, C.N. Integrating Cogeneration and Intermittent Waste-Heat Recovery in Food Processing: Microturbines vs. ORC Systems in the Coffee Roasting Industry. *Appl. Energy* **2018**, *225*, 782–796. [CrossRef]
41. Mumith, J.-A.; Makatsoris, C.; Karayiannis, T.G. Design of a Thermoacoustic Heat Engine for Low Temperature Waste Heat Recovery in Food Manufacturing: A Thermoacoustic Device for Heat Recovery. *Appl. Therm. Eng.* **2014**, *65*, 588–596. [CrossRef]
42. Amón, R.; Maulhardt, M.; Wong, T.; Kazama, D.; Simmons, C.W. Waste Heat and Water Recovery Opportunities in California Tomato Paste Processing. *Appl. Therm. Eng.* **2015**, *78*, 525–532. [CrossRef]
43. Dogbe, E.S.; Mandegari, M.; Görgens, J.F. Assessment of the Thermodynamic Performance Improvement of a Typical Sugar Mill through the Integration of Waste-Heat Recovery Technologies. *Appl. Therm. Eng.* **2019**, *158*, 113768. [CrossRef]
44. Ling-Chin, J.; Bao, H.; Ma, Z.; Taylor, W.; Paul Roskilly, A. State-of-the-Art Technologies on Low-Grade Heat Recovery and Utilization in Industry. In *Energy Conversion—Current Technologies and Future Trends*; Al-Bahadly, I.H., Ed.; IntechOpen: London, UK, 2019; ISBN 978-1-78984-904-2.
45. Thekdi, A.; Nimbalkar, S.U. Industrial Waste Heat Recovery—Potential Applications, Available Technologies and Crosscutting R&D Opportunities; 2015; Report number: ORNL/TM–2014/622, 1185778. Available online: http://www.osti.gov/servlets/purl/1185778/ (accessed on 8 October 2021).
46. Wilfert, G.L.; Huber, H.B.; Dodge, R.E.; Garrett-Price, B.A.; Fassbender, L.L.; Griffin, E.A.; Brown, D.R.; Moore, N.L. Characterization of Industrial Process Waste Heat and Input Heat Streams; 1984; Report number: PNL-4912, 6834862. Available online: http://www.osti.gov/servlets/purl/6834862/ (accessed on 8 October 2021).
47. Singh, R. *Energy in Food Processing*; Elsevier Science Publishing Company Inc.: Amsterdam, The Netherlands, 1986.
48. Wang, L. *Energy Efficiency and Management in Food Processing Facilities*; CRC Press: Boca Raton, FL, USA, 2009; ISBN 978-1-4200-6338-7.
49. National Dairy Council of Canada Guide to Energy Efficiency Opportunities in the Dairy Processing Industry 1997. Available online: https://p2infohouse.org/ref/19/18807.pdf (accessed on 8 October 2021).
50. Focus on Energy Dairy Processing Industry. Energy Best Practice Guidebook. 2006. Available online: https://focusonenergy.com/sites/default/files/dairyprocess_guidebook.pdf (accessed on 8 October 2021).
51. Woolley, E.; Luo, Y.; Simeone, A. Industrial Waste Heat Recovery: A Systematic Approach. *Sustain. Energy Technol. Assess.* **2018**, *29*, 50–59. [CrossRef]

52. Glatzl, W.; Brunner, C.; Fluch, J. GREENFOODS—Energy Efficiency in the Food and Beverage Industry. ECEEE Summer Study Proceedings. 2015. Available online: https://www.eceee.org/library/conference_proceedings/eceee_Summer_Studies/2015/2-energy-efficiency-policies-8211-how-do-we-get-it-right/greenfoods-energy-efficiency-in-the-food-and-beverage-industry/ (accessed on 14 December 2021).
53. Brunner, C.; Kulterer, K.; Fluch, J.; Glatzl, W. Greenfoods Branch Concept for Enhancing Energy Efficiency in the Food and Drink Industry. ECEEE Industrial Summer Study Proceedings. 2014. Available online: https://www.eceee.org/library/conference_proceedings/eceee_Industrial_Summer_Study/2014/2-sustainable-production-design-and-supply-chain-initiatives/greenfoods-branch-concept-for-enhancing-energy-efficiency-in-the-food-and-drink-industry/ (accessed on 14 December 2021).
54. CE-HEAT Project Decision Support System (DSS). Available online: https://www.waste-heat.eu/waste-heat-toolbox/investment-decision-support/decision-support-system (accessed on 14 December 2021).
55. Legislative Decree 4 July 2014, n°102. 2014. Available online: https://www.gazzettaufficiale.it/eli/id/2014/07/18/14G00113/sg (accessed on 28 October 2021).
56. Cervo, H.; Ferrasse, J.-H.; Descales, B.; Van Eetvelde, G. Blueprint: A Methodology Facilitating Data Exchanges to Enhance the Detection of Industrial Symbiosis Opportunities—Application to a Refinery. *Chem. Eng. Sci.* **2020**, *211*, 115254. [CrossRef]
57. Malinauskaite, J.; Jouhara, H.; Ahmad, L.; Milani, M.; Montorsi, L.; Venturelli, M. Energy Efficiency in Industry: EU and National Policies in Italy and the UK. *Energy* **2019**, *172*, 255–269. [CrossRef]
58. Kluczek, A.; Olszewski, P. Energy Audits in Industrial Processes. *J. Clean. Prod.* **2017**, *142*, 3437–3453. [CrossRef]
59. Legislative Decree 14 July 2020, n. 73. 2020. Available online: https://www.gazzettaufficiale.it/eli/id/2020/07/14/20G00093/sg (accessed on 28 October 2021).
60. Ministero dello Sviluppo Economico. Clarifications Regarding Energy Audits in Fulfillment of Article 8 of the Legislative Decree n.102/2014. 2016. Available online: https://www.mise.gov.it/images/stories/documenti/CHIARIMENTI-DIAGNOSI-14-nov-2016.pdf (accessed on 28 October 2021).
61. European Commission. *Joint Research Centre. Best Available Techniques (BAT) Reference Document for the Food, Drink and Milk Industries: Industrial Emissions Directive 2010/75/EU (Integrated Pollution Prevention and Control)*; Publications Office: Luxembourg, 2019; Available online: https://data.europa.eu/doi/10.2760/243911 (accessed on 26 October 2021).
62. Ramírez, C.A.; Patel, M.; Blok, K. From Fluid Milk to Milk Powder: Energy Use and Energy Efficiency in the European Dairy Industry. *Energy* **2006**, *31*, 1984–2004. [CrossRef]
63. Modi, A.; Prajapat, R. Pasteurization Process Energy Optimization for a Milk Dairy Plant by Energy Audit Approach. *Int. J. Sci. Technol. Res.* **2014**, *3*, 181–188.
64. Bylund, G. *Dairy Processing Handbook*; Tetra Pak Processing Systems AB: Lund, Sweden, 1995; Available online: https://dairyprocessinghandbook.tetrapak.com/ (accessed on 26 October 2021).
65. Carbon Trust. Referigeration: A Guide to Energy and Carbon Saving Opportunities. 2019. Available online: https://www.carbontrust.com/resources/refrigeration-guide (accessed on 28 October 2021).
66. Reis, C.; Nelson, E.; Armer, J.; Johnson, T.; Hirsch, A.; Doebber, I. Refrigeration Playbook. Heat Reclaim; Optimizing Heat Rejection and Refrigeration Heat Reclaim for Supermarket Energy Conservation; 2015; Report number: NREL/TP-5500-63786, 1215136. Available online: http://www.osti.gov/servlets/purl/1215136/ (accessed on 28 October 2021).
67. Istituto Nazionale di Statistica (ISTAT) ATECO (Classification of Economic Activity) 2007. Available online: https://www.istat.it/en/archivio/17959 (accessed on 28 October 2021).
68. Chaudhary, D.A.; Chaudhari, A.G. Performance Evaluation of a Plate Type (HTST) Milk Pasteurizer. *Res. J. Anim. Husb. Dairy Sci.* **2015**, *6*, 130–134. [CrossRef]
69. Hoffman, P. Simplified Balance of Line for Milk Powder Production. Processing Lines of Food Industry. 2003. Available online: https://users.fs.cvut.cz/pavel.hoffman/ (accessed on 28 October 2021).
70. Department of the Environment, Transport and the Regions. Heat Recovery from Air Compressors Heat Recovery from Air Compressors. Good Practice Guide N°238. 1998. Available online: https://www.thoriteinfo.co.uk/pdfs/good_practice_guide_238_heat_recovery_from_air_compressors.pdf (accessed on 27 October 2021).
71. Foell, W.K.; Lund, D.; Mitchell, J.W.; Ray, D.; Stevenson, R.; TenWolde, A. Low-Temperature Waste Heat Recovery in the Food and Paper Industries; 1980. Available online: https://www.osti.gov/servlets/purl/6610811 (accessed on 2 November 2021).
72. Prabhakar, P.K.; Srivastav, P.P.; Murari, K. Energy Consumption during Manufacturing of Different Dairy Products in a Commercial Dairy Plant: A Case Study. *Asia J. Dairy Food Res.* **2015**, *34*, 98. [CrossRef]
73. SpiraxSarco. The Dairy Industry Steam and Condensate Systems. 2000. Available online: https://www.spiraxsarco.com/ (accessed on 27 October 2021).
74. Quijera, J.A.; Alriols, M.G.; Labidi, J. Integration of a Solar Thermal System in a Dairy Process. *Renew. Energy* **2011**, *36*, 1843–1853. [CrossRef]
75. Nagpal, H.; Spriet, J.; Murali, M.K.; McNabola, A. Heat Recovery from Wastewater—A Review of Available Resource. *Water* **2021**, *13*, 1274. [CrossRef]

76. Brush, A.; Masanet, E.; Worrell, E. Energy Efficiency Improvement and Cost Saving Opportunities for the Dairy Processing Industry. An ENERGY STAR® Guide for Energy and Plant Managers. 2011. Available online: https://www.osti.gov/servlets/purl/1171534 (accessed on 28 October 2021).
77. Atkins, M.J.; Walmsley, M.R.W.; Walmsley, T.; Fodor, Z.; Neale, J. Minimising Energy Use in Milk Powder Production Using Process Integration Techniques. *Chem. Eng. Trans.* **2012**, *29*, 1507–1512. [CrossRef]

energies

MDPI

Article

Energy Performance of Italian Oil Refineries Based on Mandatory Energy Audits

Carlos Herce *, Chiara Martini, Marcello Salvio and Claudia Toro *

Energy Efficiency in the Economic Sectors Laboratory (DUEE-SPS-ESE), Energy Efficiency Unit Department, Italian National Agency for New Technologies, Energy and Sustainable Economic Development (ENEA), C.R. Casaccia, Via Anguillarese 301, 00123 Rome, Italy; chiara.martini@enea.it (C.M.); marcello.salvio@enea.it (M.S.)

* Correspondence: carlos.herce@enea.it (C.H.); claudia.toro@enea.it (C.T.); Tel.: +39-06-3048-3268 (C.H.); +39-06-3048-4085 (C.T.)

Abstract: Petroleum products account for the 32.3% of worldwide primary energy. There are more than 100 oil refineries in Europe that directly employ 119,000 people with a turnover of EUR 600 billion and around 1.2% to the total value added in manufacturing. Therefore, the petroleum refining sector is very important in the European economy, and its decarbonization is crucial in the energy transition. Refineries present a high degree of complexity and integration, and the continuous increase of their energy efficiency is a key topic for the sector. In this work an analysis of the energy efficiency in ten Italian refineries based on mandatory energy audits and public data is presented. The primary (0.0963 ± 0.0341 toe/t), thermal (3421.71 ± 1316.84 MJ/t), and electrical (68.20 ± 19.34 kWh/t) specific energy consumptions have been evaluated. Some insights about the impact of refined products mix (mainly driven by production of diesel fuel) and Nelson Complexity Index in energy consumption are presented. Lastly, an overview of energy performance improvement actions (EPIAs) information extracted from energy audits is presented. This work presents a first step for the benchmark of Italian refineries that should be subsequently improved.

Keywords: energy audits (EAs); specific energy consumption (SEC); energy efficiency; industry; oil refining; refineries; energy transition

Citation: Herce, C.; Martini, C.; Salvio, M.; Toro, C. Energy Performance of Italian Oil Refineries Based on Mandatory Energy Audits. *Energies* **2022**, *15*, 532. https://doi.org/10.3390/en15020532

Academic Editor: Giorgio Vilardi

Received: 12 November 2021
Accepted: 28 December 2021
Published: 12 January 2022

1. Introduction

Nowadays, fossil fuels provide more than 80% of all the energy used worldwide. The products derived from petroleum are the first primary energy source since 1970 and its consumption has been constantly growing since the end of "1980s Oil Glut" in 1983 (except during the "2008 financial global crisis" and the "2020 COVID crisis"). Oil consumption increased annually by 1.3% from 2000 to 2019 (from 154.39 EJ to 191.89 EJ). This increase is driven by non-OECD countries (mainly China and India) with an annual rate of +3.1%, meanwhile consumption in the EU has been reduced by −0.3%. However, this trend is opposite to the share of oil consumed as part of global primary energy consumption (from 39.1% to 32.3%) mainly due to the substitution of oil by coal, natural gas and renewables in power generation [1]. Since Hubbert's pioneering theory of "Peak oil" [2] the proven reserves of oil have been continuously increasing [1,3].

An oil or petroleum refinery is an industrial facility where crude oil and other feedstocks are processed into useful petroleum products. The main principle of refining is to separate and improve the hydrocarbon compounds that constitute crude oil to produce saleable products (such as gasoline, diesel fuel, petroleum naphtha, asphalt base, heating oil, kerosene, liquefied petroleum gas, jet fuel and fuel oils). A refinery includes three main process sections: separation (including the crude distillation unit [CDU]); conversion (including the gas recovery unit [GRU], hydrogen treatment unit [HTU], fluid catalytic cracking [FCC], and vacuum distillation unit [VDU]); and finishing (including catalytic reforming unit [CRU], distillate hydroforming unit [DHU], delayed coking unit [DCU], lube oil processing unit [LPU], asphalt processing unit [APU] and visbreaking). Each section is constituted by one or more process units with different configurations and operation parameters (pressure, temperature, catalyst, etc.) to perform their function [4]. In 2012, there was a worldwide total refining capacity of around 4400 million t/y, in 655 refineries (25% Asia, 20% North America and 20% Europe) [5].

In 2017, the 34.6% of global GHG emissions were produced by oil products [6], and refineries account for only 7% of all the industrial emissions in Europe [5], mainly due to combustion processes (90%) [7]. There are more than 100 oil refineries in Europe that directly employ 119,000 people with a turnover of EUR 600 billion and around 1.2% to the total value added in manufacturing [8]. Moreover, refineries are crucial for several value chains linked to energy-intensive industries, not only as fuel but also as feedstock suppliers. Therefore, the role of refineries and their decarbonization is crucial in the energy transition period from several points of view: a new hydrogen economy; carbon capture use and storage (CCUS); the circular economy; the valorisation of novel bio-feedstocks; and deep process electrification [9].

The refining sector is the main consumer of pure hydrogen worldwide [10] and it produces internally more than 1/3 of its consumption [11]. The share of internally installed production of hydrogen has tripled [12] in the past 20 years and the estimation of hydrogen-related emissions has doubled [7]. The main route of production is steam hydrocarbon reforming (more than 90% worldwide) [13,14]. Due to its extensive experience in fossil-based hydrogen, the refining sector presents a very high potential for the production of the so-called "blue hydrogen" [14–16]. This synthesis route mixes incorporate CCUS technologies in the production of hydrogen. The pure CO_2 generated during the reforming reaction is subsequently (internally or externally) used in the refinery. The "first-blue-then-green" principle proposes the use of this technology as a first step in the development of infrastructures for the massive deployment of "green hydrogen" (based 100% on renewables).

Another important aspect to consider is the production of carbon-neutral liquid fuels from a circular economy perspective. The first generation of bio-refineries based on bio-oil from energy crops [17,18] has been overtaken by the second generation of biorefineries (based on waste valorisation) [18–20], the algae-based third generation [18,19,21], or the integrated biorefineries based on bio-chemical feedstocks [22,23]. The direct electrification of refining processes presents a low potential. However, the electrification of heat and mechanical processes can be sensibly improved in order to reduce the carbon intensity of refineries [9,24].

The refineries are an excellent example of heat integration and energy efficiency in industrial processes. The European sector already applies technologies at a large scale and has increased efficiency by 13% between 1990 and 2005 [7,9]. The increase of energy efficiency in refineries is a topic that has been studied in depth due to economic and environmental related implications (see Section 2).

The purpose of this research analysis is to characterize the status of energy efficiency in Italian refineries. In order to achieve this objective an analysis based on mandatory energy audits and public data has been carried out. Firstly, the specific energy consumption (primary, thermal and electrical) in refineries as function of the refining capacity was evaluated using linear regression models. Secondly, the impact of other key parameter in refining (production slate and complexity) in energy consumption was studied. Thirdly, an overview of energy performance improvement actions (EPIAs) collected from energy audits was analysed in order to understand the implemented and potential improvements of the sector with current technologies.

Previous related research has been focused, on the one hand, on the analysis of energy efficiency refineries (mainly in the U.S.) in order to allocate the GHG emissions related to fuel transportation refining; or, on the other hand, on the analysis of technologies to reduce the energy consumption of refining. These analyses require very detailed proprietary information on the sub-processes of the refineries. Only a few studies have been focused on the analysis of energy efficiency of the refineries globally, due to the complexity of the installations.

In this work, a hybrid approach was applied with several original contributions (to the best of the knowledge of the authors). Firstly, an analysis of the overall plant was developed considering the capacity of the refineries (this variable was excluded from previous published research). Hence, the primary, thermal and electrical specific energy consumption (SEC) rates of the refineries were modelled as function of the production. Secondly, most of the analysis of SEC provides the mean value of a region or the benchmark. This work also presented the variability of the SEC (as standard deviation) for the first time, outside of the U.S. refineries. Thirdly, the analysis of EPIAs provided market-based information about the cost-effectiveness of current technologies, in order to evaluate effectively the potential and short-term scenarios for energy efficiency. Fourthly, this work is completely new for the Italian refining sector (the second country in the EU). Lastly, this study extends to refineries the general methodology developed to characterize different productive sectors from energy audits previously validated within the cement industry [25].

Section 2 of this paper is devoted to a literature review of the energy efficiency characterisation of oil refineries, with a focus on the evolution of the Italian refining market. Section 3 presents the information available from energy audits and other public sources for the analysis of energy efficiency. Section 4 estimates by means of linear regression models the primary, thermal and electrical SECs; the impact of capacity in the reliability of the models; the influence of product slate and complexity in energy consumption; and the analysis of EPIAs. Finally, in Section 5 the main remarks and the limitations of this work are discussed.

2. Literature Review

2.1. Assessment of Energy Consumption in Oil Refineries

An extensive overview of energy efficiency measures, disaggregated by process unit was developed by Worrell et al. [26] in U.S. refineries. In this work a general overview of the distribution of mass and energy flows internally to the refineries was coupled with potential EPIAs. The analysis of the implementation potential of EPIAs in the different process units was subsequently refined by Morrow III et al. in [27]. A similar work for European refineries can be found in a BREF document from the European Commission [5]. These works are very useful to understand the complexity of the refineries, to allocate energy consumptions and energy costs internally, and to classify the potential EPIAs.

The reduction of contaminants from oil products, to comply with stringent environmental quality specifications, results in an increase of energy consumption in the refineries [28]. Szklo and Schaeffer [29] studied the impact of trade-offs between local (in transportation uses) and global (in refining process) emissions of pollutants. Different options for saving energy at refineries in the study included the improvement of heat integration and waste heat recovery, fouling mitigation, advanced process control, the use of variable speed and vacuum pumps, etc.). On the other hand, alternative treatment processes are less energy intensive than hydrotreating processes (e.g., ISAL, olefin alkylation of thiophenic sulphur (OATS), oxidative desulfurization process (ODP), or catalytic distillation (CD) processes), with specific application to Brazilian market. Similar analyses have been developed in Canada (with the particularity of comparing conventional with oil sands refineries) [30], and in Sweden (focused on heat integration measures) [31,32].

The energy intensity of refineries depends on multiple factors. First, each refinery presents a unique configuration, hence the refining capacity, the integration of different units and its complexity defines the general energy consumption (generally energy consumption increases with refinery complexity). Secondly, the properties of crude oil impacts on the energy required for refining (mainly API density and sulphur content). Thirdly, the production slate and product quality (as well as the connection with other petrochemical or power plants) varies among different markets, hence the energy intensity varies with the properties of the final refined products. Lastly, the oil refining sector presents very high standards on safety and environmental issues. The related processes and devices have a non-negligible impact on energy consumption.

There are three main methodologies to evaluate the energy efficiency in refineries: the "Solomon Energy Intensity Index (EII)", the "Specific Energy Consumption" (SEC) and the "Products Method" [5]. The "Solomon EII" is the most used sectoral indicator to compare the energy intensity of mineral oil refineries [12,33]. This standard energy use index (property of Solomon Associates) is applied to benchmark the energy consumption of more than 500 refineries worldwide (including 99% of EU oil refining companies). The EII includes process unit energy standards that are individual expressions for each of the processes in the refinery and state the average standard energy consumption, and multiple confidential data from the refineries. These data are not available for all refineries and typically are considered confidential [5]. The initial value of global EII at the beginning of the use of this indicator (mid-1980's) was fixed at 100. More efficient refineries present a lower EII. The last data present a global EII of 92, that reflect an increase on energy efficiency in the refineries. The top 10% EII worldwide values were equal to or below 75 [5]. In 2005, Italian refineries presented an EII of 81 [33].

The "SEC method" calculates the ratio between the energy consumed by the refinery and the tonnes of feedstock processed [34]. It is a simple index which does not take into account the complexity of the refinery and generally represents the mean value of the sector in a region or the SEC of the best available technology (BAT). This method was applied by Worrell et al. to analyze the potential improvements of different sectors (including oil refining) in Europe, obtain the SEC for six types of oil refinery products, and present an overall typical SEC of the refinery of SEC = 0.065 toe/t [35]. This value has been recently updated to the BAT refinery in the Middle East to SEC = 0.0569 toe/t [36].

The "Products Method" takes under consideration the chemicals and energy products in the refinery, calculating an SEC benchmark per tonne of energy products produced. This indicator is subsequently normalized for all the refineries in order to give an energy consumption benchmark for each refinery compared with the overall sector [37].

It is important to note the work of Wang et al. at the Argonne National Laboratory which developed the GREET model for life-cycle analysis of vehicle technologies, transportation fuels, and other energy systems. This model was firstly applied to address the allocation of energy uses and emissions for different refinery products in a generic simplified refinery (evaluating at process unit level) [38]. This approach was subsequently applied to analyse the energy efficiency of U.S. refineries in three excellent works. In the first one 43 refineries were analysed (SEC = 0.091 $\pm$ 0.033 toe/t), which suggested that the efficiency of refineries seems to be sensitive to product slate (mainly the ratio diesel/gasoline and heavy ratio yield), crude quality (mainly API density and sulphur content), seasonal (the energy efficiency is 1% higher in winter) and regional factors, refinery configuration and complexity [39]. This analysis was subsequently refined by petroleum product (Gasoline, Diesel, Jet, RFO, LPG and Petcoke), confirming the impact of different parameters and allowing to allocate the GHG emissions intensities of different products [40]. The analysis was further extended to include 17 European refineries confirming, on the one hand, the importance of crude density (API gravity) and heavy product (HP) yields, and, on the other hand, that refineries with high complexity are more resource efficient, but more energy and GHG intensive [41]. This analysis was carried with comprehensive information on all the internal streams and mass and energy balances of the refineries without considering the impact of refinery size (only considering refineries capacities higher than 100,000 bbl/day). This method was compared with other energy content, economic value and value added models in order to allocate the GHG emissions (including the SEC) by product in European refineries [42]. This study suggests that the impact of light (hydrogen) and heavy (petcoke and fuel oil) products is crucial in the energy intensity of the refineries and its impact tends to be minimised, with a greater focus on main transportation fuels.

Nelson complexity index (NCI) is a key parameter for refineries. This index was developed in the 1960s and 1970s by W.L. Nelson in a series of articles for the Oil & Gas Journal [43] and it is still used in the annual review of refineries' complexity [44]. NCI quantifies the sophistication and capital intensity of a refinery and it is a parameter used for facility classification, cost estimation, sales price models, etc. [4]. This parameter has been included by the Argonne's group as a key parameter of energy efficiency in refineries. However, the NCI of a refinery can be obtained from different configurations, hence its importance in SEC is lower than crude and product properties [40]. Kaiser analysed in detail the primary applications of refinery complexity and as well as its limitations, providing alternative approaches to extend the applicability of the NCI [4,45]. In these works, after an extensive review of worldwide refineries, it was not possible to directly observe a correlation between complexity and throughput of the refineries and a modest correlation with conversion capacity. Hence, NCI quantitative applications must be considered with caution.

2.2. Oil Refining in Italy

Italy is the 8th largest oil importer worldwide (1.24 M barrels/day) and 2nd in EU-27 [1,6]. During 2017, the 11 Italian oil refineries processed 80.3 Mt of crude oil, which represents a refinery utilization rate of 79.6%. The 7.2% of crude oil refined was extracted in Italy (70% in the Basilicata region), therefore the oil sector is dependent on external markets. This external dependence is aligned with EU countries (the energy EU dependency rate is 61%) [46]. Despite its importance, the refining sector in Italy is presently in a contraction period. As presented in Figure 1, from mid-1980s to mid-1990s refining sector suffered a reduction in refining capacity and the decommissioning of several refineries (from 36 to 18 in Italy) due to the structural overcapacity for distillation since the "1973 First Oil Crisis" (and the subsequent "1979 Oil Shock" and "1980s Oil Glut"). The subsequent "2008 Financial Crisis" had a high impact on the refining sector. The EU refining margin fell from above to below the average margin of their competitors (U.S., Russia, Middle East and South Korea/Singapore) mainly due to the increase in energy operating costs [47]. This crisis has reduced the EU refining capacity by 10% and forced the shutdown of 5 Italian refineries from 2008 (from 16 to 11).

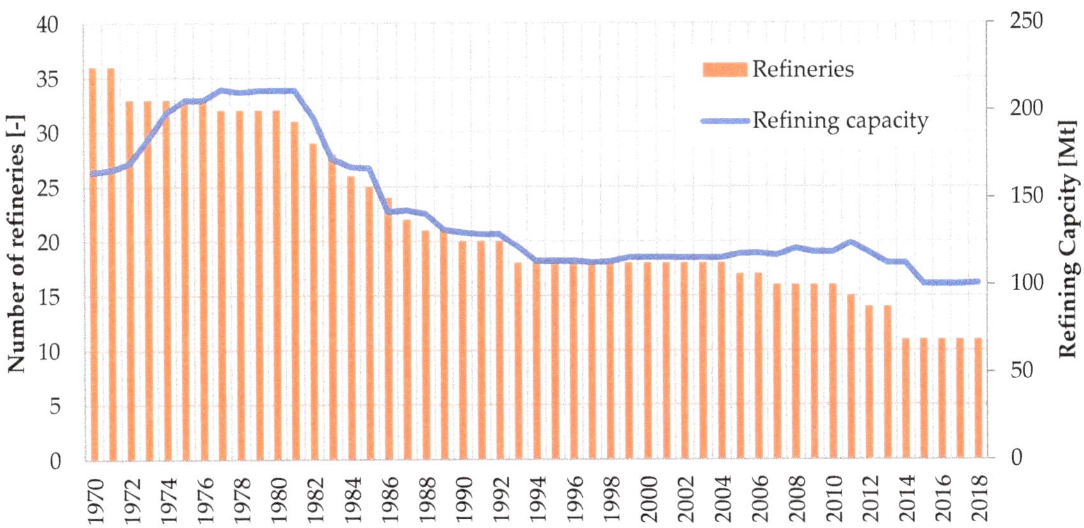

Figure 1. Italian refining capacity [Mt] and number of refineries from 1970 to 2018.

The typical product slate of OECD and Italian refineries is showed in Figure 2 [6]. On the one hand, it is important to note that 40% of Italian refinery production is diesel fuel, more than double that of gasoline. On the other hand, the ratio diesel/gasoline is almost 3:1 in OECD countries [6]. Hence, an imbalance of products is observed mainly due to internal consumption that is triple the amount of diesel compared to gasoline. This trend is aligned with EU market that exports gasoline and imports diesel [5]. The imports of Italian refinery products are 1/3 of the overall production.

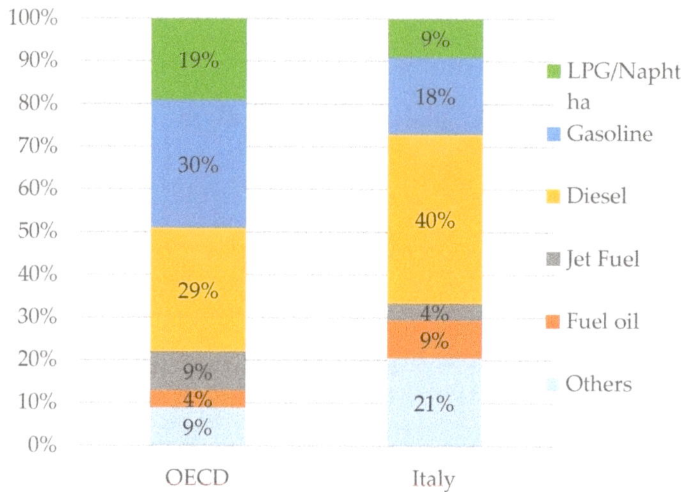

Figure 2. Distribution of products made from crude oil in OECD and Italy.

Extensive information about the refining sector in Italy is regularly published by UNEM ("Unione Energie per la Mobilità", "ex-Unione Petrolifera") and by ENI [6,48] and at the European level by CONCAWE [7,28], FuelsEurope [49] and the European Commission Joint Research Centre [5,8]. However, scientific literature about Italian refineries is relatively scarce and it is mainly focused on environmental (gas pollutant emissions [50], volatile organic compounds [51] and impact in soils [52]) and socioeconomic [53,54] assessments of refineries. Only in [55] is an overview presented of the status of implementation of BAT related to energy efficiency in some Italian industrial sites involved in the integrated pollution prevention and control (IPPC-IED) European Directives. This analysis includes 12 refineries and showed the high maturity of the Italian refining sector in terms of heat integration (except Pinch analysis), process optimization, and cogeneration.

3. Materials and Methods

According to the Italian transposition of Art.8 of European Energy Efficiency Directive, large companies and energy intensive enterprises must carry out, starting from 2015 and every four years, energy audits of their production sites [56,57]. The refining activities are highly energy intensive, and they are associated with large companies and sites with high energy consumption rates. Specifically, all companies must submit to ENEA (as national manager of energy audits database) the energy audits of all their production sites with a primary energy consumption higher than 10,000 toe [58].

In this work the energy consumption, referring to 2018 data from 10 Italian refineries, has been analysed (see Table 1). They represent the 84% of the total installed refining capacity in Italy. Two refineries have been excluded in this study due to their unique features: the ENI 2nd Generation biorefinery at Gela (the most innovative refinery in Europe, 0.75 Mt/year) [59] and the high quality bitumen ALMA refinery at Ravenna (0.55 Mt/year) [60].

Table 1. Main data of analysed refineries.

# Refinery	Capacity [Mt]	NCI	Ratio				Ref.
			LPG	Gasoline	Diesel Jet	Fuel Oil Others	
1	3.9	9.7 [2]	3.7%	17.9%	45.6%	32.8%	[61]
2	4.2	6.8 [3]	1.0%	28.3%	67.2%	3.5%	[62]
3	8.5	7.2 [3]	2.0%	29.0%	49.0%	20.0%	[63]
4	4.3	12.6 [2]	1.1%	19.0%	27.5%	52.5%	[64]
5	5.5	10.3 [3]	2.0%	16.0%	36.0%	46.0%	[65]
6	19.4	10.6 [1]	0.7%	19.5%	54.5%	25.3%	[66]
7	8.8	11.6 [2]	0.8%	16.3%	41.5%	41.3%	[67]
8	1.75	5.1 [4]	-	-	50.0%	50.0%	[68]
9	15	11.7 [1]	2.2%	32.6%	59.8%	5.4%	[69]
10	8.75	6.3 [2]	1.6%	31.6%	55.9%	10.9%	[70]

[1] Updated public data, verified in this work. [2] 2007 public data, verified in this work. [3] 2007 public data, not verified in this work. [4] Calculated in this work.

In order to ensure anonymity of the information provided by the companies only aggregated and public data are presented in this study. The capacity of different refineries is published by UNEM) [6]. Energy audits include detailed information about energy flows and consumptions inside the refinery, including exchanges of energy between units. However, the energy efficiency indicators refer to the refined crude oil, and the distribution of final products is not included. These data have been obtained from publicly available data [61–70].

The NCI of a refinery is calculated as the sum of the complexity factors of all the process units, weighted by the unit capacities relative to atmospheric distillation unit (ADU),

$$\text{NCI (Refinery)} = \sum \frac{\text{Capacity (Unit)}}{\text{Capacity (ADU)}} \cdot \text{CF(Unit)} \quad (1)$$

The complexity factors of the units are defined by the cost of the unit relative to the cost of ADU normalized on a capacity basis

$$\text{CF (Unit)} = \frac{\text{Cost (Unit)} \cdot \text{Capacity (Unit)}}{\text{Cost (ADU)} \cdot \text{Capacity (ADU)}} \quad (2)$$

The CF are standard values that depends on the processes. For example, CF of ADU is 1, CF of vacuum distillation is 2, and CF of fluid catalytic cracking is 6. Multiple CF values are listed and updated periodically. However, CF values present a non-negligible uncertainty level that drives the companies to adapt them to their specific needs [4,45,48].

Despite the extensive use of this information, the updated NCI value of most Italian refineries has not been published. Only two refineries update periodically the values of their NCI and for the rest of the refineries the last available values date back to 2007 [71]. Hence, a methodology to estimate the NCI of the analysed refineries has been developed.

Energy audits contain information about the energy flows on different units, but the capacity of each unit is not reported. Therefore, it is not possible to directly calculate the NCI. However, by cross-referencing the information from product ratios with the typical distribution of different refining processes and their relative unit capacity (see Figure 3) on conventional refineries [26,30] it has been possible to obtain a first approximation of NCIs. Subsequently, by taking the updated values as reference, some CF values have been adjusted and the NCIs have been recalculated for 7 refineries (Table 1). The averaged NCI value is equal to 9.2 with a sensible increase from past values (7.0 in 2005 and 9.0 in 2009) [5,71]. This increase is perfectly aligned with European refineries (9.2 in 2018, 8.3 in 2000) [6]. This value should be reviewed and confirmed to include the refineries excluded from the analysis and the bio-refinery units.

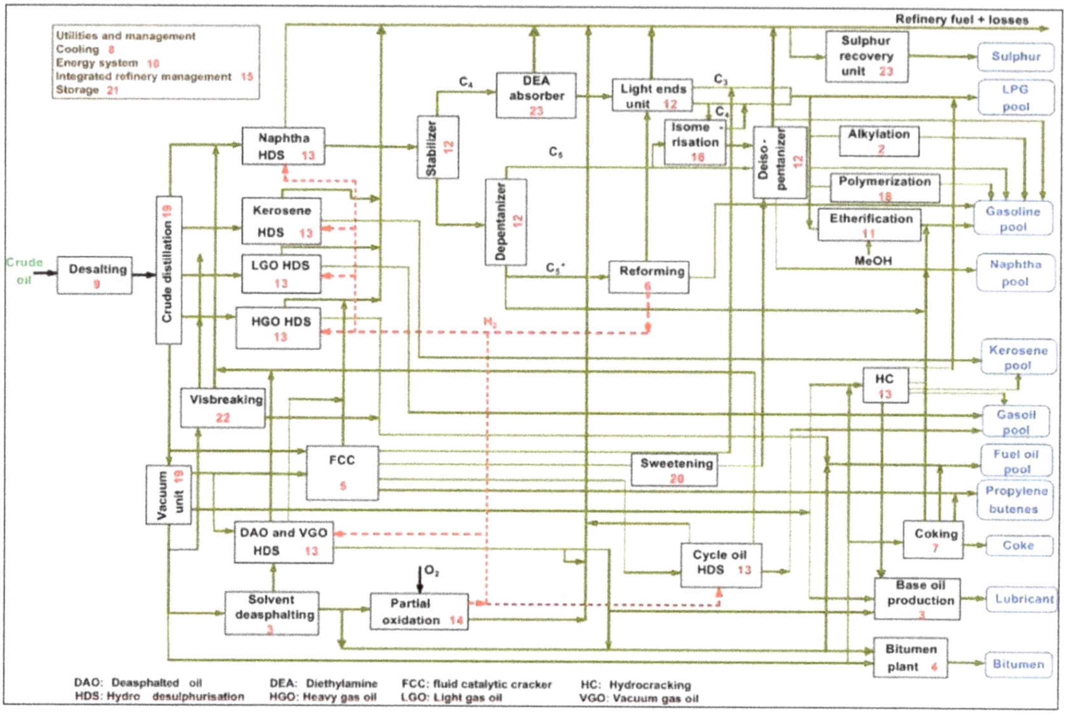

Figure 3. General scheme of a complex oil refinery [5].

Refineries present a high degree of energy integration among units and a high complexity of energy units (mainly several MW_e combined heat and power plants). In fact more than 90% of the energy of refineries is produced internally by the reuse of internal by-products [7,46,72]. The most important internal energy source is the refinery fuel gas (RFG), a mixed of different light hydrocarbons, mainly methane and ethane, with different proportions of propane, butane and hydrogen. Energy audits provide information about the uses of 10 different energy carrier flows in the units of analysed refineries with a focus on: electricity (purchased and internally generated), natural gas, auto-produced fuels, and steam (mainly internally generated in cogeneration units). The RFG accounts for nearly 65% of the total EU refinery fuel and its relative weight depends on refinery complexity [28]. The mean consumption of RFG in the analysed refineries accounts for about 49%.

It is important to highlight that the present work is focused on the analysis of energy consumption as a function of the production of oil derived fuels (GPL, gasoline, diesel/jet fuel, and fuel oil and other vacuum products). Power generation is a very important business associated to oil refineries [73], but its analysis is excluded from this work. Specifically, three integrated gasification combined cycles (IGCC) plants using refinery residues as feedstock are installed at SARAS (570 MW_e), API (280 MW_e) and ISAB (550 MW_e) refineries [55].

4. Results and Discussion

4.1. Energy Consumption Analysis

Best practices for the study of energy consumption in specific industrial sectors recommend that analyses are based on physical units of production (instead of economic data), as the information must be sufficiently disaggregated to allow for the analysis of processes and sites, and the models used in the evaluation should be general enough to be applied in different sectors [74]. Therefore, the use of the information from energy audits is ideal to define the sectoral specific energy consumption (SEC) [34]. The use of linear regression models is widely used for benchmarking analysis and energy efficiency measures [75–79] and a methodology for the characterization of productive sectors from energy audits has been developed and tested in a previous work (for a different industrial sector) [25]. Hence, the first step of the present work is to analyse the primary energy consumption (in tonnes of equivalent oil, normalized according to official conversion factors [80]) and the final electrical (in GWh) and thermal (in TJ) yearly consumptions as a function of the annual refined crude oil (in tonnes). The results of the regression are presented in Appendix A.

The linear regression between energy consumption and production presents a very high correlation ($R^2 > 0.9$), with a coefficient of correlation higher than critical Pearson correlation coefficient ($R_{crit} = 0.7079$, for a sample size $n = 10$ and $\alpha = 0.01$). Moreover, the low p-values (<0.0001) confirm that the analysis is statistically significant. However, as presented, the intercept of the regression presents a low reliability (the p-value associated with a two-tailed test "Prob > |t|" > 0.01) and a negative value. Hence, the correlation is not valid in all the crude oil refining range. As it is explained in the following, the range of validity of the correlation can be divided in three intervals of production:

- From 1.5 Mt to 3 Mt—No reliable
- From 3 Mt to 6 Mt—Medium reliability
- From 6 Mt to 15 Mt—High reliability

The SEC is defined as the ratio of energy used for refining a tonne of crude oil. Thus, SEC is calculated dividing both sides of the production function (from linear regression) and is represented by a hyperbolic function:

$$\text{SEC}\ [\text{toe}, \text{GWh}, \text{MJ}\ /\text{t}] = a\ [\text{toe}, \text{GWh}, \text{MJ}\ /\text{t}] + \frac{b\ [\text{toe}, \text{GWh}, \text{MJ}]}{x\ [\text{t}]} \quad (3)$$

where a and b respectively represent the slope and the intercept of the linear regression line. The values of primary, electrical and thermal SEC are presented in Table 2.

Table 2. Primary, electrical and thermal SEC values.

	SEC Model	Unit	Value (Mean ± SD)	Production Range [Mt]
Primary	$0.1596–3.124 \times 10^{+5}/\text{t}$	toe/t	0.0963 ± 0.0341	3–15
Electrical	$98.46–1.530 \times 10^{+8}/\text{t}$	kWh/t	68.20 ± 19.34	3–15
Thermal	$5082–1.178 \times 10^{+8}/\text{t}$	MJ/t	3421.71 ± 1316.84	3–15

The analysis of the SEC model uncertainty was developed to a significance level ($\alpha = 0.05$). Upper and lower limit curves of statistical significance (SECmodel $\pm 2\sigma$) have been defined through the propagation of the statistical error (based on the covariance matrix). The results for electrical and thermal SEC are presented in Figures 4 and 5.

It is possible to observe that there is not an economy of scale in the production, from an energy point of view. In other words, the energy consumption increases with production. This is due to the fact that there is no direct correlation between refining capacity and complexity and subsequent product slate (that define the main consumptions) of the refinery. An additional parameter that impacts on the energy consumption is the crude quality (i.e., API gravity or sulphur content) and its effect cannot be evaluated from energy audits information [41].

This analysis also shows that the SEC model presents two different areas. If production is higher than 6 Mt, energy consumption increases linearly with production and the mean and limits of the model are consistent with SEC mean values. For lower productions, the model uncertainty increases, and its accuracy decreases significantly (particularly under 3 Mt). Hence, for low productions the model is not reliable.

In the analysed energy audits, the allocation of energy flows inside the refinery is divided among the different standardized sub-processes and units [81]. The final production distribution directly depends on the presence and capacity of specific units as functions of NCI. Hence, the production slate directly impacts the energy consumption and on the subsequent allocation of GHG emissions by product [40].

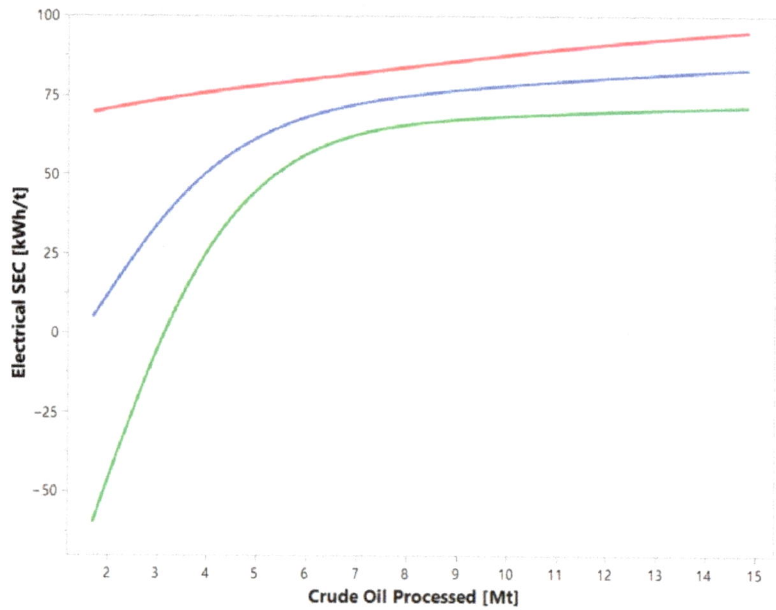

Figure 4. Electrical SEC model (blue: mean, red: upper limit, green: lower limit).

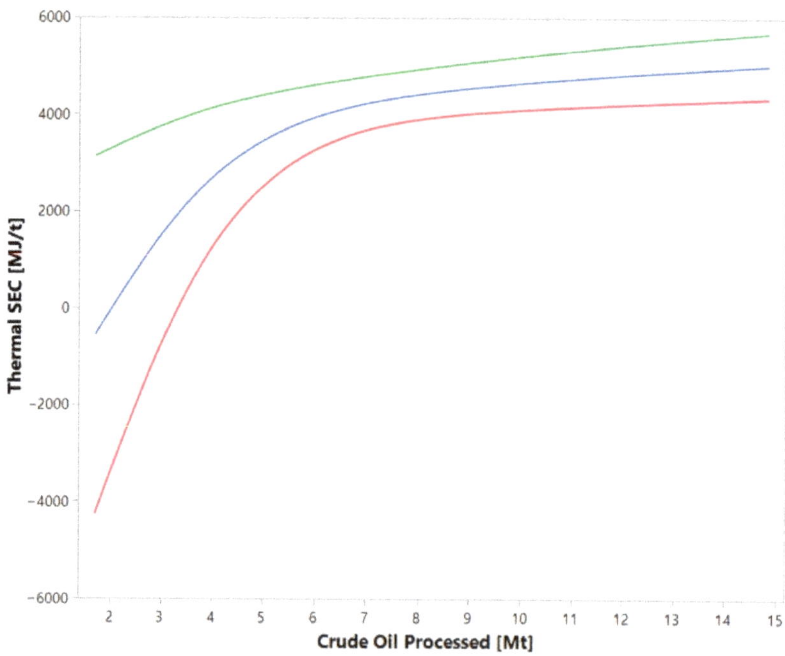

Figure 5. Thermal SEC model (blue: mean, green: upper limit, red: lower limit).

Therefore, an additional analysis based on linear regression was carried out to evaluate the correlation between primary, electrical and thermal energy consumptions and the final production in four different classes:

1. LPG: includes liquefied petroleum gas and other gaseous products (propane, propylene, etc.)
2. GASOLINE: mainly includes gasoline and virgin naptha
3. DIESEL: includes distillates mainly diesel and jet fuel. Other products are kerosene and heating oil
4. FUEL OIL & OTHERS: includes other vacuum distillation products: heavy fuel oil, petcoke, lubricating oils, waxes, asphalt, etc.

The linear correlations betweeen energy consumption and refined product classes (and their confidence intervals) are presented in Figures 6–8. The confidence intervals, which are displayed as the shaded area between linear regression and confidence curves, provide a range of values for the predicted mean for a given value of the predictor for $\alpha = 0.05$. The bands represent the uncertainty in the estimation of the true line, thus, uncertainity of the correlation increases with the confidence interval area. Table 3 summarizes the statistical regression parameters of Figures 6–8.

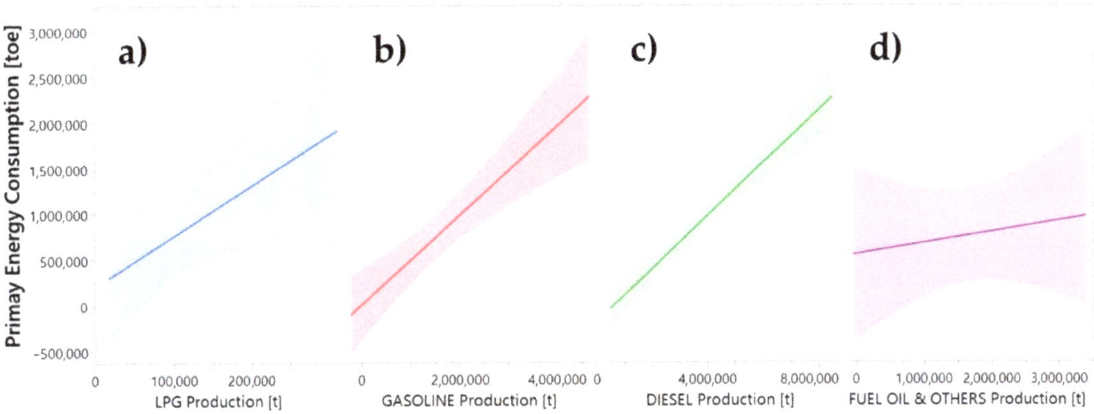

Figure 6. Primary energy consumption as function of final product class: (**a**) LPG; (**b**) gasoline; (**c**) diesel; (**d**) fuel oil & others.

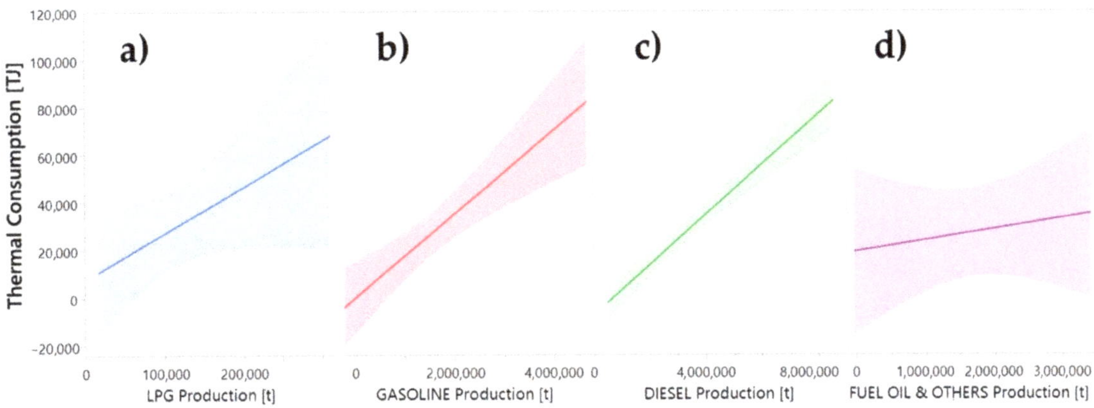

Figure 7. Thermal energy consumption as function of final product class: (**a**) LPG; (**b**) gasoline; (**c**) diesel; (**d**) fuel oil & others.

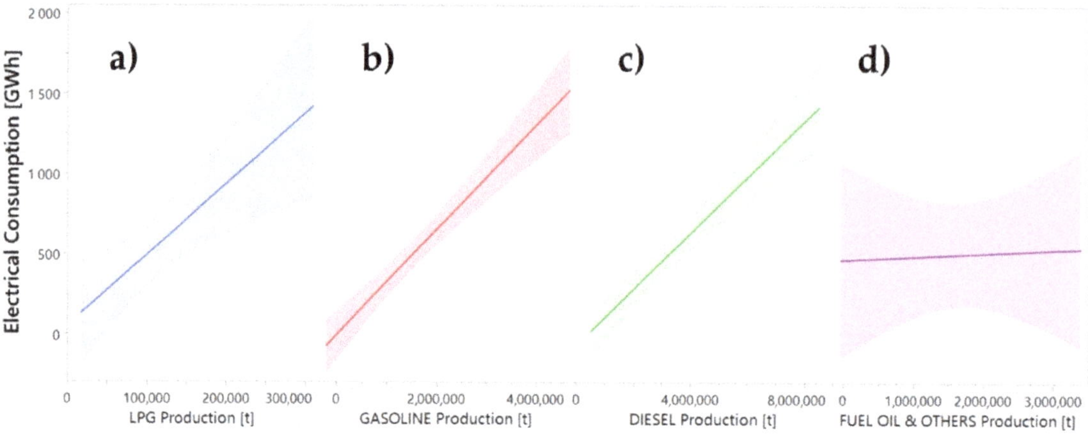

Figure 8. Electrical energy consumption as function of final product class: (**a**) LPG; (**b**) gasoline; (**c**) diesel; (**d**) fuel oil & others.

Table 3. Linear regression coefficients for energy consumption by product class.

Energy	Product Class	Energy Consumption Equation	Unit	R^2	p-Value
Primary	LPG	$203{,}020 + 5.501 \times t$	toe	0.399	0.0682
	GASOLINE	$-634.1 + 0.4875 \times t$		0.784	<0.001
	DIESEL	$-190{,}574 + 0.2888 \times t$		0.940	<0.001
	FUEL OIL & OTR	$550{,}766 + 0.1203 \times t$		0.041	0.5771
Electrical	LPG	$55.78 + 4.401 \times 10^{-3} \times t$	GWh	0.676	0.0065
	GASOLINE	$-0.739 + 3.27 \times 10^{-4} \times t$		0.914	<0.001
	DIESEL	$-65.55 + 1.74 \times 10^{-4} \times t$		0.890	<0.001
	FUEL OIL & OTR	$466.6 + 2.16 \times 10^{-5} \times t$		0.003	0.9731
Thermal	LPG	$7379 + 0.1956 \times t$	MJ	0.372	0.372
	GASOLINE	$-273.4 + 0.0176 \times t$		0.760	0.001
	DIESEL	$-7581 + 0.01057 \times t$		0.935	<0.001
	FUEL OIL & OTR	$19{,}168 + 0.00463 \times t$		0.045	0.5585

It is important to note that the three energy consumption analyses present a very high correlation with diesel production. The coefficients of determination for diesel production are: R^2(primary) = 0.940, R^2(thermal) = 0.935, and R^2(electrical) = 0.890. The energy consumption presents a high correlation with the production of gasolines ($R^2 > 0.75$) with a very high correlation with electrical consumption R^2(electrical) = 0.914. On the contrary, energy consumption presents a low correlation with LPG production and a null correlation with Fuel oil and others. Hence it is possible to hypothize that energy consumption of Italian refineries is primarly dependent on the middle distillates production and secondly from gasolines.

The main reason is linked with the relative weight of both products in the overall production. Diesel accounts for 50% of the global products, meanwhile gasoline accounts for 25%. However, it is important to note that gasoline production routes involve more processes than other products [40]. Hence the higher product-specific energy consumption ratio increases the correlation between energy consumption and gasoline products. Specifically, electricity-intensive units are mainly correlated to gasoline production (alkylation, hydrocracking and fluid catalytic cracking) [30] and this is reflected in a slightly higher correlation ($R^2_{electrical}$[Gasoline] = 0.914 vs. $R^2_{electrical}$[Diesel] = 0.890).

The importance of the diesel and gasoline ratios is in line with the literature data [38]. However this analysis is limited due to the lack of information contained in the energy audits, specifically the product slate and the mass flows in the sub-processes. Therefore it is not possible to analyse in detail the allocation of energy performance by specific product (Gasoline, Diesel, Jet, RFO, LPG and Petcoke) and to compare them with other sources [35,40,41].

Given the identified correlation between products distribution, despite the missing correspondence between NCI and capacity [4], it would be plausible to hypothesize a correlation between refineries complexity and their energy consumption and product slate as presented in [39–41]. However, not-statistically significant and very low correlations between NCI and energy consumption and production are observed (Figure 9). This result is in agreement with literature data where only a low correlation between NCI and capacity has been observed [4,45]. The lack of correlation between NCI and energy consumption is also linked to the low correlation with product slate. Usually, NCI increases with gasoline ratio, and decreases with diesel ratio. This trend is weakly observed only in diesel, showing a practically null correlation with the gasoline ratio.

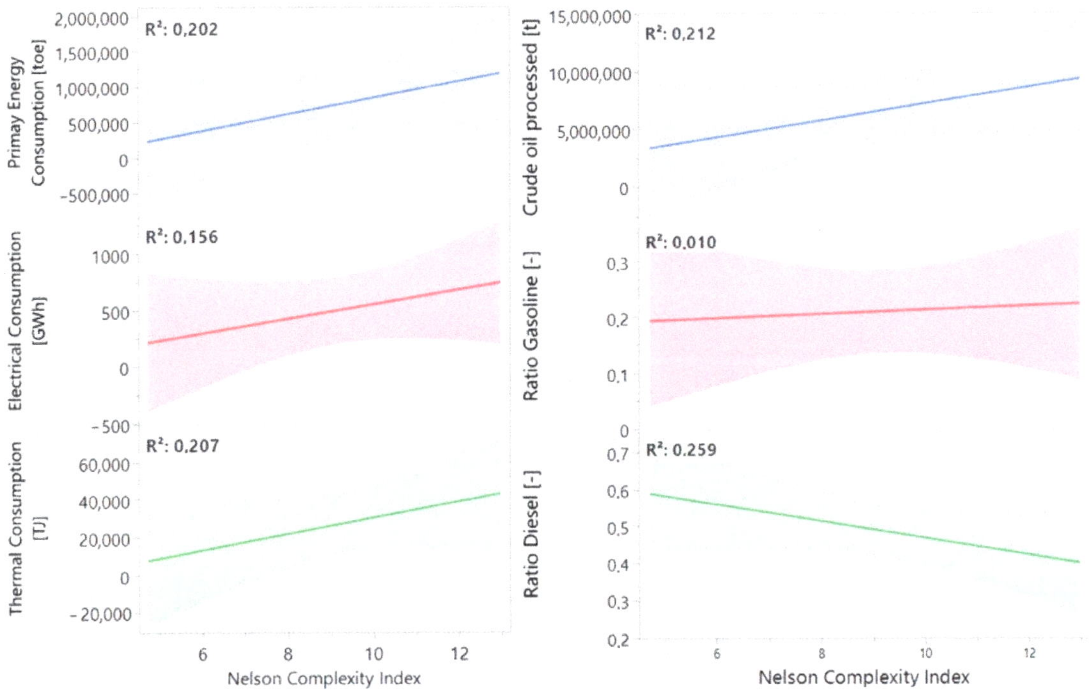

Figure 9. Primary, thermal and electrical energy consumption (**left**) and crude oil refined, gasoline and diesel ratios (**right**) as function of NCI.

It is important to note that this correlation can be also partially due to the uncertainty of the calculation of NCI. The values have been calculated from the information contained in the energy audits, but the capacity of each unit has been estimated from literature.

4.2. Energy Perfomance Improvement Actions (EPIAs) Analysis

For the ten examined refineries, energy audits also include information on energy efficiency measures implemented in the last four years, namely in the period between the last energy audit available (referred to as the December 2019 deadline) and the previous mandatory energy audit. The measures are described in terms of investments and energy savings and thus a cost effectiveness indicator can be computed, expressing the cost of saving one tonne of oil equivalent in different intervention areas.

Information on identified energy efficiency measures is also available, and on the potential savings associated with them: indeed, these measures are not yet planned, and their possible planning could be deferred in time. For these measures, a simple payback period is also available, computed without considering the access to existing incentive mechanisms for energy efficiency [82].

The energy needed for the refining process represents more than 60% of total refining costs [28] and this confirms that energy efficiency is a relevant issue also in the refining sector. Moreover, refineries are energy intensive industries and according to the Legislative Decree 73/2020 they are obliged to implement at least one of the energy efficiency measures identified in the energy audit. The analysis of implemented and planned measures shows the importance of interventions related to the production process, which are in each case very specific to the refining site examined and difficult to categorise. In fact, refineries are in general very complex sites, with different process units highly integrated with each other (Figure 3) and this implies very diversified industrial profiles.

According to the information provided in the energy audits, the refineries examined in the last four years have introduced 27 measures to improve energy efficiency. Among these measures, more than 80% have quantitative information on savings, which are equal to 44 ktoe/year of final energy and to additional 5 ktoe/year of primary energy. In the final energy savings, the main intervention category is production lines, with 30 ktoe of annual saving (66% of the total), referrings to intervention such as integration of heat recovery systems (in furnaces), flare gas recovery units, or the electrification of mechanical systems (mainly in air coolers and FCC units). This result is aligned with best available techniques as suggested by national regulations [55]. Pressure systems represent 9 ktoe (21%), followed by thermal power plant and other heat recovery systems with 5.5 ktoe (13%). The cost effectiveness indicator has the best value for production lines, around 1100 Euro/toe, followed by Pressure systems with a slightly higher value (1300 Euro/toe)

Energy audits also report 39 measures identified by the refineries analysed. Also in this case, quantitative information on savings is available for more than 80% of the measures (see Table 4). Potential savings of final energy are equal to 54 ktoe/year and potential savings of primary energy are almost negligible, since the unique measure identified in the production category from renewable sources (photovoltaic) is associated with a savings of 14 toe/year. As for implemented measures, the production lines category is associated with the majority of savings (80%, 43 ktoe/year), with interventions mainly focused on heat recovery systems and revamping of units (mainly VDU and HDS) and burners, followed by electric motors/inverters (12%, 6.3 ktoe/year), and thermal power plant and other heat recovery systems (5%, 2.5 ktoe/year). The production lines category has a good value for cost effectiveness indicator, which is around 900 Euro/toe; measures in the electric motor/inverter category have a similar value to the indicator, whereas measures in pressure systems have again a value around 1300 Euro/toe. In terms of simple payback time, the lowest value was observed in the thermal power plant and heat recovery category (lower than 2 years), followed by pressure systems and production lines (around 3 years).

Table 4. Analysis of payback time (PBT) (y), savings (toe) and investment (EUR) of identified EPIAs.

PBT Class	Number of EPIAs with Information	Saving of Final Energy (toe)	Investment (EUR)
PBT $\leq$ 1 year	8	11,563.8	5,240,000
1 < PBT $\leq$ 2 years	3	1306.0	645,000
2 < PBT $\leq$ 3 years	2	1271.8	786,000
3 < PBT $\leq$ 5 years	8	7810.6	8,917,200
5 < PBT $\leq$ 10 years	5	7612.5	10,434,000
PBT > 10 years	1	1883.4	8,000,000

The sum of implemented and identified EPIAs in the energy audits accounts for global energy savings close to 1.5% of final energy consumption of the analysed refineries. It is important to note that EPIAs usually are implemented during maintenance turnarounds of the refineries. These planned breaks in production are periodically carried out to have preventive maintenance, renovations, or upgrades. The turnarounds of the refineries take place every three or five years and for some weeks the production is stopped. Therefore, the costs are very high, and they require extensive and careful efforts in planning and coordination of the works. Hence, the analysis of EPIAs in energy audits should be integrated in the turnaround planning.

5. Conclusions

In this work the analysis of the energy performance of Italian oil refineries based on mandatory energy audits and public information was presented. For the first time primary, electrical and thermal consumptions as functions of refinery capacity have been evaluated. The analysis has been based on empirical data that present a value added for industry and academia despite their uncertainties.

A strong correlation between energy consumption and the quantity of crude oil refined has been observed. However, an analysis of SEC with production revealed that other factors have a stronger impact on the energy consumption of refineries than refining capacity. The variability and uncertainty of SEC is lower in refineries with high capacity (6–15 Mt) than in small ones (3–6 Mt). Hence the size of the plant should be considered in the calculation of the SEC.

Other key variables have been analysed. On the one hand, energy consumption is mainly driven by diesel products and, in a second order, by gasoline products with a high impact on electrical consumption. On the other hand, no correlation between the Nelson complexity index and energy consumption has been observed.

The analysis of implemented and identified EPIAs has been carried out. Despite the high degree of integration and efficiency of the refineries, most of the energy efficiency interventions are focused on the improvement and revamping of current units, with particular attention to heat integration and recovery.

This work provides important insights and updates and represents a first step for benchmarking refinery energy consumption. The analysis carried out shows that to achieve a better level of detail it will be necessary to collect additional information that is not currently contained in energy audits, such as the specific properties of crude oil, a higher detail of final products distribution, comprehensive information of mass balances by production unit integration, and a current complexity index of each refinery. Therefore, the methodology to develop more effective energy audits should be improved including information about these parameters.

The number of samples limited the statistical significance of a multiple variable linear regression analysis. However, with more details related to crude oil, product slate and sub-process mass balances, the current work could be sensibly improved. Moreover, the robustness of the models should be improved if monthly data were available.

Finally, the methodology developed should be replicated with the energy audits received every four years. The impact of the implementation of energy efficiency measures in the sector could be analysed and a detailed trend of the evolution of the refining sector with time could be studied. More detailed information gathering in the energy audits (including information about mass balances, crude and product properties and complexity) should be very useful to policymakers and improve the sectoral benchmark.

Author Contributions: C.H.—Conceptualization, Data curation, Formal analysis, Methodology, Writing—original draft. C.M.—Conceptualization, Methodology, Writing—original draft, Writing—review & editing. M.S.—Conceptualization, Methodology, Writing—review & editing. C.T.—Conceptualization, Formal analysis, Methodology, Writing—review & editing. All authors have read and agreed to the published version of the manuscript.

Funding: This work has received funding in National Electrical System Research Program (RdS PTR 2019–2021—WP1.6), "Piano Triennale della Ricerca del Sistema Elettrico Nazionale 2019–2021" implemented under programme agreements between the Italian Ministry for Economic Development and ENEA, CNR, and RSE S.p.A.

Institutional Review Board Statement: Not applicable.

Informed Consent Statement: Not applicable.

Conflicts of Interest: The authors declare no conflict of interest. The funders had no role in the design of the study; in the collection, analyses, or interpretation of data; in the writing of the manuscript, or in the decision to publish the results.

Appendix A. Bivariate Statistical Analysis

This appendix presents the main results of regression analysis for the calculation of primary, electrical, and thermal SECs, according to JMP 15 software, including details of the linear fit, summary of fit, analysis of variance and parameter estimates.

Appendix A1. Bivariate Fit of Primary Energy Consumption [toe] By Crude Oil Refined [t]

Linear Fit

Primary Energy Consumption [toe] = −312,425.5 + 0.1596493 × Crude Oil Refined [t]

Table A1. Summary of Fit.

RSquare	0.949723
RSquare Adj	0.943438
Root Mean Square Error	162,631.9
Mean of Response	750,193.5
Observations	10

Table A2. Analysis of Variance.

Source	DF	Sum of Squares	Mean Square	F Ratio
Model	1	$3.9969 \times 10^{+12}$	$3.997 \times 10^{+12}$	151.1171
Error	8	$2.1159 \times 10^{+11}$	$2.645 \times 10^{+10}$	**Prob > F**
C. Total	9	$4.2085 \times 10^{+12}$		<0.0001

Table A3. Parameter Estimates.

Term	Estimate	Std Error	t Ratio	Prob > |t|
Intercept	−312,425.5	100,583.3	−3.11	0.0145
Slope	0.1596493	0.012987	12.29	<0.0001

Appendix A2. Bivariate Fit of Electrical Consumption [GWh] By Crude Oil Refined [t]

Linear Fit

Electrical Consumption [GWh] = −152.9579 + 9.8465 × 10^{+5} × Crude Oil Refined [t]

Table A4. Summary of Fit.

RSquare	0.93824
RSquare Adj	0.93052
Root Mean Square Error	111.8482
Mean of Response	502.4196
Observations	10

Table A5. Analysis of Variance.

Source	DF	Sum of Squares	Mean Square	F Ratio
Model	1	1,520,383.5	1,520,384	121.5333
Error	8	100,080.1	12,510	**Prob > F**
C. Total	9	1,620,463.6		<0.0001

Table A6. Parameter Estimates.

Term	Estimate	Std Error	t Ratio	Prob > |t|
Intercept	−152.9579	69.17496	−2.21	0.0580
Slope	9.8465×10^{-5}	8.932×10^{-6}	11.02	<0.0001

Appendix A3. Bivariate Fit of Thermal Consumption [TJ] By Crude Oil Refined [t]
Linear Fit
Thermal Consumption [TJ] = −11,782.73 + 0.0058019 × Crude Oil Refined [t]

Table A7. Summary of Fit.

RSquare	0.93245
RSquare Adj	0.924007
Root Mean Square Error	6913.854
Mean of Response	26,834.58
Observations	10

Table A8. Analysis of Variance.

Source	DF	Sum of Squares	Mean Square	F Ratio
Model	1	5,278,785,165	$5.2788 \times 10^{+9}$	110.4316
Error	8	382,411,063	47,801,383	**Prob > F**
C. Total	9	5,661,196,227		<0.0001

Table A9. Parameter Estimates.

Term	Estimate	Std Error	t Ratio	Prob > |t|
Intercept	−11,782.73	4276.025	−2.76	0.0248
Slope	0.0058019	0.000552	10.51	<0.0001

References

1. BP. Statistical Review of World Energy 2021; BP p.l.c. Available online: https://www.bp.com/en/global/corporate/energy-economics/statistical-review-of-world-energy.html (accessed on 11 November 2021).
2. Hubbert, M.K. *Nuclear Energy and the Fossil Fuels*; Publication No. 95; Shell Development, Co.: Houston, TX, USA, 1956; pp. 1–40.
3. Nashawi, I.S.; Malallah, A.; Al-Bisharah, M. Forecasting World Crude Oil Production Using Multicyclic Hubbert Model. *Energy Fuels* **2010**, *24*, 1788–1800. [CrossRef]
4. Kaiser, M.J. A review of refinery complexity applications. *Pet. Sci.* **2017**, *14*, 167–194. [CrossRef]
5. Barthe, P.; Chaugny, M.; Roudier, S.; Delgado-Sancho, L. *Best Available Techniques (BAT) Reference Document for the Refining of Mineral Oil and Gas*; JRC: Publications Office of the European Union: Luxembourg, 2015.
6. Unione Petrolifera. Data Book 2020. Unem. Available online: https://www.unem.it/wp-content/uploads/2019/12/Data-Book-2020.pdf (accessed on 11 November 2021).
7. CONCAWE. *CO_2 Emissions from EU Refineries*; CONCAWE: Brussels, Belgium, 2008; Available online: https://www.concawe.eu/wp-content/uploads/2017/01/cr171-co2_emissions_from_refineries-2008-01728-01-e.pdf (accessed on 11 November 2021).
8. Lukach, R.; Marschinski, R.; Bakhtieva, D.; Mraz, M.; Temurshoev, U.; Eder, P.; Delgado Sancho, L. *EU Petroleum Refining Fitness Check: Impact of EU Legislation on Sectoral Economic Performance*; EUR 27262 EN; JRC Science Hub: Brussels, Belgium, 2015. [CrossRef]
9. European Commission. *Masterplan for a Competitive Transformation of EU Energy-Intensive Industries Enabling a Climate-Neutral, Circular Economy by 2050*; Publications Office of the European Union: Luxembourg, 2019. [CrossRef]
10. IEA. *The Future of Hydrogen*; International Energy Agency: Paris, France, 2019; Available online: https://www.iea.org/reports/the-future-of-hydrogen (accessed on 11 November 2021).
11. U.S. Energy Information Administration (EIA). Hydrogen for Refineries is Increasingly Provided by Industrial Suppliers. Available online: https://www.eia.gov/todayinenergy/detail.php?id=24612 (accessed on 11 November 2021).
12. Ellerington, T. Refining Fitness Check: Step 1. In Proceedings of the 11th Concawe Symposium, Brussels, Belgium, 23–24 February 2015.
13. Holladay, J.D.; Hu, J.; King, D.L.; Wang, Y. An overview of hydrogen production technologies. *Catal. Today* **2009**, *139*, 244–260. [CrossRef]

14. Baykara, S.Z. Hydrogen: A brief overview on its sources, production and environmental impact. *Int. J. Hydrogen Energy* **2018**, *43*, 10605–10614. [CrossRef]
15. Leeson, D.; Mac Dowell, N.; Shah, N.; Petit, C.; Fennell, P.S. A Techno-economic analysis and systematic review of carbon capture and storage (CCS) applied to the iron and steel, cement, oil refining and pulp and paper industries, as well as other high purity sources. *Int. J. Greenh. Gas Control* **2017**, *61*, 71–84. [CrossRef]
16. IEAGHG. *Techno-Economic Evaluation of SMR Based Standalone (Merchant) Hydrogen Plant with CCS*; IEAGHG 2017/02; IEAGHG: Cheltenham, UK, 2017; Available online: http://documents.ieaghg.org/index.php/s/HKtMncwfw2vaBxl/download (accessed on 11 November 2021).
17. Bezergianni, S.; Dimitriadis, A.; Kikhtyanin, O.; Kubička, D. Refinery co-processing of renewable feeds. *Prog. Energy Combust. Sci.* **2018**, *68*, 29–64. [CrossRef]
18. Nigam, P.S.; Singh, A. Production of liquid biofuels from renewable resources. *Prog. Energy Combust. Sci.* **2011**, *37*, 52–68. [CrossRef]
19. Ganguly, P.; Sarkhel, R.; Das, P. 2—The Second- and Third-Generation Biofuel Technologies: Comparative Perspectives. In *Sustainable Fuel Technologies Handbook*; Dutta, S., Mustansar Hussain, C., Eds.; Academic Press: Cambridge, MA, USA, 2021; pp. 29–50. ISBN 978-0-12-822989-7.
20. Hao, H.-T.N.; Karthikeyan, O.P.; Heimann, K. Bio-Refining of Carbohydrate-Rich Food Waste for Biofuels. *Energies* **2015**, *8*, 6350–6364. [CrossRef]
21. Jung, K.A.; Lim, S.-R.; Kim, Y.; Park, J.M. Potentials of macroalgae as feedstocks for biorefinery. *Bioresour. Technol.* **2013**, *135*, 182–190. [CrossRef]
22. Kohli, K.; Prajapati, R.; Sharma, B.K. Bio-Based Chemicals from Renewable Biomass for Integrated Biorefineries. *Energies* **2019**, *12*, 233. [CrossRef]
23. Ubando, A.T.; Felix, C.B.; Chen, W.-H. Biorefineries in circular bioeconomy: A comprehensive review. *Bioresour. Technol.* **2020**, *299*, 122585. [CrossRef]
24. Gangar, N.; Macchietto, S.; Markides, C.N. Recovery and Utilization of Low-Grade Waste Heat in the Oil-Refining Industry Using Heat Engines and Heat Pumps: An International Technoeconomic Comparison. *Energies* **2020**, *13*, 2560. [CrossRef]
25. Bruni, G.; De Santis, A.; Herce, C.; Leto, L.; Martini, C.; Martini, F.; Salvio, M.; Tocchetti, F.A.; Toro, C. From Energy Audit to Energy Performance Indicators (EnPI): A Methodology to Characterize Productive Sectors. The Italian Cement Industry Case Study. *Energies* **2021**, *14*, 8436. [CrossRef]
26. Worrell, E.; Galitsky, C. *Energy Efficiency Improvement and Cost Saving Opportunities for Petroleum Refineries. An ENERGY STAR® Guide for Energy and Plant Managers*; Lawrence Berkeley National Laboratory: Berkeley, CA, USA, 2005.
27. Morrow, W., III; Marano, J.; Sathaye, J.; Hasanbeigi, A.; Xu, T. *Assessment of Energy Efficiency Improvement in the United States Petroleum Refining Industry*; Lawrence Berkeley National Laboratory: Berkeley, CA, USA, 2013.
28. CONCAWE. EU Refinery Energy Systems and Efficiency. Conservation of Clean Air and Water in Europe. 2012. Available online: https://www.concawe.eu/wp-content/uploads/2017/01/rpt_12-03-2012-01520-01-e.pdf (accessed on 11 November 2021).
29. Szklo, A.; Schaeffer, R. Fuel specification, energy consumption and CO_2 emission in oil refineries. *Energy* **2007**, *32*, 1075–1092. [CrossRef]
30. Talaei, A.; Oni, A.O.; Ahiduzzaman, M.; Roychaudhuri, P.S.; Rutherford, J.; Kumar, A. Assessment of the impacts of process-level energy efficiency improvement on greenhouse gas mitigation potential in the petroleum refining sector. *Energy* **2020**, *191*, 116243. [CrossRef]
31. Marton, S.; Svensson, E.; Harvey, S. Operability and Technical Implementation Issues Related to Heat Integration Measures—Interview Study at an Oil Refinery in Sweden. *Energies* **2020**, *13*, 3478. [CrossRef]
32. Marton, S.; Svensson, E.; Subiaco, R.; Bengtsson, F.; Harvey, S. A Steam Utility Network Model for the Evaluation of Heat Integration Retrofits—A Case Study of an Oil Refinery. *J. Sustain. Dev. Energy Water Environ. Syst.* **2017**, *5*, 560–578. [CrossRef]
33. Nilsson, K.; Zetterberg, L.; Wråke, M. *Allowance Allocation and CO_2 intensity of the EU15 and Norwegian Refineries*; Report B1616; IVL Swedish Environmental Research Institute: Stockholm, Sweden, 2005; Available online: https://www.ivl.se/download/18.694ca0617a1de98f47299f/1628413669183/FULLTEXT01.pdf (accessed on 11 November 2021).
34. Lawrence, A.; Thollander, P.; Andrei, M.; Karlsson, M. Specific Energy Consumption/Use (SEC) in Energy Management for Improving Energy Efficiency in Industry: Meaning, Usage and Differences. *Energies* **2019**, *12*, 247. [CrossRef]
35. Worrell, E.; Cuelenaere, R.F.A.; Blok, K.; Turkenburg, W.C. Energy consumption by industrial processes in the European Union. *Energy* **1994**, *19*, 1113–1129. [CrossRef]
36. Rikhtegar, F.; Sadighi, S. Optimisation of energy consumption. *PTQ* **2015**, *Q2*, 1–8. Available online: http://www.eptq.com/articles/q2ppg_26032015104811.pdf (accessed on 11 November 2021).
37. Phylipsen, G.J.M.; Blok, K.; Worrell, E. International Comparisons of Energy Efficiency—Methodologies for the Manufacturing Industry. *Energy Policy* **1997**, *25*, 715–725. [CrossRef]
38. Wang, M.; Lee, H.; Molburg, J. Allocation of energy use in petroleum refineries to petroleum products. *Int. J. Life Cycle Assess.* **2004**, *9*, 34–44. [CrossRef]
39. Forman, G.S.; Divita, V.B.; Han, J.; Cai, H.; Elgowainy, A.; Wang, M.U.S. Refinery Efficiency: Impacts Analysis and Implications for Fuel Carbon Policy Implementation. *Environ. Sci. Technol.* **2014**, *48*, 7625–7633. [CrossRef]

40. Elgowainy, A.; Han, J.; Cai, H.; Wang, M.; Forman, G.S.; DiVita, V.B. Energy Efficiency and Greenhouse Gas Emission Intensity of Petroleum Products at U.S. Refineries. *Environ. Sci. Technol.* **2014**, *48*, 7612–7624. [CrossRef]
41. Han, J.; Forman, G.S.; Elgowainy, A.; Cai, H.; Wang, M.; DiVita, V.B. A comparative assessment of resource efficiency in petroleum refining. *Fuel* **2015**, *157*, 292–298. [CrossRef]
42. Moretti, C.; Moro, A.; Edwards, R.; Rocco, M.V.; Colombo, E. Analysis of standard and innovative methods for allocating upstream and refinery GHG emissions to oil products. *Appl. Energy* **2017**, *206*, 372–381. [CrossRef]
43. Nelson, W.L. *Guide to Refinery Operating Costs (Process Costimating)*; Petroleum Pub.: Tulsa, OK, USA, 1976; ISBN 9780878140732.
44. Oil & Gas Journal Research. 2021 Worldwide Refinery Survey with Complexity Analysis. Available online: https://ogjresearch.com/products/2021-worldwide-refinery-survey-with-complexity-analysis.html (accessed on 11 November 2021).
45. Kaiser, M.J. A new approach to refinery complexity factors. *Eng. Econ.* **2017**, *62*, 336–368. [CrossRef]
46. Eurostat. Energy Data—2020 Edition. 2020. Available online: https://ec.europa.eu/eurostat/web/products-statistical-books/-/KS-HB-20-001 (accessed on 11 November 2021).
47. Marschinski, R.; Lukach, R.; Barreiro-Hurle, J. Competitiveness of Energy-Intensive Industries in Europe: The Crisis of the Oil Refining Sector between 2008 and 2013. *Econ. Energy Environ. Policy* **2020**, *9*, 167–184. [CrossRef]
48. ENI. World Oil Review 2020; ENI, 2021; Volume 1. Available online: https://www.eni.com/assets/documents/eng/scenari-energetici/WORLD-OIL-REVIEW-2020-vol1.pdf (accessed on 11 November 2021).
49. FuelsEurope. *Statistical Report 2021*; FuelsEurope: Brussels, Belgium, 2021; Available online: https://www.fuelseurope.eu/wp-content/uploads/SR_FuelsEurope-_2021_FINAL-REPORT-1.pdf (accessed on 11 November 2021).
50. Bevilacqua, M.; Braglia, M. Environmental efficiency analysis for ENI oil refineries. *J. Clean. Prod.* **2002**, *10*, 85–92. [CrossRef]
51. Roveda, L.; Polvara, E.; Invernizzi, M.; Capelli, L.; Sironi, S. Definition of an Emission Factor for VOC Emitted from Italian and European Refineries. *Atmosphere* **2020**, *11*, 564. [CrossRef]
52. Duplay, J.; Semhi, K.; Mey, M.; Messina, A.; Quaranta, G.; Huber, F.; Aubert, A. Geogenic versus anthropogenic geochemical influence on trace elements contents in soils from the Milazzo Peninsula. *Geochemistry* **2014**, *74*, 691–704. [CrossRef]
53. Pagliaro, M.; Meneguzzo, F. Oil refining in Sicily: A critical perspective looking to the future. *Energy Sci. Eng.* **2020**, *8*, 566–573. [CrossRef]
54. Falcone, P.M.; Imbert, E.; Sica, E.; Morone, P. Towards a bioenergy transition in Italy? Exploring regional stakeholder perspectives towards the Gela and Porto Marghera biorefineries. *Energy Res. Soc. Sci.* **2021**, *80*, 102238. [CrossRef]
55. Mazziotti Gomez de Teran, C.; Fiore, D.; Favaroni, M.; Fardelli, A. Overview of technologies for improving energy efficiency from energy intensive industries derived from the Italian national experience in IPPC permit licensing. *J. Clean. Prod.* **2017**, *168*, 1547–1558. [CrossRef]
56. DECRETO LEGISLATIVO 14 Luglio 2020, n. 73. Attuazione Della Direttiva (UE) 2018/2002 Che Modifica la Direttiva 2012/27/UE Sull'efficienza Energetica. (20G00093) (GU Serie Generale n.175 del 14-07-2020). Available online: https://www.gazzettaufficiale.it/eli/id/2020/07/14/20G00093/sg (accessed on 11 November 2021).
57. DIRECTIVE (EU) 2018/2002 OF THE EUROPEAN PARLIAMENT AND OF THE COUNCIL of 11 December 2018 Amending Directive 2012/27/EU on Energy Efficiency 2018. Available online: https://eur-lex.europa.eu/legal-content/EN/TXT/HTML/?uri=CELEX:32018L2002&from=EN (accessed on 11 November 2021).
58. Santino, D.; Biele, E.; Salvio, M. *Guidelines for Energy Audits under Article 8 of the EED: Italy's Implementation Practices and Tools*; ENEA: Rome, Italy, 2019. Available online: https://www.efficienzaenergetica.enea.it/component/jdownloads/?task=download.send&id=377&catid=40&Itemid=101 (accessed on 11 November 2021).
59. ENI. In Gela the Most Innovative Biorefinery in Europe. Available online: https://www.eni.com/en-IT/operations/italy-gela-innovative-biorefinery.html (accessed on 11 November 2021).
60. ALMA Petroli. The Refinery. Available online: https://www.almapetroli.com/en/refinery (accessed on 11 November 2021).
61. API. Autorizzazione Integrata Ambientale (AIA)—Raffineria di Falconara Marittima—Ancona. 2017. Available online: https://va.minambiente.it/it-IT/Oggetti/Documentazione/1838/4240 (accessed on 11 November 2021).
62. ENI. Raffineria di Venezia—Dichiarazione Ambientale 2020–2022. 2020. Available online: https://www.eni.com/assets/documents/ita/attivita/mid-downstream/Dichiarazione-Ambientale-Venezia-2020.pdf (accessed on 11 November 2021).
63. ENI. Raffineria di Sannazzaro de' Burgondi—Dichiarazione Ambientale Edizione 2020. 2020. Available online: https://www.eni.com/assets/documents/ita/attivita/mid-downstream/DICHIARAZIONE-AMBIENTALE-ENI-SANNAZZARO-2020-anno-rif-2019.pdf (accessed on 11 November 2021).
64. ENI. Raffineria Livorno—Dichiarazione Ambientale Edizione 2020. 2020. Available online: https://www.eni.com/assets/documents/ita/attivita/mid-downstream/Dichiarazione-Ambientale-Livorno-2019.pdf (accessed on 11 November 2021).
65. ENI. Raffineria Taranto—Dichiarazione Ambientale Edizione 2020. 2020. Available online: https://www.eni.com/assets/documents/ita/sostenibilita/2020/Dichiarazione-Ambientale-2020-della-Raffineria-di-Taranto.pdf (accessed on 11 November 2021).
66. LUKOIL. Data Book 2020. Available online: https://www.lukoil.com/FileSystem/9/561978.xlsx?dl=1 (accessed on 11 November 2021).
67. ESSO. Autorizzazione Integrata Ambientale (AIA)—Raffineria di Augusta (ex ESSO)—Relazione di Riferimento ai Sensi del DM 272/14. *2016*. Available online: https://va.minambiente.it/it-IT/Oggetti/Documentazione/1901/4180?Testo=&RaggruppamentoID=2049#form-cercaDocumentazione (accessed on 11 November 2021).
68. IPLOM. Bilancio di Sostenibilità della Raffineria. 2020. Available online: https://iplom.it/wp-content/uploads/2020/09/Iplom_bilancio_2017_2018.pdf (accessed on 11 November 2021).

69. SARAS. Refining—Saras 2020. Available online: https://www.saras.it/en/what-we-do/refining (accessed on 11 November 2021).
70. SARPOM. Autorizzazione Integrata Ambientale (AIA)—Raffineria di Trecate. 2016. Available online: https://va.minambiente.it/it-IT/Oggetti/Documentazione/1830/4244?Testo=&RaggruppamentoID=2100#form-cercaDocumentazione (accessed on 11 November 2021).
71. Petrillo, A.W. Gruppo API—Strategia e Posizionamento di Mercato Nel Settore della Raffinazione. 2010. Available online: http://www.comitati-cittadini.org/wp-content/uploads/2012/06/api-nova-energia-slide.ppt (accessed on 11 November 2021).
72. Eurostat Physical Energy Flow Accounts (PEFA)—Key Indicators of Physical Energy Flow Accounts by NACE Rev. 2 Activity. Available online: https://ec.europa.eu/eurostat/databrowser/view/ENV_AC_PEFA04__custom_1509250/default/table?lang=en (accessed on 11 November 2021).
73. Golmohamadi, H.; Asadi, A. Integration of Joint Power-Heat Flexibility of Oil Refinery Industries to Uncertain Energy Markets. *Energies* **2020**, *13*, 4874. [CrossRef]
74. IEA. *Energy Efficiency Indicators: Essentials for Policy Making*; International Energy Agency: Paris, France, 2014.
75. Chen, Y.; Therkelsen, P. *The Effect of Linear Regression Modeling Approaches on Determining Facility Wide Energy Savings*; Lawrence Berkeley National Laboratory: Berkeley, CA, USA, 2019.
76. Therkelsen, P.; Rao, P.; Sholes, D.; Meffert, B.; Green, R.; Nimbalkar, S.U.; Mckane, A. The Value of Regression Models in Determining Industrial Energy Savings. In Proceedings of the ECEEE Industrial Efficiency 2016, Kalkscheune, Berlin, Germany, 12–14 September 2016; pp. 389–399.
77. Menghi, R.; Papetti, A.; Germani, M.; Marconi, M. Energy efficiency of manufacturing systems: A review of energy assessment methods and tools. *J. Clean. Prod.* **2019**, *240*, 118276. [CrossRef]
78. Giacone, E.; Mancò, S. Energy efficiency measurement in industrial processes. *Energy* **2012**, *38*, 331–345. [CrossRef]
79. ISO 50006:2014. *Energy Management Systems—Measuring Energy Performance Using Energy Baselines (EnB) and Energy Performance Indicators (EnPI)—General Principles and Guidance*; International Organization for Standardization: Geneva, Switzerland, 2014.
80. DUEE-SPS-ESE. Diagnosi Energetiche art 8 del D.Lgs. 102/2014. Linee Guida e Manuale Operativo. Clusterizzazione, il Rap-porto di Diagnosi ed il Piano di Monitoraggio; ENEA. 2019. Available online: https://www.efficienzaenergetica.enea.it/component/jdownloads/?task=download.send&id=170&catid=41&Itemid=101 (accessed on 11 November 2021).
81. DUEE-SPS-ESE. Unione Petrolifera. Guideline for Energy Audits in Oil Refining Sector. 2019. Available online: https://www.efficienzaenergetica.enea.it/component/jdownloads/?task=download.send&id=218&catid=36&Itemid=101 (accessed on 11 November 2021).
82. Herce, C.; Biele, E.; Martini, C.; Salvio, M.; Toro, C. Impact of Energy Monitoring and Management Systems on the Implementation and Planning of Energy Performance Improved Actions: An Empirical Analysis Based on Energy Audits in Italy. *Energies* **2021**, *14*, 4723. [CrossRef]

energies

MDPI

Article

Private Hospital Energy Performance Benchmarking Using Energy Audit Data: An Italian Case Study

Daniele Dadi [1,*], Vito Introna [1], Annalisa Santolamazza [2], Marcello Salvio [3], Chiara Martini [3], Tiberio Pastura [1] and Fabrizio Martini [3]

1 Department of Enterprise Engineering, University of Rome Tor Vergata, 00133 Rome, Italy; vito.introna@uniroma2.it (V.I.); tiberio.pastura@gmail.com (T.P.)
2 DEIM School of Engineering, University of Tuscia, 01100 Viterbo, Italy; annalisa.santolamazza@unitus.it
3 DUEE-SPS-ESE Laboratory, Italian National Agency for New Technologies, Energy and Sustainable Economic Development (ENEA), Lungotevere Thaon di Revel, 76, 00196 Rome, Italy; marcello.salvio@enea.it (M.S.); chiara.martini@enea.it (C.M.); fabrizio.martini@enea.it (F.M.)
* Correspondence: daniele.dadi@uniroma2.it

Abstract: The increased focus on energy efficiency, both at the national and international levels, has fostered the diffusion and development of specific energy consumption benchmarks for most relevant economic sectors. In this context, energy-intensive facilities, such as hospitals and health structures, represent a unique case. Indeed, despite the high energy consumption of these structures, scientific literature lacks the presence of adequate energy performance benchmarks, especially in regard to the European context. Thus, this study aimed at defining energy benchmark indicators for the Italian private healthcare sector using data collected from the Italian mandatory energy audits according to Art.8 EU Directive 27/2012. The benchmark indicators' definition was made using a methodology proposed by the Italian National Agency for New Technologies, Energy and Sustainable Economic Development (ENEA). This methodology provided the calculation of specific energy performance indicators (EnPIs) by considering the global energy consumption of the different sites and the sector's relevant variables. The results obtained were compared with those obtained from a consolidated but more complex methodology: the one envisaged by the Environmental Protection Agency. The results obtained allowed us to validate the reliability of the proposed methodology, as well as the validity and future usability of the calculated indicators. Relying on a significant database containing actual data from recent energy audits, this study was thus able to provide an up-to-date and reliable benchmark for the private healthcare sector.

Keywords: energy efficiency; EnPIs; health sector; energy audit

Citation: Dadi, D.; Introna, V.; Santolamazza, A.; Salvio, M.; Martini, C.; Pastura, T.; Martini, F. Private Hospital Energy Performance Benchmarking Using Energy Audit Data: An Italian Case Study. *Energies* **2022**, *15*, 806. https://doi.org/10.3390/en15030806

Academic Editor: Luigi Aldieri

Received: 24 November 2021
Accepted: 18 January 2022
Published: 22 January 2022

1. Introduction

In Italy, about one-third of the total energy use is attributable to the building sector. In this sense, buildings destined for hospital use are particularly significant as they are highly energy-intensive structures in addition to their social role. The average consumption in hospitals is three times higher than in the residential sector in similar climatic conditions [1]. Although these structures are intense energy users, their energy analysis and characterization have not been sufficiently investigated. Indeed, energy efficiency was not considered as one of the sector's main objectives compared with requirements such as quality of services, functionality, or patients' well-being.

Our purpose was to carry out an important first step for the energy efficiency of this relevant sector through the definition of energy performance benchmark indicators.

To achieve this objective, a large dataset that came from mandatory energy audits for several structures operating in the Italian private health sector and collected by the Italian National Agency for New Technologies, Energy and Sustainable Economic Development (ENEA) was used.

1.1. Energy Consumption in Hospitals

A hospital structure has several peculiarities from the point of view of energy consumption. Hospitals must ensure services 24 h a day, seven days a week, throughout the year. In addition to this, the structures themselves must comply with a series of constraints imposed by the regulations to ensure a high comfort level and healthiness of the environments. Despite their high complexity, hospitals have the potential to reduce consumption through the implementation of investments and interventions aimed at improving the energy efficiency of structures and systems and constraining energy waste.

In general, hospitals' energy needs consist of using electricity and heat. Electricity is used to power medical, diagnostic, and monitoring equipment, indoor and outdoor lighting, summer air conditioning, air treatment, and the operation of computerized and security systems. Thermal energy is mainly used for the heating and air conditioning of rooms, sanitary water production, sterilization, and laundry and kitchen services. In turn, the uses of electricity and heat can be classified into two categories. The first refers to hotel-type uses to guarantee the well-being of healthcare workers and patients, including indoor and outdoor lighting, summer and winter air conditioning, lifts, the preparation of domestic hot water, and laundry and kitchen activities. The second refers to the uses for surgery, treatment, and diagnosis devices, i.e., diagnostic-medical equipment and instruments for sterilization [2].

Thermal energy is the one that best lends itself to rationing interventions since, in addition to having a high impact on total energy consumption, it is mainly used for space-heating purposes. This use allows for temporary interruptions for implementing the intervention itself without compromising the well-being of the people present in the hospital. The rationing of interventions is also possible for electricity, but it is necessary to consider that significant interruptions are not allowed, as electricity is used in services of primary importance that require continuity in their supply [3].

1.2. Energy Benchmarking

Over the last few years, several studies have focused on the analysis of the energy performance of health facilities and hospitals [4–8] for different countries such as Germany [9], China [10], the United States [11], and Korea [12].

On the other hand, other studies were not limited to an energy analysis but were aimed at defining specific benchmarks for different countries under different operating conditions, such as differences in management and, above all, environmental conditions.

In this regard, the UNI CEI EN 16231: 2012 standard [13], entitled "Energy efficiency benchmarking methodology", emphasizes the importance of determining the reference indices to compare performance. This comparison can be internal to the organization, through the analysis of historical data, and external, through comparing the organization's performance with those of other organizations in the sector. Through this energy comparison, the company can become aware of its performance and invest, if necessary, in improvement programs in terms of energy efficiency.

Different benchmarking approaches were developed for the specific health sector [14–17] and at a more general level in buildings [18–22].

These works are based on different approaches ranging from the definition of energy performance indicators (EnPIs) with identification of the relevant variables [12,15,17,18,20,21] to others based on statistical linear regression models mainly using the methodology proposed by the Environmental Protection Agency (EPA), which is described in detail in the following chapters [14,19,22–24].

Although there are approaches aimed at studying the energy behavior of health buildings and attempts at benchmark definitions, we found a lack of references in the scientific literature, especially regarding the Italian or, more generally, European context. These approaches, previously mentioned, do not translate into the definition of reliable and updated benchmarks that a sector structure can use as a reference. Instead, the benchmark approach is more developed in other countries, such as the United States. However, since the structure and energy behavior are very different, they are not considered applicable to the European context.

This study aimed to define energy benchmark indicators for the Italian private health sector using a simple approach based on the calculations EnPIs following a methodology proposed by the ENEA, which has been used successfully in other contexts [25].

One of the strengths of this work is the possibility to rely on a significant database containing actual data from recent energy audits, which allowed us to obtain up-to-date and reliable results that were perfectly suited to the Italian context to which we wanted to refer.

Moreover, to discuss and validate the results obtained, these were compared with the results achieved using a consolidated methodology that required greater complexity, such as the one proposed by the EPA.

This paper is structured as follows: Section 2, i.e., Materials and Methods, describes the dataset, including the activities of data collection and preprocessing, and introduces the main step of the ENEA methodology used to determine the benchmark EnPIs and the main steps of EPA methodology used to validate the result obtained. Section 3 describes the results obtained applying the two methodologies, while Section 4 discusses the main issues encountered and compares the result obtained with the ENEA methodologies with those obtained using the methodology proposed by EPA. Finally, in Section 5, the objectives, significant results obtained, and the next steps of the research are discussed.

2. Materials and Methods

This section describes the approach used: from the methods applied to preprocess the data to obtain the final dataset to the description of the main steps of the methodological approach proposed by ENEA and the one developed by EPA, which was used to compare the results obtained.

2.1. Data Collection and Preprocessing

Directive 2012/27/EU [25] establishes that "Member States shall ensure that enterprises that are not SMEs are subject to an energy audit carried out in an independent and cost-effective manner by qualified and/or accredited experts or implemented and supervised by independent authorities under national legislation by 5 December 2015 and at least every four years from the date of the previous energy audit." For Italy, the energy audits are collected every four years by ENEA.

For the purposes of this work, the energy audits for the Italian private health sector, received by ENEA in 2019 in correspondence with the second cycle of energy audits, were analyzed in order to define benchmarks suited to the Italian context.

In order to report relevant information about their energy consumption, each organization was required to submit a summary spreadsheet with every energy audit report. Taking into account the lessons learned during the first cycle of energy audits in 2015, ENEA decided to create a summary spreadsheet to use specifically for hospitals and health facilities in order to enable the collection of more detailed information about the energy consumption of the structure.

In 2019, in reference to the NACE Q86 code (Human health activities), the number of health facilities potentially subjected to the obligation to carry out the energy audit was 328. However, for feasibility reasons multi-site health companies were allowed to carry out energy audits on a limited number of representative sites using a clustering strategy developed by ENEA. Therefore, 152 energy audits were actually received by ENEA, with a high percentage (145 audits, 95.4%) belonging to NACE code 86.1 (Hospital activities), which is why it was the only one to be considered.

Referring to the Italian economic activity classification ATECO (ATtività ECO-nomica), revised in 2007 and deriving from the European classification NACE, Table 1 reports the descriptions of the subcategories of the ATECO code 86.1 and the number of audits for each category.

Table 1. The number of audits for each subcategory of ATECO 86.1.

ATECO Code	Description	Number of Audits
86.10.10	General hospitals and nursing homes	89
86.10.20	Specialist hospitals and nursing homes	47
86.10.30	Institutes, clinics, and university polyclinics	9
86.10.40	Hospitals and long-term nursing homes	0

A first analysis of the data collected showed that a minority of organizations did not use the updated summary spreadsheet that was implemented for the health sector, but a general one belonging to the tertiary sector. Since relevant information was absent in this other type of summary spreadsheet, to conduct complete and more in-depth analyses, the sample was reduced to only the organizations that used the updated summary spreadsheet, i.e., 85. However, further analysis showed that some information collected in the files was incongruent or incomplete. Thus, as a result, a final database consisting of 58 energy audits was obtained and analyzed.

For each healthcare structure, the following information was available:

- Data of the site, or the identification of the same, the name, the city, the VAT number, the NACE code of belonging and the accreditation or not to the National Health Service (NHS);
- General details of the structure, i.e., the covered area, the health workers, the beds, and the presence or absence of the swimming pool;
- Overall consumption of electricity, heating, and cooling relating to each site;
- Consumption and data relating to two macro-areas into which it is possible to divide a hospital structure, a part for hospitalizations, and a part for diagnosis and therapy.

Figure 1 shows some of the characteristics of the final sample analyzed in terms of the sites, beds, and health workers divided by the ATECO code and in terms of accreditation to the NHS.

To complete the available data, for each structure in the database, the degree days of heating and degree days of cooling were calculated through the website Degree Days [26]. In particular, the reference temperature, based on which, the heating and cooling degree days were calculated, was set to 22 °C and 25 °C, respectively, taking into account the minimum requirements that a healthcare facility must comply with.

2.2. Data Analysis

2.2.1. ENEA Methodology

The procedure proposed by ENEA for determining the benchmark energy performance indicators ($EnPI_{bmk}$) consists of a series of steps [27]:

1. Identification of the relevant variables;
2. Calculation of the energy performance indicators (EnPI) for each site;
3. Calculation of the average energy performance indicators ($EnPI_{avg}$);
4. Definition of the $EnPI_{bmk}$;
5. Evaluation of the reliability of the $EnPI_{bmk}$.

The first step of the methodology involves the identification of the relevant variables, which are those quantifiable factors that significantly impact energy performance and routinely change (weather conditions, operating conditions, working hours, production output, etc.) [28]. The identification of these variables is usually determined by the knowledge of the energy system under analysis and is supported by the reference scientific literature. The second step involves the calculation of the energy performance indicator (EnPI) for each site in the sample considered, which is defined as the ratio between energy consumption and the representative consumption parameter (relevant variable):

$$\mathrm{EnPI}\left[\frac{\mathrm{tep}}{\mathrm{m^2, bed, etc.}}\right] = \frac{\mathrm{energy\ consumption\ [tep]}}{\mathrm{parameter\ [m^2, bed, etc.]}} \quad (1)$$

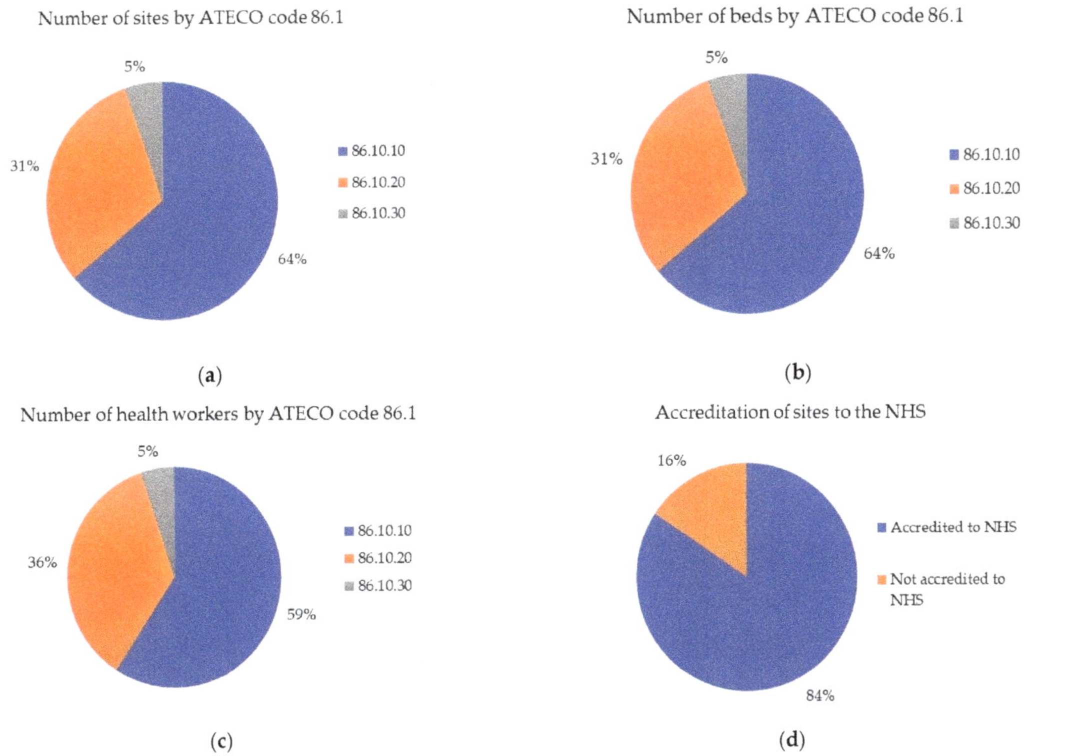

Figure 1. (**a**) Number of sites by ATECO code 86.1, (**b**) number of beds by ATECO code 86.1, (**c**) number of workers by ATECO code 86.1, and (**d**) accreditation of sites to the NHS.

Subsequently, the average energy performance indicators ($EnPI_{avg}$) are calculated, which are defined as the average of the EnPIs of the individual structures and the relative standard deviation (st.dev.), which expresses the dispersion of the data of the sample considered around the average. Therefore, the benchmark energy performance indicators are determined using the following formula:

$$EnPI_{bmk} = EnPI_{avg} \pm st.dev. \quad (2)$$

Based on the ratio value between the standard deviation and the $EnPI_{avg}$, it is possible to evaluate the reliability of the $EnPI_{bmk}$. Reliability is considered as follows:

- "High" if the ratio is less than 20%;
- "Average" if the ratio is between 20% and 60%;
- "Low" if the ratio is greater than 60%.

Figure 2 reports a schematic representation of the methodology followed.

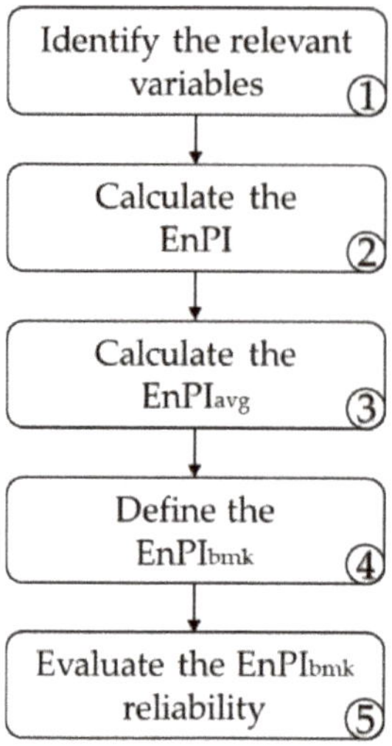

Figure 2. ENEA methodology.

2.2.2. EPA Methodology

The EPA has developed a technical methodology for evaluating the energy performance of different types of buildings; in this study, reference was made to the specific one developed for hospitals [29]. This methodology consists of a mathematical model for the definition of the energy efficiency ratio (ERR). The purpose of the methodology is to identify, through regression analysis, the key factors that determine energy consumption in order to develop a consumption forecasting model that allows for evaluating the energy performance of a hospital or, in more general terms, for a building. The procedure is divided into a sequence of phases, which have been adapted according to the information contained in the energy audits under study.

The first phase involves defining a group of structures with similar functional and operational characteristics to compare the structures themselves and overcome any technical limitations in the data. Then, it is necessary to define the variables for the regression analysis. Regarding the dependent variable, this is represented by the energy use intensity (EUI), which is equal to the total energy consumption of the site (EC) divided by the site's surface area. The independent variables, on the other hand, refer to those factors that characterize the health facility and that can impact energy consumption (X_1—health workers per square meter, X_2—beds per square meter, X_3—cooling degree days, X_4—heating degree days, and X_5—machines per square meter). Therefore, the predicted EUI is calculated as follows, with a_0, a_1, a_2, a_3, a_4, and a_5 as the parameters of the linear regression [29]:

$$\text{Predicted EUI} = a_0 + a_1X_1 + a_2X_2 + a_3X_3 + a_4X_4 + a_5X_5 \quad (3)$$

After determining the regression model for forecasting the energy use intensity, the methodology defines the energy efficiency ratio (EER) for each site as:

$$\text{EER} = \frac{\text{Actual EUI}\left[\frac{\text{tep}}{\text{m}^2.}\right]}{\text{Predicted EUI}\left[\frac{\text{tep}}{\text{m}^2.}\right]} \quad (4)$$

The numerator represents the energy consumption intensity for the specific health facility, which is calculated using measured data. In contrast, the denominator represents the expected value of the energy consumption intensity, which is calculated through the previously determined regression model using the measured values of the independent variables (X_1, X_2, X_3, X_4, X_5) for the same site as inputs. Thus, a low energy efficiency ratio indicates that the specific health facility is more efficient than the average because it uses less energy than predicted, whereas a high energy efficiency ratio indicates the opposite.

After computing the EER for each element of the sample, the results can be analyzed through a frequency distribution to highlight the differences in the energy efficiency of the sample.

Finally, by sorting the values of the EER from smallest to largest, it is possible to calculate the cumulative distribution of the EER for the sample and use regression analysis to obtain the value of the cumulative percentage as a function of the energy efficiency ratio.

In conclusion, through its mathematical formulation, the model created makes it possible to compare the energy performance of a generic health facility with those of the sample used.

3. Results

3.1. ENEA Results

We used the database defined in the previous paragraph to calculate the benchmark energy performance indicators for the private health sector. Specifically, the energy performance indicators were calculated using the energy consumption as a numerator given by the sum of the health facility's electricity, heating, and cooling energy consumptions. The denominator, instead, changed for each energy performance indicator (as shown in Table 2), using the relevant variables available in the database.

Table 2. Results of the EnPIs calculations.

Sample Sites	$EnPI_{bmk}$	$EnPI_{avg}$ ± st.dev	Reliability
ATECO 86.10.10	$EnPI_{bmk}$_ca (toe/m^2)	0.052 ± 0.023	Average
	$EnPI_{bmk}$_hw (toe/health worker)	2.101 ± 0.950	Average
	$EnPI_{bmk}$_b (toe/bed)	4.275 ± 2.593	Low
ATECO 86.10.20	$EnPI_{bmk}$_ca (toe/m^2)	0.050 ± 0.031	Low
	$EnPI_{bmk}$_hw (toe/health worker)	2.278 ± 0.875	Average
	$EnPI_{bmk}$_b (toe/bed)	7.268 ± 7.453	Low
ATECO 86.10.10 accredited to NHS	$EnPI_{bmk}$_ca (toe/m^2)	0.049 ± 0.023	Average
	$EnPI_{bmk}$_hw (toe/health worker)	1.959 ± 0.902	Average
	$EnPI_{bmk}$_b (toe/bed)	3.738 ± 2.010	Average
ATECO 86.10.20 accredited to NHS	$EnPI_{bmk}$_ca (toe/m^2)	0.057 ± 0.030	Average
	$EnPI_{bmk}$_hw (toe/health worker)	2.426 ± 0.867	Average
	$EnPI_{bmk}$_b (toe/bed)	8.546 ± 7.711	Low

The energy performance indicators were defined for the ATECO 86.10.10 (general hospitals and nursing homes) and 86.10.20 (specialist hospitals and nursing homes) codes. The ATECO 86.10.30 code (institutes, clinics, and university polyclinics) was not analyzed, as it was not significant in terms of the sample size. Moreover, the analysis was also conducted specifically for the hospitals accredited and not accredited to the NHS. Additional indicators were assessed considering a more specific part of the data available, namely, that relating to hospitalizations and diagnosis and therapy, using only the sites that had filled in the relevant fields provided within the summary file. To limit the possible distortions of energy consumption, we decided to exclude sites with a swimming pool from the sample in the analyses explained above.

3.1.1. Energy Performance Indicators: Generality of the Structure

The benchmark energy performance indicators ($EnPI_{bmk}$) were defined by relating the energy consumption to three relevant variables:

- The covered area (ca) in square meters;
- The number of health workers (hw);

- The number of beds (b).

These variables were shown to significantly impact the energy consumption for hospitals in several studies [11,15,17,18].

Table 2 shows the results of the EnPIs calculations.

For the categories identified, the indicator referring to workers always had average reliability. Good results were also obtained considering the covered area, while the worst results were obtained considering the number of beds as the relevant variable.

The same benchmark indicators were also identified only for health facilities accredited to the NHS, improving the reliability of some indicators compared with those defined considering the whole dataset.

For example, Figure 3 shows a graphical representation of one of the calculated EnPIs using the area covered in square meters as the relevant variable, showing good reliability for the ATECO 86.10.10 code.

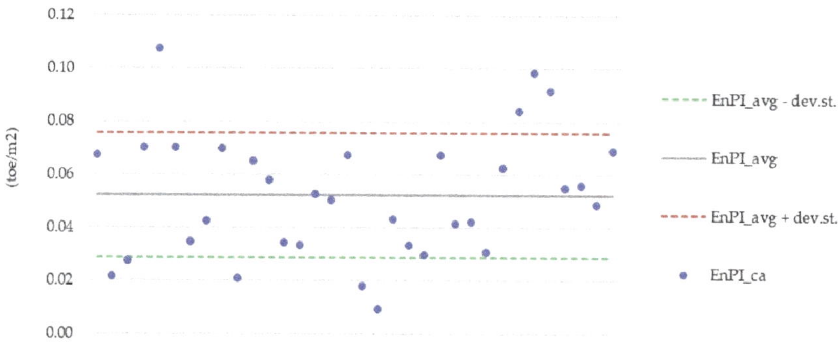

Figure 3. Graphical representation of $EnPI_{bmk\_}ca$ (toe/m^2) for the ATECO code 86.10.10.

In order to be able to differentiate the structures and conduct a more targeted analysis, a further structures subdivision was envisaged during the energy audit phase. Each structure was divided into two macro-areas: *hospitalization* and *diagnosis and therapy*. Each of these could be divided into several parts, where the results of the related analyses are given in the following paragraphs.

3.1.2. Energy Performance Indicators: Hospitalizations

The *hospitalization* macro area represented the hotel area of the health facility. We could divide the hospitalization into five specific hospital wards: overall areas of hospitalization, intensive care, day surgery, dialysis, and gyms and rehabilitation. During the energy audit, for each of the areas present within the health facility, it was possible to indicate the consumption of electricity, heating, and cooling; the number of days in the hospital; and the covered area of the relative spaces. This information was used to determine more specific EnpIs, which was useful for comparing similar structures in terms of wards.

Starting from the database and excluding the sites belonging to the ATECO 86.10.30 code and those with a swimming pool, the number of sites that provided the data requested for at least one area among the five previously listed was 24. However, these sites were different from each other in terms of the areas present within them. In the definition of the benchmark indices, this heterogeneity involved the need to consider a subset of health structures characterized in terms of the presence of the areas under analysis from time to time.

For each area, two energy performance indicators were defined. The first related the sum of the electricity consumption, heating, and cooling of the single area to the relative number of days in hospital (dh), while the second one related the sum of electricity consumption, heating, and cooling of the single area to the relative surface area (sh). Following the ENEA methodology steps defined in the previous paragraphs, it was possible to calculate the benchmark indicators and their relative reliability. Table 3 summarizes the results of the reliability evaluation for the EnPIs related to hospitalizations.

Table 3. Reliability evaluation for the EnPIs related to hospitalizations.

Hospital Ward	Reliability $EnPI_{bmk}$_dh	Reliability $EnPI_{bmk}$_sh
Overall areas of hospitalization (ward present in 24 sites)	Low	Average $EnPI_{bmk}$_sh (toe/m^2) = 0.042 ± 0.021
Intensive care (ward present in 12 sites)	Low	Low
Day surgery (ward present in 10 sites)	Low	Average $EnPI_{bmk}$_sh (toe/m^2) = 0.051 ± 0.021
Dialysis (ward present in 2 sites)	-	-
Gyms and rehabilitation (ward present in 12 sites)	Low	Low

The calculated benchmark indicators related to the surface area showed average reliability only for the overall areas of hospitalization and for the day surgery ward, while for the remaining areas, we did not find valid benchmark indicators due to the "low" reliability, both concerning the number of days hospitalization and the surface area. For the dialysis ward, it was not possible to calculate the respective indicators due to an excessively small sample.

3.1.3. Energy Performance Indicators: Diagnosis and Therapy

The *diagnosis and therapy* macro area represented the operating area of the health facility. We could divide the diagnosis and therapy into seven specific activities: operating block, sterilization, radiology and diagnostic imaging, first aid, functional and endoscopic examinations, transfusion center, and laboratory diagnostics. For each of the services provided by the health facility, among the information contained in the collected energy audits, it was possible to find the consumption of electricity, heating, and cooling; the number of services provided; and the surface areas of the spaces where the services themselves are provided.

Starting from the database defined in Section 2.1 and excluding the sites belonging to the ATECO 86.10.30 code and those with a swimming pool, the number of sites that provided the data requested for at least one of the seven activities listed was 30. However, they did not all perform the same diagnosis and therapy activities; consequently, in developing the benchmark EnPIs for each type of service provided, a subset of health facilities carrying was considered. For each activity, two energy performance indicators were defined: the first relates the sum of the electricity, thermal, and cooling energy consumption of the single activity to the relative number of services provided (ns), while the second relates the sum of the consumption of electricity, heating, and cooling of the single activity to the relative surface area (ss) where it is carried out.

Table 4 summarizes the results of the reliability evaluation for the EnPIs related to diagnosis and therapy.

All benchmark indicators calculated for dialysis showed low reliability, both for the number of services provided and the covered surface area. These results were mainly due to the high heterogeneity of the services provided within the same specific activity.

Table 4. Reliability evaluation for the EnPIs related to diagnosis and therapy.

Hospital Ward	Reliability $EnPI_{bmk\_ns}$	Reliability $EnPI_{bmk\_ss}$
Operating block (activity provided by 23 sites)	Low	Low
Sterilization (activity provided by 12 sites)	Low	Low
Radiology and diagnostic imaging (activity provided by 27 sites)	Low	Low
First aid (activity provided by 10 sites)	Low	Low
Functional and endoscopic examinations (activity provided by 25 sites)	Low	Low
Transfusion center (activity provided by 4 sites)	Low	Low
Laboratory diagnostics (activity provided by 19 sites)	Low	Low

3.2. EPA Results

Using the same starting database and following the EPA methodology described in the previous paragraphs, the first step was to define a sample of health facilities that was as homogeneous as possible. This resulted in the exclusion of 20 sites from the 58 sites initially present in the database to provide a final sample of 38 health facilities. In particular, the sites excluded were as follows:

- Those belonging to the ATECO code 86.10.30;
- Those with a swimming pool inside.

The dependent variable of the regression model was represented by the intensity of energy consumption (toe/m^2), which is equal to the ratio between the sum of the electrical, thermal, and cooling energy consumed and the covered area. For the choice of the independent variables, the data relating to both the generality of the health facility and the climatic conditions were considered, namely, the covered area, number of health workers, number of beds, heating degree days, and cooling degree days. In particular, the health workers and the beds were considered in terms of the surface density, comparing the respective values to that of the covered area. Therefore, the independent variables were as follows:

- Health workers per square meter (employee/m^2);
- Beds per square meter (bed/m^2);
- Heating degree days (°C);
- Cooling degree days (°C).

The additional independent variable "machines per square meter" mentioned in the EPA methodology was not included in the analysis since it was not among the data collected from the mandatory energy audits.

Several regression analyses were conducted to define the combination of statistically significant parameters (*p*-value lower than 0.05). After evaluating the different combinations and the presence of outliers, it was possible to define the regression model using the parameters reported in Table 5. The adjusted R^2 value was equal to 0.4677.

Table 5. Regression analysis results for Energy Use Intensity.

Variable	Results	
Dependent variable	Energy use intensity	
Observations	38	
R^2	0.5108	
Adjusted R^2	0.4677	
Standard error	0.0188	
	Coefficients	*Significance*
Intercept	0.06965	0.00004
Health workers per square meter	1.47221	0.00000
Beds per square meter	−2.70713	0.00409
Cooling degree days	−0.00015	0.02352

By analyzing the results obtained in Table 5, it is possible to make some considerations. The coefficient relating to the energy driver "health workers per square meter" was positive. In contrast, the coefficients obtained for the energy driver "beds per square meter" and "cooling degree days" were negative. All three of these coefficients were statistically significant. It is important to emphasize that the model refers to the total energy consumption (electrical, thermal, and cooling energy), and the energy consumption can have different dynamics for the structures in the dataset. For example, some relevant differences may be due to the geographical position, the main energy users, the presence of self-production systems of energy (e.g., trigeneration systems), and the daily dynamics of the sites. The energy driver "number of beds per square meter" had a negative coefficient due to a different use of the spaces among the structures: a higher amount of beds per square meter translated into a different use of the spaces, which, in turn, could lead to optimized energy consumption (e.g., air conditioning).

Since in the equation, the dependent variable is the total energy consumption of the site (EC) divided by the site's surface area (i.e., energy use intensity), the explanatory power of the site's surface area was not included in the R^2 value, altering it artificially. Thus, the EPA methodology suggests recalculating the R^2 value in terms of energy consumption (EC) [29]:

$$R^2 = 1 - \frac{\sum_{i=1}^{38}(\text{ActualEC}_i - \text{PredictedEC}_i)^2}{\sum_{i=1}^{38}(\text{ActualEC}_i - \text{ActualEC\_avg})^2} \quad (5)$$

The R^2 value thus calculated was equal to 0.8350, a more than satisfactory value.

At this point, a health facility can evaluate its energy performance by calculating the energy efficiency ratio, which is given by the ratio between the actual energy use intensity and the predicted energy use intensity, calculated through the regression model. An energy efficiency ratio value lower than one indicates that the health facility uses less energy than expected and is consequently more efficient; on the other hand, a value greater than one indicates lower efficiency.

The energy efficiency ratios of the 38 structures belonging to the sample were then calculated. Figure 4 shows the distribution of the energy performance ratios: the most energy-efficient health facility is located on the far left of the distribution, while the least efficient health facility is on the far right.

The energy efficiency ratios were sorted in ascending order, and we were able to calculate the cumulative percentage for each sample ratio. Finally, through the regression analysis, the equation of the curve was determined, which expressed the value of the cumulative percentage as a function of the energy efficiency ratio. The significance value was set at 0.05. Figure 5 graphically shows the regression performed, while Table 6 shows the results.

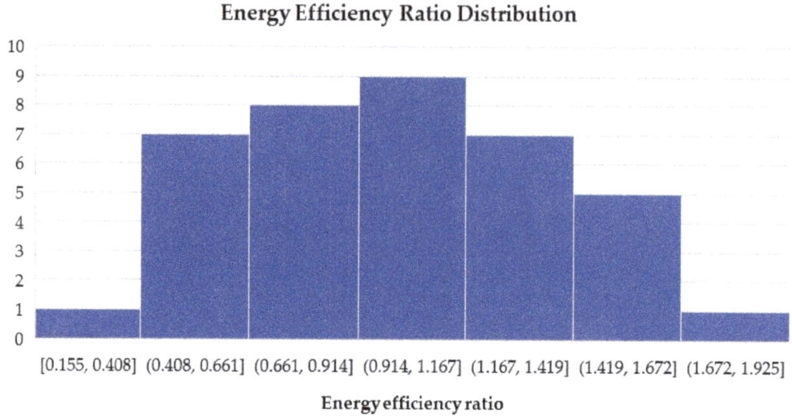

Figure 4. Distribution of the energy efficiency ratios.

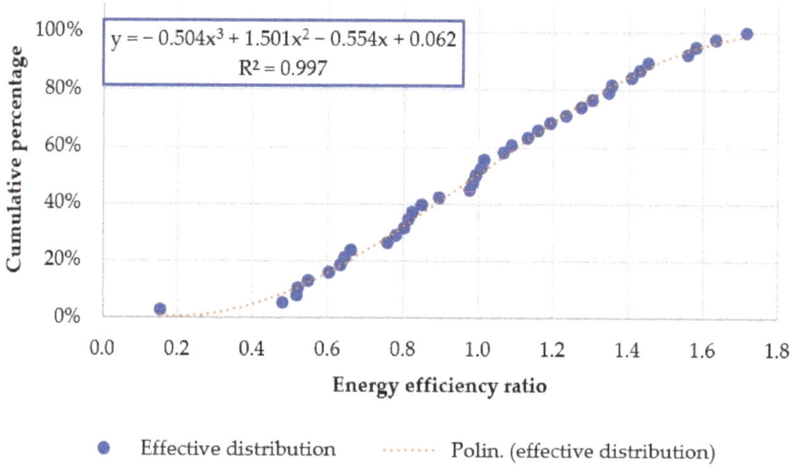

Figure 5. Cumulative distribution of the energy efficiency ratios.

Table 6. Regression analysis results for the cumulative percentage.

Variable	Results	
Dependent variable	Cumulative percentage	
Observations	38	
R^2	0.9965	
Adjusted R^2	0.9962	
Standard error	0.0180	
	Coefficients	*Significance*
Intercept	0.06177	0.04378
Health workers per square meter	−0.55418	0.00001
Beds per square meter	1.50108	0.00000
Cooling degree days	−0.50357	0.00000

After determining its energy efficiency ratio, a health facility that intends to evaluate its energy performance compared to those of the sample can calculate the corresponding cumulative percentage value and identify the percentage of sites in the sample with better or worse performances. For example, a cumulative percentage of 20% indicates that only 20% of the sample has an energy efficiency ratio equal to or lower than its own.

4. Discussion

At this point, it is possible to apply both the methodologies of ENEA and EPA and compare the two results. In particular, two structures were considered. For structure A (ATECO 86.10.10 accredited to the NHS), according to the ENEA methodology, two out of three energy performance indicators were lower than the average value of the respective benchmark indicators, indicating better performances than the average ones. The EPA methodology application resulted in an energy efficiency ratio value less than one, indicating greater efficiency. Indeed, the cumulative percentage was 18%, which meant that structure A was more efficient than 82% of the health facilities in the sample.

Table 7 shows the results obtained for structure A.

Table 7. Results of the comparison between ENEA methodology and EPA methodology for structure A.

ENEA		EPA	
ENEA methodology results	Results for structure A	EPA methodology results	Results for structure A
$EnPI_{bmk}$_ca (toe/m^2) = 0.049 ± 0.023	EnPI_ca (toe/m^2) = 0.042	EUI predicted (toe/m^2) = 0.065	EUI actual (toe/m^2) = 0.042
$EnPI_{bmk}$_hw (toe/health worker) = 1.959 ± 0.902	EnPI_hw (toe/health worker) = 1.396	EER = 0.63	-
$EnPI_{bmk}$_b (toe/bed) = 3.738 ± 2.010	EnPI_b (toe/bed) = 5.641	Cumulative percentage = 18%	-

According to the ENEA methodology, for structure B (ATECO 86.10.20 accredited to the NHS), there were two out of three energy performance indicators higher than the average value of the respective benchmark indicators, thus indicating slightly worse performances than the average ones. The EPA methodology application results in an energy efficiency ratio value greater than one, indicating lower efficiency. Indeed, the cumulative percentage was 72%, which meant that structure B was less efficient than 72% of the health facilities in the sample. Table 8 shows the results obtained for structure B.

Table 8. Results of the comparison between the ENEA methodology and EPA methodology for structure B.

ENEA		EPA	
ENEA methodology results	Results for structure B	EPA methodology results	Results for structure B
$EnPI_{bmk}$_ca (toe/m^2) = 0.057 ± 0.030	EnPI_ca (toe/m^2) = 0.063	EUI predicted (toe/m^2) = 0.051	EUI actual (toe/m^2) = 0.063
$EnPI_{bmk}$_hw (toe/health worker) = 2.426 ± 0.867	EnPI_hw (toe/health worker) = 3.231	EER = 1.24	-
$EnPI_{bmk}$_b (toe/bed) = 8.546 ± 7.711	EnPI_b (toe/bed) = 4.951	Cumulative percentage = 72%	-

Therefore, the two methods can be considered consistent from the point of view of the results.

5. Conclusions

The analyses carried out in this work made it possible to define energy performance benchmark indicators for the Italian private health sector, following a methodology developed by ENEA. One of the strengths that added value to the analysis was the possibility to rely on an extended dataset from the national mandatory energy audits, which allowed us to define an up-to-date and reliable benchmark for the private healthcare sector.

The analysis carried out concerned ATECO 86.10.10 (general hospitals and nursing homes) and 86.10.20 (specialist hospitals and nursing homes), as they represent the great majority of the sites present in the sample. The best results from the point of view of reliability were obtained for the EnPIs calculated by considering the number of health workers and the covered area as relevant variables, while the worst results were obtained considering the number of beds. However, the same benchmark indicators were also calculated only for health facilities accredited to the NHS. It seemed to improve the reliability of some indicators compared with those defined considering the whole dataset. On the other hand, no relevant results were obtained considering specific macro-areas, such as hospitalization and diagnosis and therapy.

Concerning the reliability of the indicators determined, the results appeared to be acceptable when we considered the whole dataset (Table 2). On the other hand, when we proceeded to subdivide the dataset into macro-areas (hospitalization and diagnosis and therapy, respectively in Tables 3 and 4), the calculation of the indicators provided low reliability. This low reliability was mainly attributable to a limited number of data in the various macro-categories due to the intrinsic difference of the structures and incomplete data collection.

A suitable solution involves improving the data collection phase for the next cycle of energy audits scheduled for 2023. We are confident that systematizing and simplifying the data collection phase by providing more specific and clearer indications on which parameters to report could significantly increase the reliability of the benchmark indicators.

It should be emphasized that although the subdivision of the dataset into macro-areas and macro-categories produced indicators with generally low reliability, this evaluation was of fundamental importance to identify further opportunities for improvement in view of the next mandatory scheduled audits.

In order to test the reliability of the proposed method, the results were compared with those obtained by using the EPA methodology. This test was made by comparing two different health facilities obtaining comparable results. Therefore, the two methods could be considered consistent from the point of view of results.

Moreover, using the benchmark methodology customized in this study, healthcare facilities can independently assess their energy efficiency in reference to the performance of the Italian private healthcare sector and determine how much their energy efficiency differs from the average of the sector.

Finally, given the valid results obtained in the private health sector, it could be interesting to extend the analyses carried out to the public health sector to compare the public and private health sectors.

Author Contributions: All authors contributed equally to the idea and the design of the methodology proposed and to the deployment of the research project. F.M. and V.I. were responsible for the research activities definition, coordination, and verifications. T.P. and C.M. analyzed the data. D.D. and A.S. prepared the original draft and receipted the suggestions from internal and external reviewers; M.S., F.M. and V.I. contributed to the review and editing. All authors have read and agreed to the published version of the manuscript.

Funding: This work was part of the Electrical System Research (PTR 2019–2021), implemented under Programme Agreements among the Italian Ministry for Economic Development (currently Ministry of Ecological Transition) and ENEA, CNR, and RSE S.p.A.

Institutional Review Board Statement: Not applicable.

Informed Consent Statement: Not applicable.

Data Availability Statement: Not applicable.

Conflicts of Interest: The authors declare no conflict of interest.

References

1. Grassi, W.; Testi, D.; Menchetti, E.; Della Vista, D.; Bandini, M.; Niccoli, L.; Grassini, G.L.; Fasano, G. Valutazione dei Consumi nell'edilizia Esistente e Benchmark Mediante Codici Semplificati: Analisi di Edifici Ospedalieri. Available online: https://www.enea.it/it/Ricerca_sviluppo/documenti/ricerca-di-sistema-elettrico/governance/rse117.pdf (accessed on 8 November 2021).
2. Mori, A.; Lavinia, C. Caratterizzazione Energetica Delle Strutture Sanitarie del Mezzoggiorno d'Italia. 2015. Available online: https://iris.enea.it/handle/20.500.12079/6728 (accessed on 8 November 2021).
3. Mori, A.; Martini, S.; Muzi, G. Energetic Characterization of the "G. Brotzu" Hospital Enterprise, San Michele Hospital in Cagliari, According to the Programme Agreement with the Italian Ministry of Economic Development. *Energ. Ambiente E Innov.* **2010**, *5*, 72–86.
4. García-Sanz-Calcedo, J.; López-Rodríguez, F.; Cuadros, F. Quantitative Analysis on Energy Efficiency of Health Centers According to Their Size. *Energy Build.* **2014**, *73*, 7–12. [CrossRef]
5. Christiansen, N.; Kaltschmitt, M.; Dzukowski, F. Electrical Energy Consumption and Utilization Time Analysis of Hospital Departments and Large Scale Medical Equipment. *Energy Build.* **2016**, *131*, 172–183. [CrossRef]
6. García-Sanz-Calcedo, J.; Gómez-Chaparro, M.; Sanchez-Barroso, G. Electrical and Thermal Energy in Private Hospitals: Consumption Indicators Focused on Healthcare Activity. *Sustain. Cities Soc.* **2019**, *47*, 101482. [CrossRef]
7. Sheppy, M.; Pless, S.; Kung, F. *Healthcare Energy End-Use Monitoring*; National Renewable Energy Lab: Golden, CO, USA, 2014.
8. Borges de Oliveira, K.; dos Santos, E.F.; Neto, A.F.; de Mello Santos, V.H.; de Oliveira, O.J. Guidelines for Efficient and Sustainable Energy Management in Hospital Buildings. *J. Clean. Prod.* **2021**, *11*, 129644. [CrossRef]
9. Seifert, C.; Damert, M.; Guenther, E. Environmental Management in German Hospitals—A Classification of Approaches. *Sustainability* **2020**, *12*, 4428. [CrossRef]
10. Ji, R.; Qu, S. Investigation and Evaluation of Energy Consumption Performance for Hospital Buildings in China. *Sustainability* **2019**, *11*, 1724. [CrossRef]
11. Bawaneh, K.; Ghazi Nezami, F.; Rasheduzzaman, M.; Deken, B. Energy Consumption Analysis and Characterization of Healthcare Facilities in the United States. *Energies* **2019**, *12*, 3775. [CrossRef]
12. Chung, M.; Park, H.-C. Comparison of Building Energy Demand for Hotels, Hospitals, and Offices in Korea. *Energy* **2015**, *92*, 383–393. [CrossRef]
13. *UNI CEI EN 16231: 2012*; Energy Efficiency Benchmarking Methodology; European Committee for Standardization: Brussels, Belgium, 2012.
14. Dahlan, N.Y.; Mohamed, H.; Kamaluddin, K.A.; Abd Rahman, N.M.; Reimann, G.; Chia, J.; Ilham, N.I. Energy Star Based Benchmarking Model for Malaysian Government Hospitals—A Qualitative and Quantitative Approach to Assess Energy Performances. *J. Build. Eng.* **2022**, *45*, 103460. [CrossRef]
15. González González, A.; García-Sanz-Calcedo, J.; Rodríguez Salgado, D. Evaluation of Energy Consumption in German Hospitals: Benchmarking in the Public Sector. *Energies* **2018**, *11*, 2279. [CrossRef]
16. Singer, B.C. *Hospital Energy Benchmarking Guidance—Version 1.0*; Lawrence Berkeley National Lab. (LBNL): Berkeley, CA, USA, 2009. [CrossRef]
17. Hwang, D.K.; Cho, J.; Moon, J. Moon Feasibility Study on Energy Audit and Data Driven Analysis Procedure for Building Energy Efficiency: Bench-Marking in Korean Hospital Buildings. *Energies* **2019**, *12*, 3006. [CrossRef]
18. Kim, D.W.; Kim, Y.M.; Lee, S.E. Development of an Energy Benchmarking Database Based on Cost-Effective Energy Performance Indicators: Case Study on Public Buildings in South Korea. *Energy Build.* **2019**, *191*, 104–116. [CrossRef]
19. Arjunan, P.; Poolla, K.; Miller, C. EnergyStar++: Towards More Accurate and Explanatory Building Energy Benchmarking. *Appl. Energy* **2020**, *276*, 115413. [CrossRef]
20. Ma, H.; Du, N.; Yu, S.; Lu, W.; Zhang, Z.; Deng, N.; Li, C. Analysis of Typical Public Building Energy Consumption in Northern China. *Energy Build.* **2017**, *136*, 139–150. [CrossRef]
21. de Oliveira Veloso, A.C.; Gonçalves de Souza, R.V.; dos Santos, F.N. Energy Benchmarking for Office Building Towers in Mild Temperate Climate. *Energy Build.* **2020**, *222*, 110059. [CrossRef]
22. Shang, L.; Lee, H.W.; Dermisi, S.; Choe, Y. Impact of Energy Benchmarking and Disclosure Policy on Office Buildings. *J. Clean. Prod.* **2020**, *250*, 119500. [CrossRef]
23. Estrella Guillén, E.; Samuelson, H.W.; Cedeño Laurent, J.G. Comparing Energy and Comfort Metrics for Building Benchmarking. *Energy Build.* **2019**, *205*, 109539. [CrossRef]
24. Benchmarking and Building Performance Standards Policy Toolkit. Available online: https://www.epa.gov/statelocalenergy/benchmarking-and-building-performance-standards-policy-toolkit (accessed on 8 November 2021).

25. Bruni, G.; De Santis, A.; Herce, C.; Leto, L.; Martini, C.; Martini, F.; Salvio, M.; Tocchetti, F.A.; Toro, C. From Energy Audit to Energy Performance Indicators (EnPI): A Methodology to Characterize Productive Sectors. The Italian Cement Industry Case Study. *Energies* **2021**, *14*, 8436. [CrossRef]
26. Degree Days Calculation. Available online: https://www.degreedays.net (accessed on 8 November 2021).
27. ENEA; Assoimmobiliare Benchmark Di Consumo Energetico Degli Edifici per Uffici in Italia. 2021. Available online: https://www.enea.it/it/Stampa/File/Rapporto_BenchmarkConsumiUffici_EneaAssoimmobiliare_2019.pdf (accessed on 8 November 2021).
28. *EN ISO 50001:2018*; Energy Management Systems—Requirements with Guidance for Use; ISO: Geneva, Switzerland, 2018.
29. ENERGY STAR Score for Hospitals (General Medical and Surgical). Available online: https://www.energystar.gov/buildings/tools-and-resources/energy-star-score-hospitals-general-medical-and-surgical (accessed on 8 November 2021).

MDPI
St. Alban-Anlage 66
4052 Basel
Switzerland
Tel. +41 61 683 77 34
Fax +41 61 302 89 18
www.mdpi.com

Energies Editorial Office
E-mail: energies@mdpi.com
www.mdpi.com/journal/energies

www.ingramcontent.com/pod-product-compliance
Lightning Source LLC
Chambersburg PA
CBHW040810250926
51850CB00007BA/1131
9783036549033